Naturalistic Decision Making

Expertise: Research and Applications

Robert R. Hoffman, K. Anders Ericsson, Gary Klein, Dean K. Simonton, Robert J. Sternberg, and Christopher D. Wickens, **Series Editors**

Naturalistic Decision Making

Edited by
Caroline E. Zsambok
Gary Klein
Klein Associates Inc.

LEA LAWRENCE ERLBAUM ASSOCIATES, PUBLISHERS
1997   Mahwah, New Jersey

Lawrence Erlbaum Associates, Inc., Publishers
10 Industrial Avenue
Mahwah, New Jersey 07430

Cover design by Kathryn Houghtaling

Library of Congress Cataloging-in-Publication Data

Naturalistic decision making / edited by Caroline E. Zsambok,
Gary Klein.
p. cm.
This volume represents the substance of the [Second] Naturalistic Decision Making conference, which was held in Dayton,
Ohio in June 1994.
Includes bibliographical references and indexes.
ISBN 0-8058-1873-1 (cloth : alk. paper). — ISBN 0-8058-
1874-X (pbk. : alk. paper)
1. Decision making — Congresses. 2. Problem solving —Congresses. I. Zsambok, Caroline E. II. Klein, Gary A. III. Naturalistic Decision Making Conference (2nd : 1994 : Dayton, Ohio)
BF448.N38   1996
153.8'3—dc20   96-27981
                                           CIP

Books published by Lawrence Erlbaum Associates are printed on
acid-free paper, and their bindings are chosen for strength and dura-
bility.

Printed in the United States of America
10 9 8 7 6 5 4 3 2 1

Contents

PART III: RESEARCH REPORTS

PART IV: METHODOLOGICAL AND THEORETICAL CONSIDERATIONS

Series Editor's Preface

This volume is the second to appear in this series, and it is one that will stand as representative of the standards and goals of the series. I am hopeful it will serve as a good model.

Contributors to this volume are researchers and applied practitioners, from various backgrounds including experimental psychology, ergonomics, computer science, and systems science, who have aligned themselves with a paradigm called Naturalistic Decision Making (NDM). This volume represents the substance of the Second NDM Conference, which was held in Dayton, OH, in June 1994. Domains for this applied work include health care, aviation, management, manufacturing, command and control, and decision aiding. Theoretical and methodological foundations include the study of expert–novice differences, examination of the role of recognition processes and situation assessment in problem solving, examination of hypothesis formation and testing in real-world situations, and examination of decision-making strategies in emergency situations. Both the first and the second NDM Conference were initiated, planned, and managed by Klein Associates Inc. They also secured sponsorships for the conferences. Klein Associates is, to be sure, unique. However, I should lead into the rest of this preface by saying that any laudations thrust on them are deserved, by virtue of their decades-long empirical efforts, their diverse and substantive research findings, and the important applications of their work to theory and practice. In other words, the stuff stands on its own merits.

Now, how is Klein Associates unique? Well, for one thing they have had this wacky belief that a private firm, one specializing in applications, can contribute to basic science and academia! Their links with the academy are perhaps just as strong as their links with government and industry. They have devoted considerable effort, exemplified by the NDM conferences that have brought together researchers in the academy, government, business, and industry to deal with shared, and tough, problems.

It would be safe to say that something of a revolution or paradigm shift has occurred in the area of judgment and decision making. Like all such shifts, whether modestly or boldly self-proclaimed, the new view defines itself in part in terms of what it is reacting against. In turn, this leads to counterreactions by the old guard and counterclaims of setting up straw men. Like all successful shifts, it is explicitly focused on integrating previous work and views, and on leading the field to new and broader horizons. Few private corporations can lay claim to having had such an impact.

We still see in NDM something of a promissory note, understandable given the newness of the approach and the untouched problems and challenges—ones that it, in fact, has revealed. The call to a bandwagon was clear in the edited book from the first NDM conference (Klein, Orasanu, Calderwood, & Zsambok, 1993). In the present volume, we see more of the payoff in terms of research reports on projects conducted according to the general NDM view. Here we see numerous studies of expertise in diverse domains. We see research that effectively dissolves the distinction between basic and applied cognitive or human factors research and that contributes to the affairs of society, government, and business and, at the same time, says important things to the experimental psychologist—things about proficiency in skill, in knowledge, and in reasoning.

The world could be a better place for this sort of stuff. Systems could be designed better. People could be trained better. Teams could work more effectively. Corporations could more effectively preserve and disseminate the wisdom and experience of their senior people. Emergencies could be handled better. Theories could be more realistic and relevant. More scientists would look out the windows of their ivory towers. Management might even listen more to scientists when they say: "We gotta stick some basic research in at the front end in order to do this right," or, "The best answer might not be quite the one that you want to hear," or "We really need to approach this in terms of a longer-term payoff." Or, best of all, "Hey, I can solve that problem for you."

—*Robert R. Hoffman, Series Editor*

Preface

This book has become a reality because of funding for planning and executing the 1994 Naturalistic Decision Making conference, the event which stimulated production of the chapters that appear here. Many thanks go to these sponsors of the conference: Armstrong Laboratory (Human Engineering Division), U.S. Army Research Institute for the Behavioral and Social Sciences, U.S. Army Research Laboratory (Human Research and Engineering Directorate), Crew System Ergonomics Information Analysis Center, NASA–Ames Research Center, Naval Air Warfare Center/Training Systems Division, Naval Command, Control and Ocean Surveillance Center (NRaD), and Office of Naval Research.

Many thanks also go to contributors to this volume. The editors want to acknowledge the excellent work of panel discussants who came to the conference well prepared to engage in stimulating and critical dialogue. This produced a rich foundation for chapters 3, 4, and 6 through 15, which were ably written by panel chairs after the conference had concluded. Likewise, we are appreciative of the excellent contributions from authors of research reports and theoretical or methodological papers (the conference participants and their affiliations are listed on p. xv).

Finally, we wish to acknowledge Robert Hoffman as Series Editor for his careful and insightful reviews of all chapters, and Barbara Law of Klein Associates for her tireless and exceptional contribution as production editor.

—*Caroline E. Zsambok*
—*Gary Klein*

CONFERENCE PARTICIPANTS AND AFFILIATIONS

Lee Roy Beach
University of Arizona

Herbert H. Bell
U.S.A.F. Armstrong Laboratory

Kevin B. Bennett
Wright State University

Marilyn Sue Bogner
U.S. Food & Drug Administration

LtCol James Bushman
U.S.A.F. Armstrong Laboratory

Janis A. Cannon-Bowers
Naval Air Warfare Center/Training Systems Division

Michelene S. Chi
University of Pittsburgh

Marvin S. Cohen
Cognitive Technologies, Inc.

Richard I. Cook
The Ohio State University
(currently at University of Chicago Medical Center)

Robert D. Deutsch
Evidence Based Research, Inc.

Lia Di Bello
City University of New York
(University of California, San Diego)

Hubert Dreyfus
University of California

Michael Drillings
U.S. Army Research Institute

Joseph S. Dumas
American Institutes for Research

Mica R. Endsley
Texas Tech University

Jon Fallesen
Army Research Institute, Field Unit

Rhona Flin
The Robert Gordon University

David Gaba
Stanford University and Palo Alto VAMC

Sallie Gordon
University of Idaho

Sherrie Gott
U.S.A.F. Armstrong Laboratory

Jeff Grossman
Naval Command, Control and Ocean Surveillance Center

Simon Henderson
UK Defence Research Agency

Robert R. Hoffman
Adelphi University

William C. Howell
American Psychological Association

Susan G. Hutchins
Naval Command, Control and Ocean Surveillance Center
(currently at the Naval Postgraduate School)

George L. Kaempf
Klein Associates Inc.
(currently at Pacific Bell)

Gary Klein
Klein Associates Inc.

Andrew R. Lacher
The MITRE Corporation

Zvi Lanir
Institute for Strategic Thought & Praxis

Raanan Lipshitz
University of Haifa

Colin F. Mackenzie
University of Maryland

Patricia A. Martin
Wright State University

Scott L. Martin
London House/SRA

Robert M. McIntyre
Old Dominion University

Michael D. McNeese
Armstrong Laboratory

Thomas E. Miller
Klein Associates Inc.

Christine M. Mitchell
Georgia Institute of Technology

Kathleen L. Mosier
San Jose State University Foundation
at NASA–Ames Research Center

Randall J. Mumaw
Westinghouse Science & Technology Center

William Nelson
Idaho National Engineering Laboratories

David Noble
Engineering Research Association

Judith Orasanu
NASA–Ames Research Center

Raphael Pascual
UK Defence Research Agency

Carolyn Prince
Naval Air Warfare Center/Training Systems Division

Emilie M. Roth
Westinghouse Science & Technology Center

William B. Rouse
Search Technology

Joan Ryder
CHI Systems, Inc.

Eduardo Salas
Naval Air Warfare Center/Training Systems Division

Penelope Sanderson
University of Illinois at Champaign-Urbana

Maj. John Schmitt
U.S. Marine Corps

Neal Schmitt
Michigan State University

Jan Maarten Schraagen
TNO Human Factors Research Institute

Daniel Serfaty
ALPHATECH, Inc.
(currently at APTIMA, Inc.)

James Shanteau
Kansas State University

David E. Smith
Naval Command, Control and Ocean Surveillance Center
(currently at University of California, San Diego)

Gerald F. Smith
University of Minnesota
(currently at the University of Northern Iowa)

Philip J. Smith
The Ohio State University

Janet A. Sniezek
University of Illinois

Alan F. Stokes
Florida Institute of Technology

Gunilla Sundstrom
GTE Laboratories, Inc.

Maj. Gen. Paul K. Van Riper
U.S. Marine Corps

Kim J. Vicente
University of Toronto

Wayne Waag
U.S.A.F. Armstrong Laboratory

Hugh Wood
National Fire Academy

David Woods
The Ohio State University

Yan Xiao
University of Toronto
(currently at University of Maryland School of Medicine)

Caroline E. Zsambok
Klein Associates Inc.

Part I

About Naturalistic
Decision Making

Chapter 1

Naturalistic Decision Making: Where Are We Now?

◆

Caroline E. Zsambok
Klein Associates Inc.

The short answer to the question of where we are now is: We are in very exciting times. If you are engaged in research or applications concerning Naturalistic Decision Making (NDM), you are already aware of this. If you are not an NDM researcher, you may not know about this emerging movement and the interest it is generating. For a start, over the last decade, informal estimates of funding from the U.S. Department of Defense for NDM research range from $25 million to $35 million. A multitude of projects have been funded in hopes that an NDM research perspective can improve decision support systems installed in military command-and-control suites in various aircraft, naval vessels, and land-based platforms. The U.S. Marine Corps began applying NDM research findings to improve its command and control procedures. On the civilian side, several leading companies in the nuclear power industry are beginning to realign their procedures, training, and decision support systems through the use of NDM research methods such as cognitive task analysis, and through the use of NDM-derived training methods. The aviation industry is sponsoring programs on situation awareness to improve decision making. And this is just a sampling. We are, indeed, in exciting times with NDM research.

A longer answer to the question of where we are now in NDM research can be found in the pages of this book. The purpose of this chapter is to frame the longer answer by highlighting major themes in this volume and identifying the value of pursuing this research. First, however, some background about NDM and a definition are in order.

BACKGROUND AND DEFINITION

The term NDM first appeared in 1989, when a conference was organized so that researchers who had stepped outside of the traditional decision research paradigms could discuss their findings, and then publish them in a single volume (Klein, Orasanu, Calderwood, & Zsambok, 1993). In the 1980s, these researchers had begun to study how experienced people actually make decisions in their natural environments or in simulations that preserve key aspects of their environments. In fact, the foregoing can serve as a short definition of NDM: NDM *is the way people use their experience to make decisions in field settings.*

A sampling of the types of decision makers studied by participants in that first NDM conference includes firefighters, pilots and cockpit crews, corporate executives, trouble shooters of electronic equipment, computer software designers, military commanders, and physicians.

What these researchers found is that the processes and strategies of "naturalistic" decision making differ from those revealed in traditional decision research. For example, it was discovered that in NDM, the focus in the decision event is more front-loaded, so that decision makers are more concerned about sizing up the situation and refreshing their situation awareness through feedback, rather than developing multiple options to compare to one another. In contrast, most traditional decision research has involved inexperienced people who are engaged in laboratory tasks where contextual or situational factors play a limited role. The traditional paradigm emphasizes understanding the back end of the decision event—choosing among options (Beach & Lipshitz, 1993).

Further, NDM researchers took issue with the fact that the traditional paradigm compares the quality of decisions against abstract, "rational" standards, such as multiattribute utility analysis. Rational standards and formal models of decision making may be appropriate for typical laboratory tasks, but they do not take into account effects of most contextual factors that accompany decision making in real-world settings, nor do they adequately model the adaptive characteristics of real-world behavior (Cohen, 1993). Indeed, many participants in the 1989 conference had embarked on NDM-like research because formal models lacked explanatory or predictive power in real-world settings they were studying, or because they discovered problems with the prescriptive advice from the traditional paradigm. As one of many examples, several researchers such as Jeff Grossman and Ed Salas had seen that training principles derived from formal models produced counterproductive behaviors in the military. A case in point is the military requirement of commanders to develop and compare multiple courses of action, even when time is precious and when a single option is an obvious workable choice, given the situation and context.

The identification of key contextual factors that affect the way real-world decision making occurs, in contrast to their counterparts in the traditional decision research paradigm, evolved as a major contribution of the 1989 NDM conference (Orasanu & Connolly, 1993). They are:

1. Ill-structured problems (not artificial, well-structured problems).
2. Uncertain, dynamic environments (not static, simulated situations).
3. Shifting, ill-defined, or competing goals (not clear and stable goals).
4. Action/feedback loops (not one-shot decisions).
5. Time stress (as opposed to ample time for tasks).
6. High stakes (not situations devoid of true consequences for the decision maker).
7. Multiple players (as opposed to individual decision making).
8. Organizational goals and norms (as opposed to decision making in a vacuum).

This array of *task and setting* factors is one of four defining markers for NDM research. The other three markers concern the *research participants* (*experienced* decision makers, not naive subjects); the *purpose* of the research (discovering how experienced people *actually make decisions* in context-rich environments, not how they ought to make decisions in approximation to a rational standard); and *locus of interest within the decision episode* (not just in the option selection process, but also in *situation awareness*).

Considering these four markers, a definition richer than the earlier short-hand version has evolved:

> The study of NDM asks how experienced people, working as individuals or groups in dynamic, uncertain, and often fast-paced environments, identify and assess their situation, make decisions and take actions whose consequences are meaningful to them and to the larger organization in which they operate.

This definition is too long to remember. The shorter version will do. But, this definition is important for two reasons. First, it is more comprehensive and carries more meaning than the short version. Second, it is stated in positive terms, not oppositional terms relative to what it is not. A large portion of the 1989 NDM conference was spent defining NDM in opposition to traditional research. This was an important step in the beginning of the NDM research endeavor. We needed to define our work against the backdrop of the then current paradigm—to say what NDM research is not, as a point of departure for stating what NDM research is.

Now, there is a clearer understanding of what defines NDM in positive terms. The value added by this position was visible at the second NDM conference in 1994. Much less of our time was spent clarifying who we are and what we do compared to the traditional standard. The majority of our time was spent on discussions of related research lines and what we can learn from them, on reports of new research findings, on applications of NDM research to a variety of domains, on NDM theory development, and on methods that can be used in practical applications and for pushing forward our research.

The value added by a positive definition of NDM is also reflected in this book, because it grew from the 1994 conference. Before turning to major themes in the book, two points should be clarified about the foregoing discussion and definition of NDM. First, the comprehensive definition emphasizes complex, uncertain, and unstable situations where decision makers cannot rely on routine action or thinking. But, as Rouse and Valusek (1993) pointed out, a great deal of real-world decision making involves "activities that are quite routine, with actions following well-worn patterns and observations agreeing with expectations" (p. 274). To be sure, some NDM researchers explicitly attempt to model both simple and complex decision making. For example, Rasmussen (1993) distinguished two levels of routine performance—the less conscious and habitual skill-based performance and the more conscious rule-based performance. As another example, Klein's RPD model describes simple, or routine, NDM as different from complex, or nonroutine, NDM (Klein, 1993).

The second point to be clarified is that NDM research does not mandate field studies as the only methodology. To the extent that laboratory studies can replicate the factors present in real-world decision making such that subjects take the tasks almost as seriously as they do in real life, lab studies can be considered as included under the umbrella of NDM research. Hammond (1993) argued that what matters is not where the research takes place, but what is studied. His point is that naturalistic approaches must provide testable theories of the environment that describe its formal properties and their consequences for cognitive activity. These approaches cannot remove the complexity and ambiguity present in the environment, or they will amount to removal of the object of study. As stated earlier, lab studies generally remove that complexity, but they need not. An excellent example from past NDM research is Hammond and Grassia's (1985) application of laboratory studies on interpersonal conflict and interpersonal learning to a naturalistic test within the public political domain. Many more examples are to be found in the current NDM research activities described by contributors to this volume.

THEMES IN NDM RESEARCH

This section provides a framework for answering in a more specific way the question of where we are now with NDM research. Unlike the structure represented by the table of contents, this is a framework composed of six themes that criss-cross chapters in this book. As such, it offers an organizing tool to understand where we are now in NDM research.

The Big Picture: Models and Theories

Both the development and testing of models and theories have occurred since the 1989 conference, although not to the degree that some would like (Howell, chapter 4, this volume). Five examples of development and testing contained in this volume are as follows.

First, Cohen, Freeman, and Thompson (chapter 25, this volume) describe a new model called the Recognition/Metacognition (R/M) model. This model describes a set of metacognitive skills that supplement recognitional processes in decision events involving novel situations. Among these skills are the identification of key situational assessments, checking the stories one constructs for completeness and consistency relative to these assessments, and generating alternative stories when too much conflicting information is encountered. The model was developed through critical incident interviews in two military domains.

Training based on the model has been conducted with command staff in the Army and Navy, and the results of the training are consistent with hypotheses derived from the model. Training in these metacognitive skills led participants to consider more factors in their situation assessment evaluations compared to nontrained controls. It also led them to consider less of the neutral (irrelevant) information and more of the conflicting information than did controls. Last, trained participants developed a better understanding of the situation than did controls, as evaluated by blind subject matter experts.

As a second example, Klein (chapter 27, this volume) reports the addition of a component to an earlier version of the Recognition-Primed Decision (RPD) model: the diagnostic function. Decision makers engage in diagnosis either for the purpose of evaluating a situation assessment about which they are uncertain, or to compare alternative explanations of events. The diagnostic function was added because of its evidence in new studies conducted by Klein and his colleagues, and because of its evidence in story building by jurors and others who try to diagnose situations by developing plausible explanations or stories (Pennington & Hastie, 1993). Klein also reports model testing activities in a study of skilled chess players. Findings were consistent with the model on three counts: Skilled players produced a high quality move as the first one considered; time pressure did not cripple their performance; skilled players could adopt a course of action (i.e., a chess move) without comparing and contrasting it to others.

Third, Endsley (chapter 26, this volume) describes a model of situation awareness (SA). The National Transportation Safety Board discovered that 88% of aircraft accidents in a 4-year period were due to human error in situation assessment. Endsley's model represents an initial attempt to provide a general framework for understanding processes and factors that impact SA. The model describes three levels of SA: perception, comprehension, and prediction. It also contains mechanisms for goal selection, attention to critical cues and expectancies, and action. This model is similar to other models of human performance and decision making, but it emphasizes the role of SA within the decision event.

A fourth example of model/theory development and testing concerns the work of Serfaty, MacMillan, Entin, & Entin (chapter 23, this volume). They developed a mental-model theory and framework for studying expertise in battle-command decision making. Interestingly, this framework includes among other sources Duncker's (1945) often-ignored decision theory (see Hoffman, 1994), which is remarkably compatible with the NDM perspective yet predates the NDM movement by more than 50 years.

Serfaty et al. derived a number of hypotheses from their model; these hypotheses concern the difference between experts and novices in the quality of their mental models. Serfaty et al. found that participants with higher expertise levels generated more detailed courses of action, with more contingencies, than those with lower expertise levels. Similarly, high-expertise participants were more able to focus immediately on critical unknowns and to ask diagnostic questions. The hypothesis that experts build more complex mental models was supported by five behavioral measures, including their ability to "see" what could go wrong with their plans. This finding is consistent with other NDM models described earlier.

Finally, Lipshitz and Ben Shaul (chapter 28, this volume) made a significant contribution to model and theory development by critically reviewing three NDM models described in this volume (cf. Endsley; Klein; Serfaty et al.) and incorporating relevant concepts from Neisser (1976) and Rouse and Morris (1986). Lipshitz and Ben Shaul compared findings from studies with more and less experienced sea combat personnel to predictions expected from these models. To account for these and other expert/novice differences available in the literature, they concluded that a comprehensive theory of NDM must include constructs that reference both memorial structures ("schemata") and current representations ("mental models," also called "situation assessments" by others).

These authors have developed a comprehensive model of NDM called "recognition-primed decision making as schemata-driven mental modeling." Because the model specifies the nature of influence on one another from each of five elements in the decision episode, it is possible to generate hypotheses that can be tested. This is significant because lack of testable NDM theories is a concern that has been voiced by NDM researchers. The value added by theory/model development and testing is the same as in other lines of research—this is how we can advance and refine our understanding of phenomena under scrutiny. However, theory development is of particular importance in NDM research because so much of our work is funded for applications, or to solve specific problems, rather than for basic research or comprehensive theory development. At both the 1989 and the 1994 NDM conferences, researchers raised a concern that our theoretical progress is slower than we would like, but that it *is* taking place. There is a general agreement that we should try to draw on applied project funding to cover theoretical testing whenever possible. Examples can be found in many of the contributions to this volume.

Related Lines of Research

A second theme in NDM research that cuts across the chapters that follow is that we take an interdisciplinary approach to our work. Several related lines of research are discussed, most notably expertise.

Expertise. The study of expertise is one of the most closely allied research lines to that of NDM. This is because the focus of NDM research has been to understand how people use their experience, or expertise, to make decisions.

Hubert Dreyfus (chapter 2, this volume) placed the central questions about expertise into historical perspective in his keynote address to the 1994 NDM conference. Twenty years ago, he was asked, "Should pilots get rule-based training or situation-based training?" In other words, is "skill" a set of rules? Dreyfus' answer to that question requires a look back into ancient Greek philosophy and up through Artificial Intelligence research. His summary of that answer (found in this volume) eventually tied together Heidegger's framework for distinguishing among types of skilled responses and the 5-stage model of skill acquisition derived by him and his brother, Stuart Dreyfus.

According to the model, novices initially make decisions and take action by learning a set of rules. They then gain competency by compiling those rules into more and more comprehensive and abstract rules that permit faster, more fluid decision making and acting. But, a major shift occurs at the highest levels of competency, or expertise. Experts do not rely on ever-more compiled rules to make decisions and take action. Instead, the cognitive underpinnings of how experts decide and act are qualitatively different from how novices decide and act. These are the cognitive underpinnings that NDM researchers are attempting to describe or to support through the decision aids and training they design.

Many of the studies reported in this volume, and its predecessor (Klein et al., 1993), used designs intended to reveal differences between experts and novices, which in turn were used to derive conclusions about decision processes, mental structures, and overt decision strategies involved in NDM.

NDM researchers are beginning to discover important differences in their findings about experts.

For example, Serfaty et al. (chapter 23, this volume) did not find evidence of experts being better than novices at producing an initial plan of action that is workable, as has been hypothesized by others (Klein, chapter 27, this volume). This may be due to the lower level of expertise represented in the people he studied compared to participants in the other studies, who had been working in their areas of expertise for many years. As a second example, Lipshitz and Shaul (chapter 28, this volume) found that experts collected more information than novices, not less, as has been found by others and as predicted by the RPD model. It is less obvious what the possible reasons are for this difference than in the one reported earlier. The important point is that NDM researchers are beginning to test boundary conditions of models and theories; follow-up research will help to clarify which factors operate as constraints, and under what conditions.

A healthy debate is emerging about the use of experts in NDM research. Are verbal protocols collected from experts relied on too heavily, given known difficulties in probing for expert knowledge? Howell (chapter 4, this volume) argues that these protocols should be seen as provisional and subject to external verification. If experts are a standard, then how should we deal with individual differences in the ways they understand situations, solve problems, and make decisions? Howell and others are calling for more research into experts' individual differences and the meaning we should make of these differences in terms of methodological implications and explanatory principles about the nature of expertise.

Elsewhere, Hoffman (1994) has described a virtual explosion of studies of experts in their operational settings by scientists in such fields as psychometrics, computer science, ergonomics, and military command and control. Hoffman worries that there is an uneven knowledge base among NDM researchers about this related research. Partly, this can be attributed to the cross-disciplinary nature of the NDM movement. But, voices of Hoffman and others that carry this message are important as the NDM movement attempts to mature into an effective contributor to decision research and applications.

Other Related Research Lines.   Although expertise maintains an especially strong relationship to NDM research, it is only one of many related fields. A variety of other research lines and perspectives are encountered throughout this volume. A sampling includes problem solving, ecological psychology, situation awareness, action-oriented thought, situated cognition, and process control. An integrated overview has been prepared by experienced NDM researchers concerning the relationship of NDM research to other research lines. See Beach, Chi, Klein, Smith, and Vicente (chapter 3, this volume) for this summary.

The Focus on Situation Awareness

A third common theme among subsequent chapters in this volume concerns situation awareness. If there is one thing that NDM research has done, it is the spotlighting of situation assessment (SA) processes as targets for decision research and decision aiding. A new model of SA was described earlier (Endsley, chapter 26, this volume). In addition, many of the contributors to this volume describe the importance of situation awareness. Following are a few examples.

Kaempf and Orasanu, in discussing research surrounding aviation, note that decisions in this domain have serious consequences, are often made under severe time constraints, and frequently involve ambiguous information. Here, the onus on decision makers is in their assessment of the situation: The course of action they select inextricably depends on the way they understand the situation. Yet, typical aviation training reinforces accepted procedures and minimizes sensitivity to complex conditions that can arise. Kaempf and Orasanu, in summarizing views of several NDM researchers in the aviation domain, conclude that situation awareness is important and needs to be supported through decision aids and training.

Smith and Marshall are attempting to build a "naturalistic decision aid" to support the processes familiar to the decision maker, many of which involve situation awareness. Typical aids target the decision outcome. These researchers are using a schema model as the central processing component of a new decision support system, in hopes that the system will "think" more like naturalistic decision makers.

Waag and Bell describe one of their studies that is a part of a larger effort funded by Armstrong Research Laboratory to better understand SA in fighter pilots and to design better training programs. Hutchins describes a study she performed that revealed a low percentage of matches in SA between commanders and tactical action officers and between commanders and decision aids that were designed to remedy the mismatch problem. Roth shows through illustrative examples how situation awareness (and response planning) continue to be important elements in NDM, even in cases where detailed procedures are provided, as in the nuclear power industry.

There are other examples of studies or theory development about SA from contributors of this volume. Howell raises the concern that we may have gone too far—that we are too focused on SA to the exclusion of other aspects of decision making that require study. Others disagree, believing that if NDM researchers are not the ones to focus on SA, who will be? And, they note that NDM researchers are pursuing other NDM processes, such as the research of Xiao, Milgram, and Doyle concerning preparatory actions of decision makers, Klein's research into option generation, and Cohen's research into processes involved in story generation and metacognition such as critiquing, correcting, and the "quick test."

Methodological Rigor and Advancements

One of the chief concerns raised at the 1989 NDM conference was methodological rigor in NDM research. This continues to be a topic of discussion and is a fourth theme found in this volume (see especially the chapters in Part II). There are numerous studies reported by contributors to this volume that demonstrate methodological rigor. For example, DiBello describes one of her studies in a larger program to understand the cognitive impact of technology introduced into the workplace. In this study, she trained transit workers to understand how the Material Requirements Planning (MRP) system in a remanufacturing facility "thinks," so that zero inventory and just-in-time production is supported. Through in-depth interviews that used particular cognitive probes, and through a meticulous data-analysis process, she was able to identify participants' grasp of MRP principles before and after the training. Her methods allowed her to understand how learning took place; namely, that through her "constructive" training, knowledge was reorganized rather than added to. This finding is precisely what Dreyfus meant when he described the cognitive underpinnings of expertise as qualitatively different than an accrual of ever-more-compiled rules.

As another example, Orasanu and Fischer sifted through mounds of preexisting data, using both ethnographic and analytic methods, and found six types of decisions made by cockpit crews. They also identified strategies for each decision type, and found that these strategies distinguished better from worse crews. For example, better crews monitor for more relevant cues, use more information during decision making, and do not overestimate resources. For a summary of a variety of methods available to analyze observational data, see Roth (chapter 12, this volume).

We have also produced advancements in our methods. Following are three examples discussed in this volume. First, Lipshitz's critical analysis of the panel discussion on errors clarifies a central question intimately tied to much of our methodology: What is an error? The reason this topic is so important is that the way you attempt to decrease errors via decision aids is tied to the way you conceptualize errors in the first place. If you assume an error is a decision that fails to meet a priori standards, then you design decision aids that help humans meet those standards—and you use methods designed to trap those kinds of errors. These aids run the risk of being underutilized because they generally do not match the way decision makers operate in real work settings (see also Mosier, chapter 30, this volume). If you assume that an error cannot be identified without knowing the user's (or the operator's) goals as well as the common practices and cognitions of experts, then you can design aids that will support natural means of perceiving, understanding, and acting in an operational environment.

In his chapter, Lipshitz identifies three perspectives on errors that are consistent with the NDM approach: the cognitive, interpretive, and adaptive/systemic perspectives. He also discusses methodological approaches that correspond to these perspectives. Lipshitz contrasts each perspective about what constitutes an error to a traditional standard, behavioral decision theory (BDT). His conclusion is that NDM offers a more sophisticated answer to questions of errors than does BDT. The NDM perspectives have different and more reasonable standards for identifying errors than does BDT. As such, it stands a better chance at developing useful decision aids.

A second example of methodological advancement appears in the chapter by Serfaty and his colleagues. This is a method to reliably assess battle command expertise even when there is not a universal definition of this expertise. This has particular significance to NDM researchers, because our work so heavily involves the use of experts.

A third example concerns a family of methods, collectively called *Cognitive Task Analysis* (CTA). This topic receives particularly strong coverage in this volume because it is so widely used in NDM applications. Roth (chapter 12, this volume) goes so far as to state that CTA is a prerequisite to decision aid design. Schraagen (chapter 22, this volume) offers a particularly strong case for CTA in his discussion of a successfully implemented decision support system for the Royal Netherlands Navy that was based on results of a cognitive task analysis.

Because of its wide use, one entire panel session at the 1994 NDM conference was devoted to discussing CTA. Gordon and Gill's summary of that session states that CTA is a "natural" for NDM research, because the purpose of a *cognitive* task analysis (as opposed to *behavioral* task analysis) is to understand the decision requirements that lie behind an experienced person's job or task performance. Most CTA methods involve lengthy data collection and analysis procedures. Gordon and Gill discuss the pressing need to develop more streamlined, and less time-consuming, CTA methods. Recently, the Navy Personnel Research and Development Center funded a research project headed by Militello and Klein to develop streamlined alternatives and to validate their usefulness. We look forward to publication of those results.

Applications of NDM Research

Another prevalent theme in this book concerns the ways NDM research and theory are applied in particular domains or in technologies that cross domains. Domains include healthcare (Bogner), command-and-control environments (Drillings & Serfaty; Pascual & Henderson), aviation (Kaempf & Orasanu; Orasanu & Fischer; Stokes, Kemper, & Kite; Waag & Bell), and business and industry (Schmitt; G. Smith). Technologies include system design, training, cognitive task analysis, and process control, each of which is discussed throughout the chapters in this book. Concerning system design, one particularly innovative application of NDM theory to the reuse of software design is described by Mitchell, Morris, Ockerman, and Potter. Training receives focused attention in Cannon-Bowers and Bell, and in Cohen, Freeman, and Thompson, regarding individual decision makers. The training of teams is the focus of both Zsambok and Salas, Cannon-Bowers, and Johnston. A summary of applications discussed in some of these chapters and from other sources is offered in Klein's introductory chapter to Part II.

The value added by NDM research is perhaps most clearly displayed in these chapters. Most of the impetus behind the NDM research endeavor has been to address real-world problems. Whatever our successes have been, they are most notably captured in our applications.

What We Need

A final theme that cuts across chapters in this book is identification of what we need most in NDM research and applications. A sampling of these needs follows. Gerald Smith discusses the need for a problem-centered approach, like that used in research on expertise, as a way to characterize real-world problems. Roth and others discuss the need for ecological interfaces in decision aids—aids that are more supportive of human decision processes than of prescribed decision strategies. Howell calls for a functional taxonomy of tasks, more theory building, and long-term evaluation of applications. Zsambok and Schmitt echo the latter, and emphasize the need for more funding to accomplish this level of evaluation. Mosier, Roth, and DiBello all highlight the need to design training that teaches users how decision supports systems think. Gordon and Gill, as well as Roth, want to see new CTA methods that are less time-intensive than those in current use. And, Miller and Woods offer a comprehensive list of needs and difficulties faced by system designers who are attempting to bring an NDM approach to their work.

FINAL REMARKS

The long answer to where we are now in NDM lies in the pages of this book. The themes identified earlier provide a framework for discovering the answer, and the examples anticipate the content of the answer. However, other contributors to this book discuss equally important issues—issues that do not happen to fall neatly within this framework. For example, Bogner and Schmitt both discuss the inadequacy of previous descriptions of contextual factors common to NDM research settings. Bogner adds important insights about the impact of especially high stakes when patients must make their own life-and-death decisions. Schmitt discusses the effects of diffused decision making and extremely long action/feedback loops on organizational decision making.

One additional element to the answer of where we are now in NDM research concerns our ability to communicate our insights, questions, and research findings. Journal articles are one way. However, journals dedicated to decision research seem to have been reluctant to publish findings based on the types of studies typically funded by sponsors of NDM research. Conferences and resultant books are another way, and we have been fortunate in securing funding for two such conference-book cycles in the recent past. But, conferences and books are long in the making, and future funding is never assured.

Fortunately, another avenue has opened up. The Human Factors and Ergonomics Society (HFES) has formed a new technical group, called the Cognitive Engineering and Decision Making Technical Group. The group sponsors sessions about NDM-related topics at the annual HFES conference, and produces a newsletter. A web page was also recently established, and an electonic listserver is currently in operation.

We look forward to the by-products of this development: ease of access by those new to the NDM endeavor; identification of research topics; development of dissertation topics (a crucial step for assuring continuing growth in NDM research); more timely communication among NDM researchers; and the evolution of an organized body to shape the lessons learned—a body whose strength in numbers can influence research agendas and funding priorities of the future.

REFERENCES

Beach, L. R., & Lipshitz, R. (1993). Why classical theory is an inappropriate standard for evaluating and aiding most human decision making. In G. A. Klein, J. Orasanu, R. Calderwood, & C. E. Zsambok (Eds.), *Decision making in action: Models and methods* (pp. 21–35). Norwood, NJ: Ablex.

Duncker, K. (1945). On problem solving. *Psychological Monographs, 58* (5, Whole No. 270).

Cohen, M. S. (1993). The naturalistic basis of decision biases. In G. A. Klein, J. Orasanu, R. Calderwood, & C. E. Zsambok (Eds.), *Decision making in action: Models and methods* (pp. 51–99). Norwood, NJ: Ablex.

Hammond, K. R. (1993). Naturalistic decision making from a Brunswikian viewpoint: Its past, present, future. In G. A. Klein, J. Orasanu, R. Calderwood, & C. E. Zsambok (Eds.), *Decision making in action: Models and methods* (pp. 205– 227). Norwood, NJ: Ablex.

Hammond, K. R., & Grassia, J. (1985). The cognitive side of conflict: From theory to resolution of policy disputes. In S. Oskamp (Ed.), *Applied social psychology annual: Vol. 6. International conflict and national public policy issues* (pp. 233–254). Beverly Hills, CA: Sage.

Hoffman, R. R. (1994). *A review of "naturalistic decision-making" research on the critical decision method of knowledge elicitation and the recognition priming model of decision-making, with a focus on implications for military proficiency* (Working report). Garden City, NY: Adelphi University.

Klein, G. A. (1993). A recognition-primed decision (RPD) model of rapid decision making. In G. A. Klein, J. Orasanu, R. Calderwood, & C. E. Zsambok (Eds.), *Decision making in action: Models and methods* (pp. 138–147). Norwood, NJ: Ablex.

Klein, G. A., Orasanu, J., Calderwood, R., & Zsambok, C. E. (1993). *Decision making in action: Models and methods.* Norwood, NJ: Ablex.

Neisser, U. (1976). *Cognition and reality.* San Francisco: Freeman.

Orasanu, J., & Connolly, T. (1993). The reinvention of decision making. In G. Klein, J. Orasanu, R. Calderwood, & C. E. Zsambok (Eds.), *Decision making in action: Models and methods.* (pp. 3–20). Norwood, NJ: Ablex.

Pennington, N., & Hastie, R. (1993). A theory of explanation-based decision making. In G. Klein, J. Orasanu, R. Calderwood, & C. E. Zsambok (Eds.), *Decision making in action: Models and methods* (pp. 188–201). Norwood, NJ: Ablex.

Rasmussen, J. (1993). Deciding and doing: Decision making in natural contexts. In. G. Klein, J. Orasanu, R. Calderwood, & C. E. Zsambok (Eds.), *Decision making in action: Models and methods* (pp. 158–171). Norwood, NJ: Ablex.

Rouse, W. B., & Morris, N. M. (1986). On looking into the black box: Prospects and limits on the search for mental models. *Psychological Bulletin, 100*(3), 349–363.

Rouse, W. B., & Valusek, J. (1993). Evolutionary design of systems to support decision making. In G. A. Klein, J. Orasanu, R. Calderwood, & C. E. Zsambok (Eds.), *Decision making in action: Models and methods* (pp. 270–286). Norwood, NJ: Ablex.

Chapter 2

Intuitive, Deliberative, and Calculative Models of Expert Performance

◆

Hubert L. Dreyfus
University of California

I would not be here today talking about naturalistic decision making and models of expertise if it were not for Gary Klein. Not just for the obvious reason that he invited me to give this talk, but rather because he got me involved in thinking about expertise in the first place. About 20 years ago Gary was called for advice by an Air Force Captain named Jack Thorpe. Thorpe was a psychologist battling his superiors over whether pilots should be trained to cope with emergencies by memorizing a set of rules as they had always done, or be given what he called *situational* emergency training. Gary suggested Thorpe read my recently published book, *What Computers Can't Do*, and I soon received a call from the Air Force saying they wanted to sponsor me to do research on skill acquisition. Because I had no knowledge of pilots or of skill acquisition I suggested we involve my brother, Stuart, who taught Operations Research at Berkeley and who at least had begun to question the adequacy of formal models of decision making.

Captain Thorpe agreed to a meeting. Thorpe suspected that skills were learned by responding to many situations, but the older officers in charge of emergency training thought that all which was needed for emergency coping was to write out the appropriate rules in bold-faced type and require that pilots memorize them. The problem for Jack Thorpe was that the psychology books on skill, influenced by the dominant information-processing model of the mind, took skill to be the result of compiled rules and so supported his superiors. Stuart and I agreed to think about the problem.

I went back to my philosophy texts to study what the great thinkers of the past said about skills and discovered that they too supported the bold-face approach. Plato (trans. 1948) described one of Socrates' earliest dialogues aimed at understanding moral experts' skills. Socrates asked Euthyphro, a religious prophet, to tell him how to recognize piety: "I want to know what is characteristic of piety ... to use a standard whereby to judge your actions and those of other men" (p. 7). Instead of revealing his piety-recognizing heuristic, however, Euthyphro gave him *examples* from his field of expertise. Socrates ran into the same problem with craftsmen, poets and, to his horror, even statesmen. None could articulate the rules underlying his expertise. Socrates therefore concluded that none of these experts knew anything and he did not know anything either. Not a promising start for Western philosophy.

Fortunately, Plato admired Socrates and came to his defense. Experts had, Plato said, learned the principles that generated expert behavior in another life, but they had forgotten them. Plato argued that in domains such as mathematics and ethics, where we have a priori knowledge, we must assume that the experts apply explicit, context-free rules or theories they somehow learned before they entered the everyday world. Once learned, such theories control the thinker's mind whether he is conscious of them or not. Thus, thanks to Plato, that branch of the philosophical tradition that descends from Socrates, to Descartes, to Leibniz, to Kant, to Piaget, to AI takes it for granted that understanding a domain consists in having a *theory* of that domain. A theory formulates the relationships between objective, *context-free* elements (simples, primitives, features, attributes, factors, data points, cues, etc.) in terms of abstract principles (covering laws, rules, programs, etc.).

Plato's account did not apply to everyday skills but only to domains in which there is a priori knowledge. However, at the beginning of modernity the success of theory in the natural sciences suggested that in *any* orderly domain there must be some set of context-free elements and some abstract relations between those elements which accounts for the order of that domain and that the interiorization of this theory accounts for man's ability to act intelligently in it.

Building on Plato, Descartes assumed that all understanding consisted in forming and manipulating appropriate mental representations, that these representations could be analyzed into primitives, and that all phenomena could be understood as complex combinations of these primitive elements. Moreover, at about the same time—around 1600—Hobbes (1958) assumed that the elements were formal elements related by purely syntactic operations, so that reasoning could be reduced to calculation. "When a man *reasons*, he does nothing else but conceive a sum total from addition of parcels," Hobbes wrote, "for REASON ... is nothing but reckoning" (p. 45). Leibniz (1951), working out the classical idea of mathesis—the formalization of everything—sought a grant to develop a universal symbol system, so that "we can assign to every object its determined characteristic number" (p. 18). According to Leibniz, in understanding we analyze concepts into more simple elements. Leibniz envisaged "a kind of alphabet of human thoughts" (p. 20) whose "characters must show, when they are used in demonstrations, some kind of connection, grouping, and order which are also found in the objects" (p. 10). Finally Leibniz boldly generalized the rationalist account of understanding to all forms of intelligent activity, even everyday practice.

So I found that traditional philosophers would not help Jack Thorpe, but Stuart and I decided to carry on anyway. Our hope now lay in Stuart, because he had never read a word of philosophy and was as uncorrupted by Plato as one could hope to be in our rationalistic culture. However, there was a further problem; we knew nothing about planes or pilots. We solved that one by thinking about cars and drivers. Doing the phenomenology of his own driving skill, Stuart gradually arrived at the five-stage model of skill acquisition, which is reviewed briefly here.

STAGE 1: NOVICE

Normally, the instruction process begins with the instructor decomposing the task environment into context-free features that the beginner can recognize without benefit of experience in the task domain. The beginner is then given rules for determining actions on the basis of these features, like a computer following a program.

For purposes of illustration, let us consider two variations: a bodily or motor skill and an intellectual skill. Student automobile drivers learn to recognize such interpretation-free features as speed (indicated by the speedometer) and rules such as "shift to second when the speedometer needle points to 10 miles an hour."

Novice chess players learn a numerical value for each type of piece regardless of its position, and a rule: "Always exchange if the total value of pieces captured exceeds the value of pieces lost." They also learn that when no advantageous exchanges can be found, center control should be sought, and they are given a rule defining center squares and one for calculating extent of control. Most beginners are notoriously slow players, as they attempt to remember all these rules and their priorities.

STAGE 2: ADVANCED BEGINNER

As novices gain experience actually coping with real situations, they begin to note, or an instructor points out, perspicuous examples of meaningful additional aspects of the situation. After seeing a sufficient number of examples, students learn to recognize them. Instructional maxims now can refer to these new *situational aspects*, recognized on the basis of experience, as well as to the objectively defined *nonsituational features* recognizable by the novice. Advanced beginner drivers use (situational) engine sounds as well as (nonsituational) speed in their gear-shifting rules. They shift down when the motor sounds like it is straining.

With experience, chess beginners learn to recognize overextended positions and how to avoid them. Similarly, they begin to recognize such situational aspects of positions as a weakened king's side or a strong pawn structure despite the lack of precise and universally valid definitional rules.

STAGE 3: COMPETENCE

With more experience, the number of potentially relevant elements of a real-world situation that the learner is able to recognize becomes overwhelming. At this point, because a sense of what is important in any particular situation is missing, performance becomes nerve-wracking and exhausting, and the student might well wonder how anybody ever masters the skill.

To cope with this problem and to achieve competence, people learn, through instruction or experience, to adopt a hierarchical perspective. First they must devise a plan, or choose a perspective, that then determines which elements of the situation are to be treated as important and which ones can be ignored. Thus, restricting themselves to only a few of the vast number of possibly relevant features and aspects, decision making becomes easier.

The competent performer thus seeks new rules and reasoning procedures to decide on a plan or perspective. However, these rules are not as easily come by as the rules given beginners in texts and lectures. The problem is that there are a vast number of different situations that the learner may encounter, many differing from each other in subtle ways. There are, in fact, more situations than can be named or precisely defined so no one can prepare for the learner a list of what to do in each possible situation. Competent performers, therefore, have to decide for themselves what plan to choose without being sure that it will be appropriate in the particular situation. Now, coping becomes frightening rather than exhausting, and the learner feels great responsibility for his or her actions. Prior to this stage, if the learned rules did not work out, the performer could rationalize that he or she had not been given good enough rules. Now he or she feels remorse if things do not work out. Of course, often at this stage, things do work out well, and the performer experiences elation.

A competent driver leaving the freeway on a curved off ramp, after taking into account speed, surface condition, criticality of time, an so forth, may decide the car is going too fast. The driver then has to decide whether to let up on the accelerator, or step on the brake. The driver is happy to get through the curve without mishap and is shaken if the car begins to go into a skid.

The class A chess player, here classed as competent, may decide after studying a position that his or her opponent has weakened king's defenses so that an attack against the king is a viable goal. If the attack is chosen, chess players ignore features involving weaknesses in their own position created by the attack. Successful plans induce euphoria, whereas mistakes are felt in the pit of the stomach.

As the competent performer becomes more and more emotionally involved in his or her tasks, it becomes increasingly difficult to draw back and to adopt the *detached* rule-following stance of the beginner. Although it might seem that this involvement would interfere with detached rule testing and so would inhibit further skill development, in fact just the opposite seems to be the case. If the detached rule-following stance of the novice and advanced beginner is replaced by involvement, one is set for further advancement, whereas resistance to the frightening acceptance of risk and responsibility can lead to stagnation and ultimately to boredom and regression.

STAGE 4: PROFICIENT

Proficiency seems to develop if, and only if, experience is assimilated in this theoretical way and the performer's theory of the skill, as represented by rules and principles, is gradually replaced by situational discriminations accompanied by associated responses. Thus intuitive behavior gradually replaces reasoned responses. As the brain of the performer acquires the ability to discriminate between a variety of situations entered into with concern and involvement, plans are intuitively evoked and certain aspects stand out as important without the learner standing back and choosing those plans or deciding to adopt that perspective. Action becomes easier and less stressful as the learner simply sees what needs to be achieved rather than deciding, by a calculative procedure, which of several possible alternatives should be selected. There is less doubt that what one is trying to accomplish is appropriate when the goal is simply obvious rather than the winner of a complex competition. In fact, at the moment of involved intuitive response there can be no doubt, because doubt comes only with detached evaluation of performance.

Remember that the involved, experienced performer sees goals and salient facts, but not what to do to achieve these goals. This is inevitable because there are far fewer ways of seeing what is going on than there are ways of responding. Proficient performers simply have not yet had enough experience with the wide variety of possible responses to each of the situations they can now discriminate to have rendered the best response automatic. For this reason, proficient performers, seeing the goal and the important features of the situation, must still *decide* what to do. To decide, they fall back on detached, rule-based determination of actions.

The proficient driver, approaching a curve on a rainy day, may realize intuitively that the car is going dangerously fast. The driver then consciously decides whether to apply the brakes or merely to reduce pressure by some selected amount on the accelerator. Valuable moments may be lost while a decision is consciously chosen, or time pressure may lead to a less that optimal choice. However, this driver is certainly more likely to negotiate the curve safely than the competent driver who spends additional time *deciding*, based on speed, angle of curvature, and felt gravitational forces, that the car's speed is excessive.

Proficient chess players, who are classed as masters, can recognize a large repertoire of types of positions. Recognizing almost immediately and without conscious effort the sense of a position, they set about calculating the move that best achieves the goal. They may, for example, know that they should attack, but they must deliberate about how best to do so.

STAGE 5: EXPERTISE

The proficient performer, immersed in the world of his skillful activity, *sees* what needs to be done, but *decides* how to do it. The expert not only knows what needs to be achieved, based on mature and practiced situational discrimination, but also knows how to achieve the goal. A more subtle and refined discrimination ability is what distinguishes the expert from the proficient performer. The expert distinguishes, among situations all seen as similar with respect to plan or perspective, those requiring one action from those demanding another. With enough experience with a variety of situations, all seen from the same perspective but requiring different tactical decisions, the proficient performer gradually decomposes this class of situations into subclasses, each of which share the same decision, single action, or tactic. This allows the immediate intuitive response to each situation which is characteristic of expertise.

The expert chess player, classed as an international grandmaster, in most situations experiences a compelling sense of the issue and the best move. Excellent chess players can play at the rate of 5 to 10 seconds a move and even faster without any serious degradation in performance. At this speed they must depend almost entirely on intuition and hardly at all on analysis and comparison of alternatives. For such expert performance, the number of classes of discriminable situations, built up on the basis of experience, is immense: A master chess player can distinguish roughly 50,000 types of positions.

Automobile driving probably involves the ability to discriminate a similar number of typical situations. The expert driver, generally without any awareness, not only knows by feel and familiarity when slowing down on an off ramp is required; the foot performs the appropriate action without the driver having to calculate and compare alternatives. It seems that a beginner makes inferences using rules and facts just like a heuristically programmed computer, but that with talent and a great deal of involved experience the beginner develops into an expert who intuitively sees what to do without applying rules.

As you all know, everyday, skilled decision making has been systematically overlooked in laboratory studies that study decision making outside natural context in which the decision maker has experience-based expertise. Such studies force the subject to behave in nonskillful ways and so enforce the traditional account of deliberation as applying rules to situations defined in terms of context-free features.

Once Stuart had worked out the five stages using his driving skills as his example, we just changed car to plane and driver to pilot and wrote a report for the Air Force. It was called "The Psychic Boom: Flying Beyond the Thought Barrier." To our amazement Captain Thorpe was delighted, and we realized that our skill model was generalizable. So we wrote a book entitled *Mind Over Machine* (Dreyfus & Dreyfus, 1986), applying our model to issues in management, education, and the law.

Our timing could not have been better. Edward Feigenbaum had just published his book *The Fifth Generation: Artificial Intelligence and Japan's Computer Challenge to the World* (Feigenbaum & McCorduck, 1983), and expert systems were all the rage. Feigenbaum claimed that, thanks to expert systems, we would soon have "access to machine intelligence—faster, deeper, and better than human intelligence" (p. 236). Amazed that he and his AI colleagues could have such confidence, I was led to return again to philosophy and to examine the assumptions underlying this incredible faith.

I found that the argument that AI could be based on the use of rules and features had first been formulated by Allen Newell and Herbert Simon in 1957 (Newell & Simon, 1957/1990). They realized that the strings of bits manipulated by a digital computer could be used as symbols to stand for anything—numbers, of course, but also features of the real world. Moreover, programs could be used as rules to represent relations between these symbols, so that the system could infer further facts about the represented objects and their relations.

This way of looking at computers became the basis of a way of looking at minds. Newell and Simon hypothesized that the human brain and the digital computer, although totally different in structure and mechanism, had, at the appropriate level of abstraction, a common functional description. At this level, both the human brain and the appropriately programmed digital computer could be seen as two different instantiations of a single species of device—one that generated intelligent behavior by manipulating symbols by means of formal rules.

Again philosophy had paved the way for the seeming self-evidence of this representational view of mind. In 1922 Ludwig Wittgenstein had published his *Tractatus Logico-Philosophicus* (1922/1981) working out in details Hobbes' syntactic, representational view of the relation of the mind to reality. He defined the world as the totality of logically independent atomic facts and argued that these atomic facts and their logical relations were represented in the mind. AI can be thought of as the attempt to find Wittgenstein's atomic facts and logical relations in the subject (man or computer) that mirror the primitive objects and their relations which make up the world. Newell and Simon's physical symbol system hypothesis in effect turns the Wittgensteinian vision—which is itself the culmination of the classical atomist-rationalist philosophical tradition—into an empirical claim, and bases a research program on it.

It is one of the ironies of intellectual history that Wittgenstein's devastating attack on his own *Tractatus*, his *Philosophical Investigations*, was published in 1953, just as AI took over the atomistic tradition he was attacking. After writing the *Tractatus*, Wittgenstein (1975) spent years doing what he called "phenomenology"—looking in vain for the atomic facts his theory required. He ended by abandoning his *Tractatus* and all atomistic, rationalistic philosophy. He argued that the analysis of everyday situations into facts and rules is itself only meaningful in some context and for some purpose. Thus the elements chosen already reflect the goals and purposes for which they are carved out. When we try to find ultimate context-free, purpose-free features, as we must if we are going to find what corresponds to the primitive symbols in a computer, we are in effect trying to free aspects of our experience of just that pragmatic organization which makes it possible to use them intelligibly in coping with everyday problems.

Why were the AI researchers and expert system builders so confident, then? Their argument always came back to the claim that the brain must be operating like a rule-following physical symbol system because no one had any idea of any other way of producing intelligent responses. Newell and Simon (1990) admitted that "The principal body of evidence for the symbol-system hypothesis...is negative evidence: the absence of specific competing hypotheses as to how intelligent activity might be accomplished—whether by man or by machine" (p. 118).

Feigenbaum (cited in Leibniz, 1951) likewise assumed the "What else could it be" argument when he said, "When we learned how to tie our shoes, we had to think very hard about the steps involved…. Now that we've tied many shoes over our lifetime, that knowledge is 'compiled,' to use the computing term for it: it no longer needs our conscious attention." (p. 48) On this Platonic view, the rules must be functioning in the expert's mind whether he is conscious of them or not. How else could one account for the fact that the expert can perform the task? After all, we *can* still tie our shoes, even though we no longer can say how we do it. So the rules must have been compiled. However, the claim that the expert is using compiled rules is like claiming that because as children we once needed training wheels to ride bicycles, as accomplished bicyclists we must be using *invisible* training wheels.

The phenomenon of skill acquisition had shown Stuart and me that the expert does not seem to compile the rules he or she uses as a beginner. Rather, the expert seemed to put them aside just like training wheels. Our view was that, when badgered by a Socrates or an expert systems builder, experts can remember the rules they once used as graduate students, but these rules no more make possible expert performance than would retrieving the training wheels from the attic make for expert bike riding.

Happily, there are now models of computer learning using neural networks that show that the brain may well switch from the symbolic, rule-following mode of representation to a more holistic nonsymbolic mode as the learner moves from novice to expert, just as the phenomenology suggests.

If multilayered networks succeed in fulfilling their promise, researchers will have to give up the conviction held from Hobbes to early Wittgenstein that the only way to produce intelligent behavior is to mirror the world with a formal theory in the mind. Indeed, they may have to give up the more basic intuition at the source of philosophy that there must be a theory of every aspect of reality; that is, there must be elements and principles in terms of which one can account for the intelligibility of any domain. Neural networks may show that Wittgenstein was right in thinking that we can behave intelligently in the world without having a theory of that world, and that a theory of everyday life may well be unobtainable.

Once one has abandoned the philosophical approach of classical AI and accepted the atheoretical claim of neural net modeling, one question remains: How much of everyday intelligence can such a network be expected to capture? Network modelers point out that their networks are good at learning and pattern recognition. Classical AI researchers respond that networks have so far had difficulty dealing with step-wise problem solving. One might think, using the metaphor of the right and left brain, that perhaps the brain uses each strategy when it is appropriate. The problem would then be how to combine them. One cannot just switch back and forth, for, as the gestaltists saw, the pragmatic background plays a crucial role in determining relevance even in everyday logic and problem solving, for experts in any field, even logic, grasp operations in terms of their functional similarities.

It is even premature to consider combining the two approaches, because so far neither has accomplished enough to be on solid ground. Neural network modeling may simply be getting a deserved chance to fail as did the symbolic approach. Network researchers have yet to solve the problem of appropriate generalization. Neural net modelers agree that for a net to be intelligent it must be able to cluster; that is, given sufficient examples of inputs associated with one particular output, it should associate further inputs of the same type with that same output. Nets should also generalize. That is, after learning from experience successful responses to various situations, a new, similar but not identical, situation should evoke a new, similar but not identical, associated and successful response. The question arises, however: What counts as the similar? The designer of the net has a specific definition in mind and counts it a success if the net generalizes to other instances of this type. When the net produces an unexpected association, however, can one say it has failed to generalize? One could equally well say that the net has all along been acting on a different definition of the type in question and that the difference has just been revealed.

A net must share size, architecture, initial connections, configuration, and socialization with the human brain if it is to share our sense of appropriate similarity. If it is to learn from its own "experiences" to make associations that are human-like rather than be taught to make associations which have been specified by its trainer, it must also share our sense of appropriateness of outputs, and this means it must share our needs, desires, and emotions and have a human-like body with the same physical movements, abilities, and possible injuries. Neural networks have none of these capacities. Nevertheless, such networks show something very important—our phenomenology of expertise as an intuitive response that does not require compiled rules can be implemented in a brain. We now know what else the causal basis of intelligence could be.

However, if all that we can say about expertise is that it is a direct, intuitive response to situations, and if the brain's underpinning is not the implementation of a theory, but is just a lot of unanalyzable synaptic connections, what is left for philosophers and psychologists to study? It is as if where there once were three possible levels of investigation, there would now be only two. The three were: (a) the phenomenological level investigated by phenomenologists and gestalt psychologists; (b) the information-processing level investigated by rationalist philosophers, cognitive psychologists, and decision theorists; and (c) the brain level studied by neuroscience. Our model of expertise and the possibility of neural networks working seems to have wiped out the second level, but there must be more to the story.

Fortunately, there is. The philosopher Martin Heidegger provides us with a way to see that there are several kinds of skilled response to a situation each with its own phenomenology and its own appropriate mode. Heidegger (1927/1962) distinguished three kinds of coping activities. In our normal everyday coping we deal with *ready-to-hand* equipment without any thought at all. This skillful dealing may be general and routine, but, as our skill model makes clear, it can be as specific and subtle as the response of a chess grandmaster to a complex chess position. This is the kind of expert behavior Stuart and I have been studying.

At the other extreme are the sorts of decisions modeled by decision theory or implemented in expert systems. These systems as we have already noted operate with context-free features and formal rules. Heidegger called the way of being of these elements *present-at-hand*. The things in a person's world approach in present-at-hand situations are so unfamiliar that no experience is relevant. We have seen that the novice experiences the world this way when learning a new skill in a completely strange domain. Of course, we are never confronted with *totally* strange domains, but we might suppose that astronauts have to begin their satellite repairing skills in something close to this context-free and rule-like way. Decisions that have no precedents present the same sort of problem. For example, one might have had to resort to decision analysis in deciding where to place the first nuclear power plant. In addition, changing political, technological, and economic contexts make precedents unreliable, so decision analysis is still going to be helpful.

However, these cases of involved transparent coping and detached logical analysis are the extremes. Heidegger noted that things do not always go smoothly, but the only alternative to intuitive response is not detached analysis. Rather, there is a middle kind of coping. For example, when a piece of equipment is missing or when the situation is otherwise abnormal we have to stop and think. Heidegger said we have to deal with the *unready-to-hand*. We then act deliberately rather than responding intuitively, but we do not resort to abstract, rational deliberation. Unlike the case of dealing with the present-at-hand, such deliberate action takes place in a familiar context. Usually when experts have to make such decisions they are in a situation in which they have already had a great deal of experience. The expert, however, is not able to react intuitively, either because the situation is in some way unusual, or because of the great risk and responsibility involved, or because he has to be able to justify his decision to his clients or superiors.

This, I take it, is the kind of expert decision making studied under the title *naturalistic decision making* (NDM). I understand that researchers looking at actual cases of decision making have found that in dealing with difficult, real-life situations in a domain in which they are already experts, the experts do not generate multiple courses of action, weigh the pros and cons of each, and then select the best line of action according to some evaluation function. Rather, the experts draw on their context-based, intuitive understanding, but check and refine it to deal with the problematic situation.

Thus, whereas most expert performance is ongoing and nonreflective, the best of experts, when time permits, think before they act. Normally, however, they do not think about their rules for choosing goals or their reasons for choosing possible actions, because if they did, they would regress to the competent level. Rather, they reflect on the goal or perspective that seems evident to them and on the action that seems appropriate to achieving that goal. We investigated this issue a bit during our Air Force research and touched on it in our book *Mind Over Machine* (1986). We call the kind of inferential reasoning exhibited by the novice, advanced beginner, and competent performer as they apply and improve their theories and rules "calculative rationality," and what experts exhibit when they have time "deliberative rationality." Deliberative rationality is detached, reasoned observation of one's intuitive, practice-based behavior with an eye to challenging and perhaps improving intuition without replacing it by the purely theory-based action of the novice, advanced beginner, or competent performer.

Work on Naturalistic Decision Making, by studying how decision making works in complex, uncertain, unstable situations such as emergencies, where experts do not have enough experience to have an immediate, intuitive response, complements the work that Stuart and I have done on transparent, intuitive coping. In such special situations the decision makers must assess the situation and generate plans of actions. It is important to understand how this is done so that the decision makers can draw on their expert understanding rather than falling back on formal models which cause the decision makers to loose touch with their expert intuition altogether. Stuart and I have gone as far as phenomenology can go in describing some of the ways experts can respond deliberately to problematic situations. We would now like to learn from those who have carried out empirical research in this area.

REFERENCES

Dreyfus, H. L., & Dreyfus, S. (1986). *Mind over machine: The powers of human intuition and expertise in the era of the computer.* New York: The Free Press.

Feigenbaum, E., & McCorduck, P. (1983). *The fifth generation: Artificial intelligence and Japan's computer challenge to the world.* Reading, MA: Addison-Wesley.

Heidegger, M. (1962). *Being and time* (M. Macquarrie & E. Robinson, Trans.) New York: Harper and Row. (Original work published 1927)

Hobbes, T. (1958). *Leviathan.* New York: Library of Liberal Arts.

Leibniz, G. W. von. (1951). *Selections* (P. Weiner, Ed.). (pp. 17–25). New York: Scribner's.

Newell, A., & Simon, H. A. (1990). Computer science as empirical enquiry: Symbols and search. In M. Boden (Ed.), *The philosophy of artificial intelligence* (pp. 105–132). Oxford, England: Oxford University Press. (Original work published 1957)

Plato. (1948). *Euthyphro, VIII* (F. J. Church, Trans.). New York: Library of Liberal Arts.

Wittgenstein, L. (1953). *Philosophical investigations.* Oxford, England: Basil Blackwell.

Wittgenstein, L. (1975). *Philosophical remarks.* Chicago: University of Chicago Press.

Wittgenstein, L. (1981). *Tractatus logico-philosophicus* (C. K. Ogden, Trans.). London: Routledge. (Original work published 1922)

Chapter 3

Naturalistic Decision Making and Related Research Lines

◆

Lee Roy Beach
University of Arizona

Michelene Chi
University of Pittsburgh

Gary Klein
Klein Associates Inc.

Philip Smith
Seville Training Systems

Kim Vicente
University of Toronto

This panel addressed the diversity of ideas about what constitutes naturalistic decision making (NDM). The danger for a movement that regards itself as new is that it may try to embrace everything that is not old. This guarantees a lack of consensus about what the movement is and why it exists.

In an effort to demarcate the boundaries of NDM, the panel members were asked to define NDM either by providing a distinct definition or by providing a working definition in terms of relevant research issues. Gary

Klein cited Caroline Zsambok's (chapter 1, this volume; Orasanu & Connolly, 1993) list of themes encompassed by NDM, and offered the Recognition-Primed Decision model (Klein, 1989) as an exemplar. Kim Vicente continued the discussion by outlining the ties between NDM and other areas of cognitive research. Philip Smith argued that the unique aspects of NDM call for appropriately unique research tacts, and in doing so further delineates NDM. Michelene Chi explored expertise, which is a central issue in NDM, with particular emphasis on decision contexts. Finally, Lee Roy Beach illustrated the research potential of NDM models using an extension of Image Theory (Beach, 1993) for examining decision making in organizations.

In Klein's view, NDM focuses on how people use their knowledge and experience to assess complex and uncertain conditions and take action. His colleague, Caroline Zsambok (this volume), identified four themes of NDM: (a) Task and setting involve ill-structured problems, uncertain, dynamic environments, shifting ill-defined or competing goals, action/feedback loops, time pressure, high stakes, multiple players, and organizational goals and norms. (b) Subjects are experienced participants. (c) Locus of interest includes situation awareness, diagnosis, and plan generation rather than fixating on the moment of choice. In addition, (d) the purpose of research is to describe the strategies people use rather than prescribing the strategies they ought to use. In short, we study NDM because we want to know how reasonably experienced people actually make judgments and decisions in realistic settings.

The Recognition-Primed Decision (RPD) model provides an example of NDM. Although not synonymous with NDM and certainly limited in what it attempts to explain, the research on RPD demonstrates that people can make good decisions without having to perform extensive analysis: People use experience to recognize problems that they have previously encountered and for which they already know solutions; people use their experience to form mental simulations that suggest solutions. This strategy allows them to quickly make difficult decisions, rather than having to decompose situations into basic elements and perform analyses and calculations on the elements. The model departs most sharply from the majority of classical models of decision making in its attempt to trace the use of experience.

If we want to study problem solving, situation awareness, and mental simulation, and not just the moment of choice, why use the term *Naturalistic Decision Making*, which seems to evoke the decision event and the moment of choice? Why not a broader term such as *ecological psychology* (except that that tends to exclude mental events) or *organizational decision making* (except that that tends to include artificial tasks performed by naive subjects) or *naturalistic problem solving* (except for its connections to Artificial Intelligence)? Why not something neutral like G. Smith's (1994) term, *action-oriented thought* or *situated cognition*? Several people have argued that NDM is just a historical accident stemming from decision research in the mid-1980s; there is some truth in this. Even though the term NDM has shortcomings, so do each of the alternatives. None of them fully covers the range of concerns and the focus of NDM.

Moreover, NDM has a strong appeal to the customer. The military services, paramilitary organizations such as firefighters, and even Fortune 500 companies seem to resonate more to NDM than they might to ecological psychology or the others cited earlier. If we want to support these applied communities it is essential that we speak a language with which they can feel comfortable. They want to make better decisions under the conditions that Orasanu and Connolly (1993) described (i.e., ill-defined goals, incomplete information, etc.). We may feel that in order to reach this goal they may need to apply insights from ecological psychology or situated cognition, but these are not the terms in which they are framing their goals and their problems. The purpose of this panel, therefore, ceased to be one of drawing boundaries (which had been its charge), but one of tracing connections. We had a sense that we are in a fortunate position of working at a time when we can make rapid theoretical progress and, concurrently, find valuable applications. The NDM research field is attractive to many of us because it is grounded in the messy conditions that require our help the most. We are unlikely to turn away from these messy conditions simply because they do not offer the opportunity to conduct tightly controlled, fully factorial experiments. Instead, most of us become more interested the greater the number of complications! This probably is the most important feature of the NDM viewpoint.

It can, of course, be argued that, strictly speaking, there is no such thing as "naturalistic decision making." Vicente (1994) advanced this view, citing field studies conducted in complex work environments. Invariably, such studies conclude that decision making in the classic sense (i.e., explicitly deliberating between possible alternatives and then selecting a course of action) is rarely observed (Klein, 1989). Furthermore, there are strong interactions among information processing activities that have traditionally been studied in isolation by experimental psychologists; diagnosis and action are as intertwined as perception and decision making. Thoroughly understanding behavior in naturalistic contexts requires understanding of perception, decision making, action, problem solving, attention, planning, metacognition, and team/social processes. None plays a privileged role because all are intertwined. Indeed, human behavior in natural contexts is not solely, or even primarily, about decision making per se.

The fact that this meeting is about NDM is, in Vicente's opinion, a historical accident. It results from adopting the traditional judgment and decision making (J/DM) literature as a reference point. However, one could just as well select, for example, traditional accounts of problem solving. This would lead one to coin the term *naturalistic problem solving*, but that too would be arbitrary and potentially misleading, because it too oversimplifies what is going on.

Why does this matter? Is this just academic nit-picking? None of us thinks so. Adopting the J/DM literature as a referent makes NDM sound new, when in fact adoption of a different referent might not do so. The study of human behavior in naturalistic contexts is not new; there is a significant body of research, particularly in Europe (e.g., Rasmussen, 1974; Rasmussen & Jensen, 1974) and particularly in the domain of process control (e.g., Edwards & Lee, 1974), dating back many years. Perhaps the NDM claim for novelty is somewhat exaggerated. Because others have been preoccupied by similar issues for quite some time, the NDM community has much to gain by examining this earlier body of work.

Research in naturalistic environments offers important benefits. First, such studies offer insights that cannot easily be acquired with reductionistic research. Second, such studies are comparatively rare, especially in North America. Third, from the applied viewpoint, such studies are essential for sound research design (See Hoffman & Deffenbacher, 1993; Vicente, 1994). Fourth, from a basic viewpoint, such studies provide hope for results that generalize beyond the laboratory. Vicente especially encouraged researchers to continue to investigate human behavior in naturalistic settings, but also to become familiar with the valuable lessons learned by others who have rushed in where academics fear to tread (cf. Rasmussen, 1988).

Panelist Philip Smith agreed with Vicente that there are many reasons why naturalistic research on decision making and problem solving should be pursued. Among them is the potential to understand how people perform in "real" settings in order to have an impact on "real" problems.

Unfortunately, research in such settings tends to produce large amounts of initially unstructured data, and it is not only difficult to separate the wheat from the chaff, it is difficult to know what to make of the wheat. Smith proposed four general methods to help with this problem:

First, one should *use multiple research paradigms*. It is generally believed that progress in science requires collection of converging evidence to ensure the validity of conclusions and models. The basis for this belief is the assumption that research methods and resultant inferences can be flawed due to the presence of unrecognized confounds or incorrect assumptions underlying the analysis. This is of particular concern when sample sizes are small, as often is the case in NDM studies, because it is easy for the researcher to "see" patterns in what really are random data.

Second, one should use *existing models as research vehicles*. Many researchers and practitioners are interested in NDM for particular application areas. One of the most effective ways to examine these areas is through use of existing models of cognition and perception in designing tasks or in directing data analysis. These models aid in detection of patterns, because it is easier to identify predefined patterns than to discover unknown patterns.

Third, one should use *domain-specific models*. Just as is the case with general models of human performance, domain-specific models serve an important attention-focusing function. When looking at performance in a particular application area, there often are strategies and knowledge specific to that domain. Such models often are developed in an iterative fashion, explaining patterns in the initially collected data and indicating areas for future data analysis or data collection.

Finally, we should use *symbolic models*. It has long been recognized that the development of symbolic models forces greater precision in researchers' thinking, helping to identify implicit assumptions and unasked questions. The tools provided by software development can be thought of as representations that can increase salience of important issues.

Studies of NDM seem to offer considerable potential for producing insights into human performance that are of practical value in guiding system design. This potential is based in large part on the assumption that new insights can be gained by studying performance in the rich context of naturalistic settings. Smith also had a caution: This very richness can result in discovering false patterns or, of equal concern, can result in missing important patterns or factors that influence performance. Consequently, NDM research always must deliberately consider its methodological foundations.

The decision-making events of interest to NDM researchers call on decision makers to use a broad range of contextual knowledge and expertise in both situation analysis and problem solving in order to arrive at a decision. Michelene Chi addressed this theme by examining how novices learn as they progress toward expertise. To understand this process it is necessary to understand the demands and limits set by that context. First, context provides examples of the conditions that call for actions, the range of permissible actions, and the consequences of actions. It provides opportunities to develop tacit knowledge about subtle features of the situation such as who has information that can be shared, with whom collaboration is possible, and about group goals, values, and standards (culture). Context provides the opportunity not only to learn what to attend to, but also what to filter out as irrelevant. Finally, context suggests short cuts and rules of thumb for doing tasks (which seldom are taught in classrooms) and allows the becoming-expert to acquire knowledge that is specific to a job or job site.

The importance of this to NDM is that an increased focus on decision makers working in familiar contexts doing familiar tasks must not overlook the importance of the context in shaping decision making. Indeed, the study of contextual demands and limits may well be as important as studying decision processes themselves.

Finally, Lee Roy Beach followed Gary Klein's example, arguing that the ability to use an NDM model to do interesting research demonstrates that the viewpoint is viable. In this research, an NDM model, called Image Theory, which has been extensively tested (Beach, 1993) as a model of individual decision making, has been generalized to organizational decision making.

For individuals, Image Theory posits three knowledge structures (images) that influence individual decision making: the decision maker's enduring beliefs and values, the decision maker's agenda of goals, and the decision maker's ongoing plans that are being implemented in an effort to attain the goals on the goal agenda. Selected options must not seriously violate beliefs or values and must not significantly interfere with implementation of plans that are aimed at goal achievement.

In adapting Image Theory for organizations, a parallel is drawn between an individual's beliefs and values and an organization's culture, between an individual's goal agenda and an organization's vision, and between an individual's plans and an organization's strategic plan. That is, organizational culture is the shared beliefs and values of the organization's members. Vision is the shared goal agenda that those members hold for the organization. The strategic plan (however formal or informal) is the shared blueprint for realizing that vision. As is the case for individuals, decision options are evaluated on the basis of how well they complement the organization's culture, vision, and strategic plan.

Research (Weatherly & Beach, 1996) shows that when managers try to introduce decisions that are incompatible with the culture, the organization's members refuse to endorse them, predict that they will fail, and may balk at attempts to implement them. Moreover, the greater the incongruity between salient aspects of the culture and how organization members think the culture ought to be, the lower their job satisfaction, the lower their commitment to the organization, and the greater their intent to leave.

The implications of this work is that NDM, as represented in this case by Image Theory, invites one to view decision making in a broader context than does traditional decision theory. Together, the various approaches to understanding decision making discussed by these panelists provide a clear picture of an emerging area of theoretical, laboratory, and real-world work that is coming to be recognized as Naturalistic Decision Making.

REFERENCES

Beach, L. R. (1993). Broadening the definition of decision making: The role of prechoice screening of options. *Psychological Science, 4*, 215–220.

Edwards, E., & Lee, F. P. (1974). *The human operator in process control.* London: Taylor & Francis.

Hoffman, R. R., & Deffenbacher, K. A. (1993). An analysis of the relations between basic and applied psychology. *Ecological Psychology, 5*, 315–352.

Klein, G. A. (1989). Recognition-primed decisions. In W. B. Rouse (Ed.), *Advances in man–machine systems research* (Vol. 5, pp. 47–92). Greenwich, CT: JAI.

Rasmussen, J. (1974). *The human data processor as a system component: Bits and pieces of a model* (Riso–M–1722). Roskilde, Denmark: Danish Atomic Energy Commission, Research Establishment Riso.

Rasmussen, J. (1988). Cognitive engineering: A new professional. In L. P. Goodstein, H. B. Andersen, & S. E. Olsen (Eds.), *Tasks, errors, and mental models: A festschrift to celebrate the 60th birthday of Professor Jens Rasmussen* (pp. 17–36). London: Taylor & Francis.

Rasmussen, J., & Jensen, A. (1974). Mental procedures in real-life tasks: A case study of electronic trouble shooting. *Ergonomics, 17,* 293–307.

Orasanu, J., & Connolly, T. (1993). The reinvention of decision making. In G. A. Klein, J. Orasanu, R. Calderwood, & C. E. Zsambok (Eds.), *Decision making in action: Models and methods* (pp. 3–20). Norwood, NJ: Ablex.

Smith, G. F. (1994). Managerial problem solving. In *Encyclopedia of library and informational science, Vol. 53* (Supplement 16, pp. 210–236). New York: Marcel Dekker.

Vicente, K. J. (1994). A pragmatic conception of basic and applied research: Commentary on Hoffman and Deffenbacher (1993). *Ecological Psychology, 6,* 65–81.

Weatherly, K. A., & Beach, L. R. (1996). Organizational culture and decision making. In L. R. Beach (Ed.), *Decision making in the workplace: A unified perspective* (pp. 117–132). Mahwah, NJ: Lawrence Erlbaum Associates.

Chapter 4

Progress, Prospects, and Problems in NDM: A Global View

◆

William C. Howell
American Psychological Association

BACKGROUND

Consistent with the conference on which it is based and the one that preceded it, this book constitutes a status report on an endeavor known as *naturalistic decision making* (NDM). The ensuing chapters are written, by and large, by individuals whose work is giving definition and direction to this endeavor. Most of them are applied scientists who found the substantial existing literature on human judgment and decision making unsatisfactory for addressing their particular system design or training needs and went looking for a different approach. Each arrived at NDM by a somewhat different route and sees it in somewhat different terms. Like the proverbial elephant, therefore, NDM is the composite of a number of narrower perspectives.

By contrast, the present chapter represents an attempt to draw a rough sketch of the entire elephant along with some background on where it came from and speculation about where it is headed. It is written by someone whose role in the conference, as in the NDM endeavor, has been more that of observer than active participant. It reflects observations drawn larqely from a concluding session at the conference in which four major presenters—James Shanteau, Judith Orasanu, Marvin Cohen, and Jeff Grossman—and the entire audience engaged in a discussion of the "big picture" issues.

The discussion was guided by four preselected topics: progress over the years since the last NDM conference, burning current issues, long-term future prospects, and a reflection on how particular paradigms are shaping the NDM endeavor. Because they are clearly interdependent, however, and they were interwoven throughout the discussion, no attempt is made to preserve them separately in the following overview. Rather, the presentation is organized around more specific nodes that emerged in the course of the session; and, like the content it seeks to organize, it is undoubtedly colored to an extent by the biases of this particular observer.

WHAT IS NDM?

The word *endeavor* is used consistently and intentionally with reference to NDM in the preceding paragraphs, the purpose being to enter the present section with a properly neutral mind-set. Other terms that have been applied to it such as *field, area, specialty, theory, paradigm, approach, movement,* and *cult* all carry surplus meaning—meaning, incidentally, that was clearly intended by the particular user and not without affective overtones. By contrast, an "endeavor" is something that can be described in fairly objective and-unemotional terms—it could be good or a total waste. Certainly no one would question that a growing number of people have been engaging in a growing number of activities that they and others have chosen to call NDM. And when they get together to talk about it, or they write about it, they have some collective sense of what they mean: One might go so far as to say they share the rudiments of a mental model. By this definition, NDM is clearly an "endeavor."

However, has it become more than that, and if so, what? Directly or indirectly, much of the discussion during the session and throuqhout the conference touched on this issue. Although no clear consensus emerged, there was considerable reflection on how NDM fits into the larger scheme of judgment and decision research and relates to other research areas. This line of discussion sheds a great deal of light on the definition problem.

Since its inception, NDM has carried revolutionary connotations. Those responsible for inventing the term defined it primarily by contrast with "traditional" decision theory and research. Gary Klein's fabled lists depicting NDM as complex, relevant, realistic, and useful as distinct from the largely simplistic, irrelevant, unrealistic, and useless profile ascribed to the existing "brand X" set the tone. He was joined quickly by other dissidents, and NDM became a code word for protest against all the alleged conceptual, methodo-logical, and practical shortcomings of the prevailing order.

Although undeniably a direct route to change, revolution carries with it some inevitable baggage—all of which is evident here. It promotes an adversarial posture, exaggeration of differences between and similarities within competing camps, and the unrealistic hope that the new order can cure all the ills of the old. Even if a revolution succeeds, there remains the daunting task of governing captured territory and making peace with all the natives and former adversaries.

It seems that over the past few years NDM has indeed established a beachhead and is now struggling with the troublesome consequences. Among those invited to the conference, for example, were "traditionalists" such as Shanteau, Beach, and Howell, who pointed out that concerns about task complexity, realism, relevance, and so on are far from new; in fact, they were central to the pioneering efforts of Hammond, Einhorn, Slovic, and others several decades ago. These latter efforts, incidently, produced both theoretical and practical advances that NDM revolutionaries have tended to overlook in their zeal to distance NDM from Brand X. Hammond's "cognitive continuum theory," for example, sought to explain the effects of complexity along with a number of other task variables within a coherent conceptual framework (Hammond, McClelland, & Mumpower, 1980). However, although one finds occasional reference to the notion in the NDM literature, and Hammond participated in the first NDM conference, the impact of this important conceptualization on the actual NDM research has been minimal.

Several speakers and considerable discussion focused on the need for NDM to move beyond the constraining influence of such revolutionary, definition-by-contrast thinking. Theory development was seen as an important step in this direction, but note was taken of the fact that little progress has been made toward theory building since the last conference—a failure made all the more obvious by the prominent attention accorded it at that meeting. Although there have been some efforts to elaborate existing models, and these were duly reported at the present gathering, the NDM movement as a whole has seemingly not heeded its own call for construction of a broader, stronger theoretical foundation. Of course, some would argue that it was the call, rather than the lack of a more enthusiastic response to it, that was misguided, and that the research produced during the interconference interval proves it.

In addition, the position was expressed and generally endorsed that NDM can strengthen itself and the cause to which it is committed (improving real-world decisions) by enlisting the cooperation of more "traditional" decision researchers rather than driving the wedge of contrast still deeper. It would seem, therefore, that the climate for a constructive rapprochement with the judgment/decision-making mainstream has improved since the last meeting. It remains to be seen what will come of this conciliatory attitude.

In the course of wrestling with the identity question, relations of NDM to research areas outside the traditional decision domain were also explored. Among the more noteworthy are those sharing with NDM a focus on cognition, expertise, and task complexity (including ambiguity and team aspects). Far from taking an adversarial stance, NDM advocates have recognized from the outset that their future is intimately linked to progress in these areas. After all, the whole thrust of NDM is to draw on an understanding of how real experts conceptualize and deal with complex cognitive tasks. Therefore, anything that advances such understanding contributes directly to the potential of NDM to enhance system performance.

Progress in each of these areas was updated and all the well-recognized problems associated with them revisited: the relative inaccessibility of expert knowledge, the controversial nature of "mental models," the adequacy of "expert models" as criteria, and so on. One observation emerging from these discussions was that NDM, perhaps due to the prominence of the RPD model and the neglect of predecisional processes by classical theorists, has become preoccupied with *situation assessment* (SA) and all but ignores subsequent stages of the decision process. Clearly not *all* realistic decisions involve simple, relatively automatic pattern recognition; *problem solving* is an important research area that deserves greater attention from the NDM community than it has thus far received, although it was represented at this conference. If, as some suggest, the current emphasis on SA is merely a phase in the NDM evolution—an attempt to address an obvious and pressing knowledge gap—then it is perhaps time to consider progressing to the next phase.

In summary, it would seem that NDM can best be described at this point as a *movement*—one founded on a revolutionary philosophy that turned out perhaps not so revolutionary, directed against a common enemy that turned out not so common and perhaps not even an enemy—that is currently undergoing a modest identity crisis. It has not achieved the status of a distinct field or area of specialization as some of its proponents might have liked, but its impact on the broader field of judgment and decision-making research has been undeniable—and its relation to research being carried on under the banners of basic cognition, expertise, task analysis, and others is extremely close. It is so close, in fact, that some discussants see them as indistinguishable. Grossman, for example, is still searching for the unique essence of NDM!

WHAT IMPACT HAS NDM HAD?

Clearly, one of the principal influences has been to attract a number of new researchers to the decision field. As Orasanu so aptly put it at the conference, NDM:

> legitimized and gave courage to a lot of researchers who felt that their work was on the fringe of respectability in the decision-making world. It legitimized their study of real people making real decisions in their natural or complex environments.

Some, of course, might regard Shanteau's livestock judges and weather forecasters as "real people making real decisions," and his work on them, as I recall, was hardly considered "on the fringe" when it was published even without the benefit of NDM. Whether special validation of such work by NDM was necessary is thus debatable but beside the point. The fact that common research issues were being faced by a cadre of individuals who saw themselves as isolated from the mainstream and were drawn together under the NDM flag has had a stimulating effect. They now have a forum for the exchange of ideas and knowledge as these two NDM conferences have so vividly demonstrated. And such interactions inevitably lead to innovation, expanded research vistas, and progress, all of which are apparent in a comparison of the two conferences.

The full impact of NDM, however, has not been limited to its own emerging community of researchers. It seems to be attracting at least some attention in the mainstream judgment and decision literature as well, and more importantly, in the world of applications. Admittedly, this second conference was intentionally focused on applications, but even so, the expansion of the domain into which tools such as "storyboarding" and the recognition-primed decision (RPD) technique have gravitated is impressive. Once concentrated within the national defense and public safety communities, NDM-based applications are now showing up in medical, process control, manufacturing, aviation, software design, and a host of other contexts.

It is still too early to judge what the extent of the practical contribution will be in any of these application areas. Most of the presentations seemed in the vein of illustrating what *could* be done rather than offering evidence of proven *accomplishment*. Moreover, it is difficult in many applications to isolate the unique NDM contribution because the particular approach often involves techniques such as cognitive task analysis, simulation, or modeling of expertise that could well have developed without it. In any case, the philosophy underlying NDM has definitely struck a responsive chord, and the attention that it has aroused has served to promote a number of-related activities. Collectively, these activities promise to make a noticeable difference in the quality of real-world decision making in the long term, if they have not already.

WHAT ARE NDM'S MAIN ACHIEVEMENTS?

Each of the discussants considered NDM's contributions—potential or realized—in terms of theory, methodology, and applications. As noted earlier, the consensus was that not much had happened on the theory front since the last meeting. It is certainly true, however, that theoretical developments in related areas have influenced NDM thinking, and refinements in some of the core constructs have been made based on empirical findings. Several papers illustrating such work were in fact presented at the meeting. Thus, although it would be hard to find a significant body of theory to which NDM could lay exclusive claim, the interval between conferences has not been totally devoid of theoretical endeavor.

In the methodology area, the picture seems a bit brighter. Refinements on and adaptations of tools such as RPD have seen wide application. Various techniques for dealing with one or another facet of NDM, such as situation awareness measurement, were described and discussed. The "toolkit," as Orasanu put it, "is really expanding."

Among the more prominent trends in methodology is the increased reliance on simulations. This class of "tools" is certainly not new, and the current trend represents an attempt to bring task realism into the laboratory. Unfortunately, many of the difficulties that have plagued simulation research since its inception are still unresolved. For example, how much fidelity is enough, and how does one know which features to simulate? How much of the task specificity inherent in high fidelity can one sacrifice in the interest of obtaining principles that generalize? Questions such as these become extremely important when the task domain is knowledge intensive, as is the case for most of those in which NDM has an interest.

Shanteau addressed these issues in a call for NDM to broaden the range of tasks and expertise investigated. One aim of such an expansion would be to derive a taxonomy within which apparent task similarities and differences might be organized. Research could then seek to validate and refine the distinctions based on performance.

In a related point, Shanteau reiterated a concernover NDM's growing willingness to accept at face value the word and conceptualization of "experts" as criteria for system performance. Experts have been shown to be quite suboptimal—indeed little better than novices—in many decision domains where their performance could be checked against "ground truth" or some other objective index. It is important to study and rely on expertise, but not without consideration of how experts typically fare in the task domain of interest. Cohen seemed to concur in this reservation and noted the further complication that experts, even when behaving expertly, often do so in very different ways. Now *there* is a criterion problem worth worrying about! The issue here is not whether the behavior of experts is worth capturing and, in many cases, modeling. Rather, it is about the relative emphasis placed on understanding and evaluating such behavior versus merely assuming its superiority because someone has declared its source an "expert." The latter assumption, which seems to have gained dominance within the NDM community just as it has within the AI domain, is not without risks.

Nor can it be safely assumed that the verbal protocols on which such heavy reliance is placed in trying to capture expertise are uniformly either reliable or valid. It has long been recognized that whatever unique procedural knowledge experts may have tends to be automated, and hence is relatively inaccessible to direct introspection and self-report. Rather than accepting verbal protocols as "ground truth," therefore, Shanteau argued for regarding them as provisional and subject to external verification. Increasingly, it seems, NDM is slipping into the habit of taking them at face value—a habit that could wind up leading to badly informed training or aiding applications.

Applications, everyone agreed, is where NDM has made the most solid gains, and where most of the participants would probably look for future growth. Perhaps it was just the program emphasis that made it seem so visible, but the enthusiasm for NDM tools shown by those engaged in implementing them was palpable.

Doubtless many of these programs will be declared successful, and a few may even manage to provide some evidence to prove it (a rare event in applied circles). Either way, testimonials will breed further applications and growth will continue. The question then becomes: Where will the incentive to innovate—to support the R&D necessary to move from "good enough" to "better and better"—come from? NDM has made excellent use of its thinly laid foundation in theory, method, and scientific data. That foundation must be reinforced if the structure is to be built higher in the future rather than just wider. Some contend that applications will, in time, spawn a particularly salient body of theory by identifying the *real* problems to which theorists and researchers should direct their attention. Perhaps. Unfortunately, practical, imperatives usually favor the "D" over the "R," and the search for true understanding—the goal of most theory-driven research—gets lost in the shuffle.

WHAT ARE NDM'S MAIN LIMITATIONS?

Most of the areas where NDM is still deficient have already been mentioned in passing. They deserve elaboration here, however, both to do justice to the attention accorded them in the discussion, and to reinforce the message that the NDM community still has plenty of work ahead of it.

First, the paucity of ground-breaking theory discussed at length in the last conference report is still regarded by many as a primary weakness. In all likelihood, progress on this front will come chiefly through concepts developed elsewhere that are adapted to the NDM environment. It is hard to fault the wisdom of this emphasis. Indeed, a point raised earlier in this chapter and throughout the conference was that alliances with the maligned "traditional" decision establishment, as well as researchers in the problem-solving and other cognitive domains, should be strengthened.

Second, although considerable progress has been made toward expanding the task domains in which NDM is being *applied*, the same could not be said for the *research*. Given the narrowness of expertise in real-world settings, the wide range of decision task requirements, and the heavy relevance on simulation in research, generalization is achieved only through broad task representation. A comprehensive functional taxonomy needs to be developed and refined through validatin research. This, too, was called for at the first conference but has not materialized due in large part to practical constraints. Research cannot proceed without funding, so unless those who support new application areas are willing to fund associated research efforts, the necessary ingredients for a meaningful taxonomy cannot be generated. Thus sponsors of NDM applications should be "educated" to the long-term value of correlated research wherever the opportunity presents itself.

Third, progress in both research and application has been hampered to an extent by the tendency for NDM to coalesce around paradigms that seem to work, without proper regard for boundary conditions (beyond which they do not) or alternative strategies applicable to those conditions. Cohen, for example, cited the need to explore decision strategies that are *not* merely a matter of automatic recognition; strategies such as those involving *controlled* situation assessment and action selection. "When," he asked, "does the decision maker rely on RPD, and when is a more attention-demanding strategy called for?" Or, as Grossman put it, "What conditions cause people to shift from one strategy to another?" It would seem that task conditions are a determining factor, but if so, what are those conditions? Perhaps a serious review of Ken Hammond's work (e.g., Hammond et al., 1980) is in order!

Fourth, the matter of individual differences surfaced a number of times, primarily in the context of expertise, decision strategies, and training or aiding. The NDM approach rests heavily on the assumption that there is an expert model for every task and that training or aiding should be geared toward enabling less proficient operators to function closer to that mode. However, this assumption may not always be valid. There may be different cognitive routes to "expert performance," and as we have seen, experts may not always perform so expertly. More attention needs to be given to these variations both in developing criteria for system performance and in setting the parameters for aiding and training.

A fifth and final area of concern focuses directly on interventions. As noted earlier, the major growth area for NDM has been in applications, and training and aiding represent the principal intervention strategies. Unfortunately, these applications seem to be proceeding without comparable growth in understanding of how best to train or aid, or the kind of built-in evaluation that would enable such understanding to develop. Neither the operational nor training communities have traditionally shown much interest in the long-term payoff of evaluation research, let alone research on more fundamental training and aiding issues. If something seems to hold promise for bringing about incremental improvement over the current situation, the pressure to adopt it prematurely is intense. Such, unfortunately, is the apparent direction in which NDM is heading. If so, its contribution to the general improvement of decision systems may asymptote very quickly if, indeed, it has not already done so. Deeper understanding is necessary in a number of basic areas if NDM is to avoid the fate of so many other promising movements that have deteriorated into passing fads.

CONCLUDING COMMENT

Looking at the collective opinions expressed during the final session of the conference leads this particular observer to the following summary conclusion about the state of NDM. It is a movement that has made great progress over the past few years in recruiting a constituency that identifies strongly with it. Its unique characteristics are not yet defined very well except through a rather counterproductive and dubious contrast with "traditional decision theory," and its unique contribution to the decision world has so far been mostly in applications based on a fairly limited set of theoretical concepts and paradigms.

Nevertheless, it has stimulated renewed interest in a facet of the decision process that has been largely neglected in the judgment and decision literature: the structuring of ill-structured problems. This is the facet that precedes the point where most "traditional" decision theory and research enters the picture. In emphasizing this "situation assessment" (or "situation awareness" however, NDM may itself be neglecting critical facets of the total process and thereby limiting its usefulness.

In the view of the present observer, the future holds a number of challenges for NDM, not the least of which is heeding its own call for a more concerted effort toward theory building. Without it, and the deeper understanding that it can bring to both applications and research, there is a very real danger that NDM will fall short of its potential—and aspirations. To develop a substantive theoretical foundation, the movement will need to abandon its revolutionary zeal and somewhat parochial outlook in search of links to the existing theory and research in judgment/decision making, problem solving, simulation, and various other domains that bear on human cognition. In the process, it may lose some of its identity. However, premature adoption and overselling (or, more accurately, overeager buying) of provisional NDM tools constitute a far greater long-term threat.

REFERENCE

Hammond, K. R., McClelland, G. H., & Mumpower, J. (1980). *Human judgment and decision making: Theories, methods, and procedures.* New York: Praeger.

Part II

Applications of NDM
Research:
Perspectives from Panels
of Applied Researchers

Chapter 5

An Overview of Naturalistic Decision Making Applications

◆

Gary Klein
Klein Associates Inc.

If the naturalistic decision making (NDM) movement does not offer direct implications for improving decision making, then we must judge it a failure no matter how reasonable the definition and features. It may be early to expect an extensive track record of NDM applications, but we should at least be able to predict what these applications are likely to be.

In order to understand why the NDM framework should generate useful applications, there is a paradox that needs to be considered: The initial impetus behind the NDM movement was to *describe* what people do, whereas the motivation behind traditional decision research was to *improve* the way people make decisions. Therefore, NDM research should have *less* applied value than the traditional research, rather than more.

The traditional decision research tried to identify a Rational Choice method (generate a range of options, identify evaluation criteria, evaluate each option on each criterion, calculate the results, and select the option with the highest score) that could help people make better decisions. These steps are a general strategy intended to prescribe better methods than people ordinarily use. The classical approaches to decision making are centered around application. They are general because they try to improve process, regardless of content area.

A second traditional approach to decision research involves the demonstration that subjects in laboratory experiments often show biases because of the way they use heuristics, such as shortcuts in reasoning. This heuristics and biases approach appears to have applied potential. If we can describe these biases we can take steps to train people to overcome them, or to build decision aids to detect and alert the operators to decision biases, or to take other steps. These are also improvements that should generalize across different content areas.

In contrast, NDM research tries to describe the strategies proficient decision makers are using, and does not yet have any central claims about what might lead to implications for improving decision quality. It may not make sense, for example, to train naturalistic strategies such as those contained in the RPD model (see Klein, chapter 27, this volume), because the model describes what people already do.

THE RATIONALE FOR APPLYING NDM

There are several reasons for expecting NDM research to result in applications that will improve decision quality more than the traditional approaches to decision making:

- Classical methods do not apply in many naturalistic settings.
- Experienced decision makers can be used as standards for performance.
- Naturalistic Decision Making tries to build on the strategies people use.
- Experience lets people generate reasonable courses of action.
- Situation awareness may be more critical than deliberating about alternative courses of action.
- Decision requirements are context specific.

Classical Methods Do Not Apply in Many Naturalistic Settings

Attempts to apply the context-free Rational Choice strategies have largely met with failure (Means, Salas, Crandall, & Jacobs, 1993; Zakay & Wooler, 1984). General strategies may be weak strategies, because a one-size-fits-all strategy might not fit any specific setting very well. The constraints of naturalistic settings, such as time pressure and ambiguous information, typically make it impractical or impossible to apply methods for Rational Choice. Some studies have even shown higher performance from subjects using unsystematic strategies than subjects trained and directed to use Rational Choice strategies (e.g., Driskell, Salas, & Hall, 1994). There are conditions under which a Rational Choice method would be useful. Hammond and Adelman (1976) described a successful use of a Rational Choice strategy for selecting a type of bullet for use by the Denver Police Department, to satisfy the needs of police officers and citizens. Time pressure was low, expertise was low, the problem was stable, and different stakeholders were involved.

However, in many cases people wrestle with choices that have little consequence. Decision makers will find it hardest to make choices as the differences between the choices diminish, but as the differences diminish the implications of the choice also diminish.

Turning to the studies of heuristics and biases, studies are now showing that experienced decision makers do not show the types of biases found under the restricted laboratory conditions (e.g., Christensen-Szalanski, 1986; Fraser, Smith, & Smith, 1992; Gigerenzer, 1987; Lopes, 1981). At present, it is unclear how much of an impact these heuristics have on less-experienced decision makers.

Experienced Decision Makers Can Be Used as Standards for Performance

Often, the strategies and cues used by experienced decision makers can serve as criteria against which to measure the actions of novices. Therefore, even in situations where researchers cannot define the statistical or logical optimal strategy, they can use the processes found in experienced decision makers as performance standards.

Naturalistic Decision Making Tries to Build on the Strategies People Already Use

It does not attempt to totally replace these strategies.

People With Experience Can Use Their Experience to Generate a Reasonable Course of Action as the First One Considered

This assertion is based on the result described by Klein, Wolf, Militello, and Zsambok (1995) on chess players, showing the high quality of the first move they considered. Similarly, Stokes, Belger, and Zhang (1990) showed that even when they had to generate several options, experienced pilots generally selected the first option they considered. When this assumption holds, such as, in domains where people can build up experience and where time pressure and other constraints make it impractical to search for the best option, then we should not be encouraging decision makers to generate and evaluate large sets of options. In domains where critical decisions are made by inexperienced people, then our task is to boost the skill level.

Situation Awareness Appears to Be More Critical Than Deliberating About Alternative Courses of Action

Kaempf, Klein, Thordsen, and Wolf (1996) studied U.S. Navy officers who had been involved in anti-air operations, and found that most decisions concerned the nature of the situation. For those decisions about adopting a course of action, fewer than 5% involved comparisons between alternatives. Therefore, in naturalistic settings we will obtain a greater payback by helping people size up situations, either by training them to recognize cues and patterns, or by designing management information systems that help them quickly get a sense of the big picture.

Decision Requirements Are Context Specific

The decision requirements of a task refer to the key decisions and how they are made. Decision requirements include the judgments and decisions required by the task, and the cues and information required to make the decisions. For key decisions, the analyses can describe the reasons why these are difficult, along with the cues, patterns, inferences, and strategies used by proficient personnel to overcome the difficulties.

The advantage of using decision requirements is that they do not try to prescribe a generic method for making decisions. Instead, the approach is to look at the ways that experienced decision makers reason within their own domains. Rather than searching for general methods, practitioners can search for the decision requirements of the specific situation. What matters is not just *how* people think (their strategies) but also *what they think about* (the content).

Decision requirements can become the target of interventions. If a battle command function is studied and shows that inexperienced commanders have difficulty interpreting the intent of enemy forces, or judging where to place reserve units, or setting priorities for air assets, these would become the decision requirements for a better training program. If the studies identify the factors that make these judgments and decisions difficult, these would be highlighted for training programs (e.g., in designing scenarios) and decision support systems (e.g., for organizing human–system interfaces). If we can learn how proficient commanders work around the difficulties, we can use this information for training and for system design as well. Such an approach would have higher face validity for users, who would see the immediate benefits of the interventions. It would also have greater likelihood for transferring to the operational environment, because the intervention would be oriented around the barriers and difficulties that users encounter in the field.

TOOLS AND TECHNIQUES FOR USING DECISION REQUIREMENTS

Given the assumptions described earlier, the NDM framework may have application to several areas, as listed in Table 5.1. In covering the topics presented in Table 5.1, the greatest emphasis is on the decision-centered

TABLE 5.1

Applications of NDM Research

1. Decision-centered training of individuals

2. Decision-centered training of teams

3. Decision-centered designs

4. Additional forms of application
   Personnel selection
   Organizational re-engineering
   Market research

training of individuals. The other applications in Table 5.1 are also important, although they are not covered in the same level of detail.

Decision-Centered Training of Individuals

This refers to the use of decision requirements in designing training programs and training devices. (See chapter 10, this volume, by Cannon-Bowers and Bell, for a more in-depth treatment of individual decision training.) For the past 20 years, systems approaches to training have been dominant. The objective of these systems approaches is to decompose complex tasks into basic elements, specifying the initiating and terminating conditions for each element or subtask, and measuring the success of training by the achievement of performance criteria. However, systems approaches are intended for procedural tasks (e.g., setting up equipment) and not for judgment and decision tasks. The potential of NDM work is to provide guidance for exactly those skills where systems approaches are insufficient.

We can distinguish between the decision skills that can be trained (Table 5.2) and the methods for training them (Table 5.3). Table 5.2 presents types of decision *skills* that would be emphasized within an NDM framework. None of these skills is generic. Each must be trained within the context of the operational setting in which it is needed.

Situation Awareness, Pattern Matching, and Cue Learning. The ARI Research Unit at Ft. Leavenworth is sponsoring applied research by Marvin Cohen (chapter 25, this volume) on the training of situation awareness. This training includes strategies for considering alternative hypotheses and explanations of what is happening, so that the soldiers become less likely to fixate on a single explanation. The situation-awareness training is aimed at helping officers to judge when they are explaining away too much so that they can begin to search for different explanations.

Typical Cases and Anomalies. Proficient decision makers have had so many experiences that they have learned to recognize typical cases. The

TABLE 5.2
Decision Skills That Can Be Trained

- Situation awareness, pattern matching, and cue learning
- Typical cases and anomalies
- Mental models
- Time horizons
- Managing uncertainty and time pressure

TABLE 5.3
Methods for Training Individuals to Make Better Decisions

- Design of training scenarios
- Cognitive feedback within the after-action review (AAR)
- Cognitive modeling and expert/novice contrasts
- Test and evaluation techniques
- On-the-job training (OJT) methods
- Training device specification

different interventions listed in Table 5.3 can be organized to speed up the ability to recognize typical cases and patterns. At the same time, the training should make it easier to detect anomalies, details that do not quite fit. These anomalies are often the early warnings to begin taking precautions or preparing for contingencies.

Mental Models.   Many NDM researchers, such as Cannon-Bowers, Salas, and Orasanu, are investigating methods for teaching mental models, to help trainees see situations the way experts do. These may be mental models of the task, the equipment, the teamwork, or other considerations. The goal of training mental models is to provide a stronger basis for appraising situations.

Time Horizon.   One finding of NDM research is that more proficient decision makers can see further ahead into the future in planning their actions. Therefore, we can establish as a training requirement the design of scenarios, exercises, and feedback around chains of events to enable trainees to anticipate more effectively.

Managing Uncertainty and Time Pressure.   The U.S. Marine Corps has applied the ideas of naturalistic decision making to develop and initiate training that calls for rapid tactical situation assessment and reactions under uncertainty. The Marine Corps uses low fidelity tactical decision games to simulate combat conditions. The goal of this training is to develop meta-cognitive skills so that trainees can overcome the desire for more information and gain an ability to better manage the time cycle.

After reviewing some of the skills that would be addressed in training, the next issue is what sorts of *methods* would be promoted by the NDM framework. Several types of interventions are listed in Table 5.3. Each intervention is a way of helping to strengthen the skills listed in Table 5.2.

Design of Training Scenarios.   Decision requirements can tell us which judgments and decisions need to be emphasized and which contextual factors create the most difficulty. This information can be used to guide the development of scenarios. Often, people develop training scenarios to present the right level of difficulty, without having clear objectives about the nature of the challenges. By using decision requirements, scenario developers can get the structure they need. We must remember that *Practice ≠ Training.* Simply providing decison makers with an opportunity to practice does not *necessarily* translate into better and more meaningful training. Often, increased practice will translate into skill development. However, there have been too many instances in which exercises were poorly conceived and did not take advantage of important opportunities to train decision skills. Sometimes, exercises are designed so carelessly that they teach the wrong thing, namely, habits that may be dysfunctional in the workplace. Hopefully, naturalistic decision strategies can be used to systematically shape scenarios.

Cognitive Feedback.   Currently, feedback sessions concentrate on the specific actions taken. This is important, but it may miss the opportunity to use the sessions to teach why the mistakes were made so that the decision makers can learn what they are doing wrong. Often, a limited amount of time is budgeted for the exercises, leaving little opportunity for the process feedback. Because of this, trainees can learn poor habits during the exercise, and get a chance to practice these poor habits, without ever learning that they are doing things wrong. If the debrief covered cues and strategies, then the effect of the training exercise could be greatly leveraged at little cost. We would not want to see additional training in separate decision making modules—it is best to embed such training in the context of the exercises already being run.

Cognitive Modeling and Expert/Novice Contrasts.   Sometimes it is possible to present trainees with information about how experienced personnel make certain types of decisions. This can let the trainees see what is possible, what cues and relationships the experts notice, and how the experts differ from the trainees in the way they size up situations. Materials can include contrasts between the way experts and novices frame the same situations.

Test and Evaluation Techniques.   When a clear set of decision requirements has been developed, it can be used to establish criteria for measuring the cognitive performance of trainees. The decision requirements can provide a basis for assessing the ability of trainees to rapidly detect anomalies, to respond at the appropriate time horizon, to notice that expectancies have been violated and that events did not occur.

Decision-Centered Training of Teams

The NDM framework has been useful for focusing on team decision making (see Zsambok, chapter 11, this volume). For example, Prince, Chidester, Bowers, and Cannon-Bowers (1992) have developed several well-accepted team training programs for the Navy. Helmreich's Cockpit Resource Management (CRM) program has been expanded in its use in commercial aviation (e.g., Wiener, Kanki, & Helmreich, 1993) and is now referred to as Crew Resource Management as it is used in other domains. Zsambok, Klein, Kyne, and Klinger (1992) have developed a team decision training module that has been institutionalized at the Industrial College of the Armed Forces.

The interaction between naturalistic decision making work and teams is fairly subtle, and is best understood by looking at what is *not* happening, rather than at what is being addressed. Conventional decision research emphasizes a Rational Choice model for individuals, and therefore would direct attention to the use of a rational choice model (i.e., multiattribute utility analysis) for teams. Under a traditional framework, we should be trying to help teams by finding ways for them to work together to generate more options to be considered, by helping them identify more dimensions for evaluating these options, by assisting them to properly weight the evaluation dimensions, and by supporting them in rating the options and synthesizing the ratings across team members. Traditional decision researchers have studied ways of providing all of these types of aid, including brainstorming (to increase the set of options and evaluation dimensions), and delphi and nominal group techniques (to improve the synthesis of weights and ratings).

However, none of the NDM researchers whose work is discussed in the chapter on improving team decision making is looking at any of these types of methods or functions. The view of team decision making directly parallels the view of individual decision making found in accounts such as the Recognition-Primed Decision model. Situation awareness is a critical function for teams, and much of the work aims at using the inputs of team members to build a better situation awareness, and ways to encourage shared situation awareness among the team members. The use of situation awareness to focus attention on critical cues is important for team members, and one of the challenges is to find ways to selectively share information so that information overload is avoided. Expectancies are important for coordinating team activities, and for enabling all team members to notice discrepancies and anomalies and thereby to notify the team that the shared situation awareness may be inaccurate. Situation awareness includes the description of goals, and one of the important functions within team decision making is for the leader to communicate intent so that the other members can make their own decisions about how to carry out directions and how to improvise. Situation awareness should provide a sense of appropriate actions, and NDM researchers are finding that in many settings it does not make sense to generate alternative courses of action, despite doctrinal recommendations. Finally, mental simulation is an important strategy for evaluating courses of action in teams as well as in individuals. The collective experience base of the team can be effectively utilized by having the members review how the course of action is intended to be carried out, so that pitfalls can be spotted along with opportunities for improvement. Viewed in this way, the naturalistic approach to team decision making is radically different than traditional approaches.

Decision-Centered Design

Decision-centered design refers to the use of decision requirements in designing information management systems and human system interfaces. Currently, the dominant methods are a *systems-centered design* approach (focus on the technology, not on the user's needs and abilities) and a *data-centered design* approach (identify all the relevant information and pack it into the displays).

In contrast, the NDM approach identifies the decision requirements in a given job or task and uses these in the conceptual design stage, to help guide the process. The impact of decision-centered design has been shown in several projects. For example, Klinger et al. (1993) redesigned the AWACS Weapons Director's interface. The new interface resulted in a performance improvement of approximately 20% after only 4.5 hours of practice, compared to the performance of the same Weapons Directors using their regular interface on which they had received more than 1,500 hours of practice. Another successful design project (Miller, Pyle, & Shore, 1993) was for a decision support system for weaponeers.

Vicente and Rasmussen (1990) have described ways of using ecological design principles. Woods and Roth (1980) and Smith, McCoy, and Layton (1993) have shown the benefits of using design approaches based on cognitive and decision making needs, as opposed to data-centered or system-centered strategies. Mitchell (chapter 29, this volume) describes the use of a Recognition-Primed Decision model to achieve a more effective system design.

Additional Forms of Application

There are several other possible types of applications for using decision requirements. Because these have received less attention than training or system design, or because they have a more narrow focus of applicability, I just list them briefly.

Personnel selection could use decision requirements to identify the important types of judgments and decisions that a worker must be able to make, and design selection instruments to evaluate the applicants' abilities to handle these types of judgments and decisions. Some selection methods do attempt to capture these skills, but there may be benefits in trying to do so more explicitly.

Organizational reengineering refers to attempts to modify organizational structure, usually to reduce management staff positions and become more efficient and responsive. One of the core ideas of reengineering is that the extra layers of management can actually reduce performance, so it is possible to reduce costs and improve performance by shifting responsibilities and even cutting staff. One way to direct efforts at reengineering is to identify the organization's decision requirements and use these as a starting point for the redesign. The new structure can be built to handle these decision requirements.

Market researchers can study the naturalistic decision strategies of consumers. The market research community primarily relies on structured surveys and interviews, but also has begun to adopt the use of focus groups. Therefore, in-depth Cognitive Task Analysis interviews would constitute a sharp departure from standard procedures. The outputs of these studies appear to be very useful for formulating marketing and development plans.

CONCLUSIONS

This chapter has identified a range of methods for improving decision making. The basis for many of these methods is the use of decision requirements. I have tried to explain the advantages of shifting from generic, context-free, normative models of decision making to a perspective that centers around the critical judgments and decisions within a specific domain. I have also listed a range of interventions that are consistent with NDM tenets. Researchers have made important strides during the past 5 years in learning how to apply the NDM framework.

ACKNOWLEDGMENT

This chapter was based upon a technical report prepared for the U.S. Department of the Army, PO #DASW01–94–M–9906.

REFERENCES

Christensen-Szalanski, J. J. J. (1986). Improving the practical utility of judgment research. In B. Brehmer, H. Jungermann, P. Lourens, & G. Sevón (Eds.), *New directions for research in judgment and decision making* (pp. 383–410). New York: North-Holland.

Driskell, J. E., Salas, E., & Hall, J. K. (1994). *The effect of vigilant and hyper vigilant decision training on performance.* Paper presented at the annual meeting of the Society of Industrial and Organizational Psychology, Nashville, TN.

Fraser, J. M., Smith, P. J., & Smith, J. W. (1992). A catalog of errors. *International Journal of Man–Machine Studies, 37,* 265–307.

Gigerenzer, G. (1987). Survival of the fittest probabilist: Brunswik, Thurstone, and the two disciplines of psychology. In L. Kruger, G. Gigerenzer, & M. S. Morgan (Eds.), *A probabilistic revolution: Ideas in the sciences* (Vol. 2, pp. 49–72). Cambridge, MA: MIT Press.

Hammond, K. R., & Adelman, L. (1976). Science, values, and human judgment. *Science, 194,* 389–396.

Kaempf, G. L., Klein, G. A., Thordsen, M. L., & Wolf, S. (1996). Decision making in complex command-and-control environments. *Human Factors, 38*(2), 220–231.

Klein, G., Wolf, S., Militello, L., & Zsambok, C. (1995). Characteristics of skilled option generation in chess. *Organizational behavior and human decision processes.* San Diego, CA: Academic Press.

Klinger, D. W., Andriole, S. J., Militello, L. G., Adelman, L., Klein, G., & Gomes, M. E. (1993). *Designing for performance: A cognitive systems engineering approach to modifying an AWACS human–computer interface* (AL/CF–TR–1993–0093). Wright–Patterson AFB, OH: Department of the Air Force, Armstrong Laboratory, Air Force Materiel Command.

Lopes, L. L. (1981). Decision making in the shortrun. *Journal of Experimental Psychology: Human Learning and Memory, 1,* 377–385.

Means, B., Salas, E., Crandall, B., & Jacobs, O. (1993). Training decision makers for the real world. In G. A. Klein, J. Orasanu, R. Calderwood, & C. E. Zsambok (Eds.), *Decision making in action: Models and methods* (pp. 306–326). Norwood, NJ: Ablex.

Miller, T. E., Pyle, D. M., & Shore, J. S. (1993). Development of a prototype decision support system for munitions effects assessment. *Proceedings of the special session held concurrently with the Sixth International Symposium on Interaction of Nonnuclear Munitions with Structures* (pp. 1–2). Vicksburg, MS: U. S. Army Engineer Waterways Experiment Station.

Prince, C., Chidester, T. R., Bowers, C., & Cannon-Bowers, J. (1992). Aircrew coordination—Achieving teamwork in the cockpit. In R. W. Swezey & E. Salas (Eds.), *Teams: Their training and performance* (pp. 329–353). Norwood, NJ: Ablex.

Smith, P. J., McCoy, E., & Layton, C. (1993). Design-induced error in flight planning. *Proceedings of the Human Factors and Ergonomics Society 37th annual meeting* (pp. 1091–1095). Santa Monica, CA: HFES.

Stokes, A., Belger, A., & Zhang, K. (1990). Investigation of factors comprising a model of pilot decision making: Part II. In *Anxiety and cognitive strategies in expert and novice aviators* (Tech. Rep. No. ARL–90–8/SCEEE–90–2). Urbana–Champaign, IL: Institute of Aviation.

Vicente, K., & Rasmussen, J. (1990). The ecology of human–machine systems II: Mediating "direct perception" in complex work domains. *Ecological Psychology, 2*(3), 207–249.

Wiener, E. L., Kanki, B. G., & Helmreich, R. L. (Eds.). (1993). *Cockpit resource management.* San Diego: Academic Press.

Woods, D. D., & Roth, E. M. (1988). Cognitive systems engineering. In M. Helander (Ed.), *Handbook of human–computer interaction* (pp. 1–41). Amsterdam: Elsevier.

Zakay, D., & Wooler, S. (1984). Time pressure, training, and decision effectiveness. *Ergonomics, 27,* 273–284.

Zsambok, C. E., Klein, G., Kyne, M. M., & Klinger, D. W. (1992). *Advanced team decision making: A developmental model.* Fairborn, OH: Klein Associates Inc. (Prepared under Contract MDA903–90–C–0117 for U.S. Army Research Institute for the Behavioral and Social Sciences)

Chapter 6

Naturalistic Decision Making in Health Care*

◆

Marilyn Sue Bogner
U.S. Food and Drug Administration

This chapter presents a synopsis of current thinking among experienced professionals about current and future applications of naturalistic decision making (NDM) to health care. These researchers brought both clinical experience and applied research to their consideration of the topic at the 1994 conference on Naturalistic Decision Making: David Gaba, MD; Colin Mackenzie, MD; Patricia Martin, RN, PhD; David Woods, PhD; and the present author (a PhD).

Decision making in health care is a legitimate domain for NDM research because it has the characteristics that identify an NDM setting (Orasanu & Connolly, 1993). Woods cautioned that considering and studying decision making by domain is artificial, perhaps counterproductive, and certainly nonparsimonious. Some on the panel agreed with this cautionary note, suggesting that progress in understanding NDM processes, strategies, and decision aiding applications or technologies might be better served if we look across domains rather than within domains. Others felt a balance is warranted where NDM researchers look both within and across domains.

The balance of addressing issues both within and across domains was one of the goals of the NDM conference and the present volume. This chapter is a discussion of factors and issues that bear on applications of NDM

---

*The opinions expressed are solely the author's and do not represent those of the Food and Drug Administration or the Federal Government.

research, both current and future, in the health care domain. In addition, examples of applications or technologies from other domains are discussed as they relate to health care.

BACKGROUND

The amount of money spent on health care in the United States exceeds that of any industry except the Federal government. In spite of that, research in vital areas in health care—such as consumers' decisions to access care and the health care professionals' decisions as to what care to provide—has received paltry funding when compared to funding for decision-related research in aviation and nuclear power. Gaba believes one reason for the lack of significant funding is regulation of the practice of medicine essentially is nonexistent.

The practice of medicine is regulated to the extent that health care providers pass State Boards for licensure. As Gaba pointed out, after licensure, professional activities of health care providers are not regulated; the devices and drugs which they use are, however. Approximately 30,000 physician anesthesiologists and about the same number of nurse anesthetists are practicing as unregulated independent businesses. For those businesses, the absence of regulation decreases the impetus for funding performance-related research.

Another factor in the lack of significant research funding is the absence of health care disasters in which many people die at one time. Disasters such as plane crashes or impending disasters that threaten a large population as in the Three Mile Island nuclear power incident receive extensive media coverage. That coverage incites public concern, which sparks Congressional interest. Such interest often is manifest in funding for research to reduce the likelihood of similar events occurring. In health care, adverse events (incidents of inappropriate treatment) claim peoples' lives one at a time. Of the people who die as a result of adverse events, most are sick or old or both. For those populations, death often is not a particularly noteworthy or unexpected occurrence.

The importance of studying decision making in health care, including medical care decision making, is attested by the 18-year existence of the Society for Medical Decision Making. Research published in the Society's journal reflects the traditional methods of studying decision making, that is, using students as decision-making surrogates for health care professionals or patients. In discussing their research on decisions for breast cancer treatment, Chapman, Elstein, and Hughes (1995) noted that it is essential to question whether findings such as theirs would be evident in women in the age groups most at risk for breast cancer or in actual breast cancer patients. This underscores the importance of decision-making research in the real world.

Health care is emerging as a domain for actual NDM research. Within a NDM context, mention has been made of various aspects of decision making in medicine such as in medical diagnosis (Pennington & Hastie, 1993) and selecting outcomes of importance (Christensen-Szalanski, 1993).

The value of NDM research for health care decision making was emphasized by Martin in her discussion of an NDM study of nurses in a neonatal intensive care unit (ICU; Crandall & Getchell-Reiter, 1993). The nurses were asked to identify difficult decision periods such as generalized infections. The NDM techniques employed in obtaining that information resulted in information sharing. The information was synthesized so that problems could be more readily recognized. This problem-centered information is valuable to the nurses in their decisions regarding the infants in their care. The availability of such information is particularly important because health care professionals often do not have the time to discuss their work with each other.

FACTORS IN HEALTH CARE DECISION MAKING

In discussing health care decision making, the focus becomes medical decision making by health care professionals. Martin provided a cogent warning against such a focus. She believes that research in health care decision making can be useful in understanding the often overlooked, but very important decisions that consumers make regarding their health care. This includes decisions not only about when and how to access the health care system, but also decisions about questioning and complying or not complying with the prescribed regimen. An understanding of those decisions can help target efforts to assist people to take care of their health and by so doing, decrease the need for extensive, expensive treatment and associated personal costs to the patients and to society.

The time-stress characteristic of NDM settings is an important factor in health care decision making. Mackenzie described his research findings that during emergency room (ER) surgery compared with elective surgery, care providers tend to shed tasks as time pressure increases. Gaba discussed time pressure that occurs in the operating room (OR) and the ICU, which often is the result of a crisis situation in a patient's condition compounded by the providers' heavy workload. Both Mackenzie and Gaba spoke of fatigue caused by lack of sleep or an excessive cognitive workload as a pervasive, time-related factor that has a deleterious effect on decision making. This occurs for lay health care providers as well as health care professionals.

A stress-inducing factor that particularly is inherent in the three high-activity, technology-intensive, high-cost settings of the ER, OR, and ICU is the concern for cost savings. This results in production pressure—pressure to do more and be more productive in a given time period. Gaba stated that production pressure causes stress that comes not from the patients' condition as in time pressure, but from sources such as surgeons who demand that anesthesiologists complete their part of a case and start another so that the surgeons' schedule can be maximally filled. Under these conditions, Gaba, Howard, and Jump (1994) found that anesthesiologists have less time to consider alternatives and will tend to cut corners. An example of cutting corners is being overly reliant on technology. Technology, although offering information quickly, provides less information than experienced clinicians can obtain by using their skills, according to Gaba et al. (1994).

The emphasis on reducing health care costs introduces a factor that impacts health care decisions for the consumer, lay health care provider, and health care professional in more ways than the issue of access to care. Because time is money and money is linked with productivity, production pressure compounds time pressure with the resultant impact on the decision making of clinicians. The cost savings emphasis is for hospitals to perform more procedures and care for more people using the same personnel or using technology to do more in less time. This assumes that technology is a viable and useful health care team member. The emphasis on cost is based on the belief that short-term cost cutting results in long-term savings—a belief that may not be substantiated. This impacts the consumers and lay health care providers because patients are discharged to home health care at an earlier stage in their recovery than previously.

Decisions made by professional and lay health care providers can be influenced by the medical devices they use. Woods feels that many devices, particularly computerized devices, can be characterized as clumsy automation; that is, they provide poor feedback, require awkward actions for use, or involve excessive and unnecessary controls. The speed with which technology in medical devices changes—sometimes in less than a year—and the rapid introduction of medical devices into use differ from equipment in other domains. For example, technology in an airplane's cockpit may change appreciably every 10 to 20 years and the introduction of that cockpit into use may occur over several years. Thus, more effort can be directed to reducing the clumsiness of that automation than for the time-sensitive development of medical devices.

Another medical device issue is that the same type of device may differ significantly within a given institution. For a commonly used device such as infusion pumps (computer-based devices that measure and deliver medication), the newest technologically sophisticated model may be in some departments, whereas other departments have older models. Clinicians are expected to effectively use an unfamiliar device, often under time if not production pressure. They must make decisions about the manner in which the diverse devices operate and develop ways to accommodate incompatible accessories.

Woods believes that medical devices afford tremendous opportunities for latent failures or failures that manifest themselves in human performance. These errors happen from high-end medical applications, as in costly institutional care, through the various levels of the health care system, to home health care. In the case of home care, technologically complex medical devices are put in the home to be used by poorly trained, often elderly, sometimes ill, and usually stressed and fatigued lay care givers. Woods proposed that studies of decisions involving medical devices, such as infusion devices, throughout the health care system would be helpful in reducing the incidence of adverse outcomes of injuries and deaths. Such studies could identify adverse event inducing aspects of devices that were designed for specific institutional uses such as the OR, yet are used throughout the hospital and even put into home care situations.

Much of the onus of personal responsibility triggered by focusing on adverse events per se, might be avoided by exploring precursor events (Woods, 1994). Information about precursor events is vital to reducing error. As Woods discussed, a precursor event to committing a mode error (e.g., not understanding that a switch would cause different activities than assumed) was the precipitating factor in a plane crash. The pilot did not know the plane was descending at an abnormal rate. Understanding precursor events allows interruption of the cascade of errors that accompanies adverse outcomes. Based on data from such events, remedial actions were taken by the aviation industry to reduce the possibility of mode error. With technology so pervasive in health care, it is probable that analogous incidents occur frequently with adverse outcomes.

An example of a system precursor event offered by Mackenzie was the difficulty in obtaining information about whether a breathing tube is actually in the trachea (windpipe) so the patient can breathe, or if it is erroneously placed in the esophagus, which leads to the stomach. This is particularly a problem in anesthesia in the ER because the resuscitation bag used in transporting the patient has no meter to measure the exhaled carbon dioxide—information indicating that the tube is placed appropriately. The meter is attached to the ventilator and it may take as long as several minutes to attach the patient to that device and obtain the necessary information. If the tube is in the esophagus during that time, an adverse outcome of death or a persistent vegetative state may occur. This is an example of a problem in equipment design.

Other people are a factor in health care decision making. Although a variety of people may be present with a lay provider in a home care setting, the major impact of other people on decision making occurs for the professional clinician. Peer pressure in multiple provider settings such as the OR and ER impinges on decision making. Mackenzie and Gaba agreed that this especially is so with the delivery of anesthesia. In anesthesia, the patient monitors as well as the anesthesiologists' actions are in relatively clear view of other people so that inappropriate actions are readily observable. This has not been the case with surgeons, who may be the only ones able to observe their actions within a relatively small incision in traditional open surgery. The situation is changing with the increasing use of laparoscopic surgical techniques. In that surgery, a miniature video camera and surgical instruments are inserted at the surgical site via small incisions. Video images of the surgeons' actions are projected on monitors to guide their actions. These images can be seen by others in the OR.

Another aspect of peer pressure is the lack of clear command structure in the OR and ER. Surgeons often feel that the patient is their responsibility, hence that they are in command. Anesthesiologists also feel the patient is their responsibility and that they are in command. This often is a source of problems in decision making. When a crisis occurs, there is a question as to the locus of command. Patient care can be and sometimes is briefly subordinated to resolving that question.

Both Mackenzie and Gaba believe that the lack of or inappropriate communication affects medical decision making. Each individual in the OR or ER is trained to perform certain tasks essentially independently, yet they are expected to behave as members of a crew. Anesthesia care providers must interact not only with those in the anesthesia "crew" and but also with those in other "crews." Communication among the several crews often is difficult.

As there can be distortion in communication, distortion also can occur in the perception of events and the perceivers' role in the events. Mackenzie spoke of his study of videotapes (Mackenzie, Craig, Parr, Horst, & Level One Trauma Anesthesia Simulation Group, 1994) in which of the 11 untoward events that occurred, only 3 were reported in self-evaluation. People who provide care in the ER may be so involved in their tasks that unless an event is of considerable magnitude, it may not be perceived; hence it is not reported and valuable data are lost.

The intercity ERs, Mackenzie pointed out, are very rich settings in which to study medical decision making because of the conditions of providing care to severely injured patients, injuries that often compromise care, and extreme time constraints. In such a setting, approximately 40% of the trauma patients die in the first hour not from adverse events, but from the severity of the injuries—circumstances that challenge decision-making processes.

Another factor in health care decision making is information about adverse events. Reduction in the incidence of medical adverse events can occur only if there are viable data through which to understand and study how health care is delivered. High reliability organizations, Woods noted, welcome information about critical incidents; that is, adverse events, and organizational factors that contribute to them. The current litigious attitude toward health care discourages people from reporting problems and keeps health care from becoming a high-reliability organization. The current trends in health care are in the opposite direction, away from encouraging information gathering. Medical care providers have no safe forum for discussing issues, problems, and possible solutions. Having a "black box" in the OR similar to that in a plane's cockpit might be useful in collecting data; however, experience with the aviation industry indicates that such devices tend to create "sterile cockpits" where, in order to avoid incrimination, communication is reduced or absent.

Because knowledge of adverse events in health care may result in litigation, it is not surprising that there is a reluctance to report these occurrences. The number of one-at-a-time deaths from adverse events, however, is nontrivial. It has been estimated that there are approximately 100,000 preventable deaths annually from adverse events in hospitalized patients in the United States (Leape, 1994). This is twice the annual highway death rate.

Serious injuries as well as deaths incurred by medical treatment of hospitalized patients were identified through a retrospective chart review by the Harvard Medical Practice Study (HMPS; Brennan et al., 1991). Not only do those adverse outcomes have profound personal consequences for patients and their families, they also have an economic impact on society. An extrapolation of the findings of the HMPS (Brennan et al., 1991; Johnson et al., 1992) and the determination that 70% of the problems identified in that study were preventable (Leape, 1994) indicate an esti-mated cost of $25 billion for the consequences of preventable adverse events in hospitalized patients for the year of the Harvard study (Bogner, 1994). Gaba feels that $25 billion is a mere pittance compared with the overall cost of health care in this country. In an absolute sense, however, $25 billion is a significant amount of money. The magnitude of the problem should stimulate significant research funding.

It is not by chance that the preceding discussion of factors in health care decision making is in terms of correcting a problem. The orientation of health care and health care decision making is to correct a health problem, to restore the individual to a previous state of health, to facilitate homeostasis. This is contrasted to decision making in other domains such as business, in which the decision maker has a vision of the direction to develop the organization.

The orientation of health care also indicates a reactive mode for health care decision making. Health care providers are presented with a set of symptoms to which they must respond. Their responses are a series of decisions to determine: which of the presenting symptoms fits a tentative diagnosis, what tests should be conducted to aid the diagnosis, a course of action, and the medication and treatments to execute the course of action either in part or in its entirety.

An additional aspect of the reactive characteristic of health care is the influence of environmental factors on the decision maker. A central factors is time. Time stress is a characteristic of NDM situations; however, in some domains, such as business, the locus of the time stress is, for the most part, internal to the decision maker. To achieve their goals, business decision makers may develop strategies to manipulate aspects of the environment to create advantageous conditions for action. There is time for negotiation in many instances. In health care, the locus of the time stress most often is outside the decision maker. Both lay and professional health care providers must react to change in the patients' condition, must make decisions in a hurry and without a full complement of information.

Another factor present in many situations where NDM is prevalent is high stakes (Orasanu & Connolly, 1993). As Chapman et al. (1995) described, the personal consequences of decisions in health care frequently are greater than in other domains. When a health care decision affects the decision makers' personal health, characteristics of that decision are likely to differ from those operating when the decision maker considers the health of a family member. The characteristics of both of those decisions probably are not comparable to those in making decisions regarding treatment of a patient.

The range of personal consequences of a decision varies across domains of decision making. Business decisions impact the financial status of the company or industry and perhaps even the safety of the employees, but rarely is the physical well-being of the decision makers directly affected. In aviation, the decisions of pilots when flying have personal consequences as well as consequences for the passengers; that is, when the plane crashes, pilots as well as the passengers may die. This is not the case for decisions by air-traffic controllers. Their decisions impact the fate of the planes under their control, but not their own personal, physical well-being. Thus, the target of the decision to be made and the consequences of the decision on the decision maker are factors to be considered in studying the consequences of decision making.

In conclusion, there are some idiosyncratic characteristics of decision making in health care that should be considered by NDM researchers in addition to those that are typical across many domains. Health care decision making has a: restorative orientation, reactive approach, nonnegotiable time stress, and often major personal consequences. Those characteristics describe a unique "footprint" of factors. The factors that are inherent in the health care domain challenge traditional decision making research and underscore the importance of this domain for research by the NDM community.

REFERENCES

Bogner, M. S. (1994). Human error in medicine: A frontier for change. In M. S. Bogner (Ed.), *Human error in medicine* (pp. 373–393). Hillsdale, NJ: Lawrence Erlbaum Associates.

Brennen, T. A., Leape, L. L., Laird, N. M., Herbert, L., Localio, A. R., Lawthers, A. G., Newhouse, J. P., Weiler, P. C., & Hiatt, H. H. (1991). Incidence of adverse events and negligence in hospitalized patients. *New England Journal of Medicine, 324*(6), 370–376.

Chapman, G. B., Elstein, A. S., & Hughes, K. K.(1995). Effects of patient education on decisions about breast cancer treatments: A preliminary report. *Medical Decision Making, 15*(3), 231–240.

Christensen-Szalanski, J. J. J. (1993). A comment on applying experimental findings of cognitive biases to naturalistic environments. In G. A. Klein, J. Orasanu, R. Calderwood, & C. E. Zsambok (Eds.), *Decision making in action: Models and methods* (pp. 252–264). Norwood, NJ: Ablex.

Crandall, B., & Getchell-Reiter, K. (1993). Critical decision method: A technique for eliciting concrete assessment indicators from the intuition of NICU nurses. *Advances in Nursing Science, 16*(1), 42–51.

Gaba, D. M, Howard, S. K., & Jump, B. (1994). Production pressure in the work environment. *Anesthesiology, 81*(2), 488–500.

Johnson, W. G., Brennan, T. A., Newhouse, J. P., Leape, L. L., Lawthers, A. G., Hiatt, H. H., & Weiler, P. C. (1992). The economic consequences of medical injuries. *Journal of the American Medical Association, 267*(18), 2487–2492.

Leape, L. L. (1994). The preventability of medical injury. In M. S. Bogner (Ed.), *Human error in medicine* (pp. 13–25). Hillsdale, NJ: Lawrence Erlbaum Associates.

Mackenzie, C. F., Craig, G. R., Parr, M. J., Horst, R. L., & Level One Trauma Anesthesia Simulation Group. (1994). Video analysis of two emergency tracheal intubations identifies flawed decision-making. *Anesthesiology, 81*(3), 763–771.

Orasanu, J., & Connolly, T. (1993). The reinvention of decision making. In G. A. Klein, J. Orasanu, R. Calderwood, & C. E. Zsambok (Eds.), *Decision making in action: Models and methods* (pp. 3–20). Norwood, NJ: Ablex.

Pennington, N., & Hastie, R. (1993). A theory of explanation-based decision making. In G. A. Klein, J. Orasanu, R. Calderwood, & C. E. Zsambok (Eds.), *Decision making in action: Models and methods* (pp. 188–204). Norwood, NJ: Ablex.

Woods, D. D. (1994). Cognitive demands and activities in dynamic fault management: Abductive reasoning and disturbance management. In N. Stanton (Ed.), *The human factors of alarm design* (pp. 63–92). London: Taylor & Francis.

Chapter 7

Naturalistic Decision Making in Command and Control

◆

Michael Drillings
U.S. Army Research Institute

Daniel Serfaty[1]
APTIMA, Inc.

The command of armed forces in battle is a complex and difficult skill. Commanders must make decisions under quite difficult circumstances that may have enormous consequences. Command and control (C^2) is the term that describes the job of the battle commander. C^2 is characterized by ill-structured problems, changing conditions, high stakes, and time demands. The field of naturalistic decision making (NDM) has some of its roots in the military's need to better understand the human dimensions of C^2. Naturalistic decision making research has investigated the cognitive processes that underlie how individuals actually make decisions and how they use their experience and training under demanding conditions. The research has resulted in a change in how psychologists, and ultimately battle commanders, view the process of decision making.

Our main goal is to specify the relationship between NDM and C^2. This chapter is based partly on a panel discussion including, in addition to the authors: Jon Fallesen of the U.S. Army Research Institute; Simon Henderson of the Defense Research Agency, U.K.; Jeff Grossman of the Naval

[1]Formerly with ALPHATECH, Inc.

Command, Control and Ocean Surveillance Center; and James Bushman of the U.S. Air Force Armstrong Laboratory.

Joint Pub. 1–02 (1994) of the Joint Chiefs of Staff provides a comprehensive definition of the process of C^2—one that encompasses a variety of interpretations:

> The exercise of authority and direction by a properly designated commander over assigned forces in the accomplishment of the mission. Command and Control functions are performed through an arrangement of personnel, equipment, communications, facilities, and procedures employed by a commander in planning, directing, coordinating, and controlling forces and operations in the accomplishment of the mission. (p. 78)

As the definition implies, C^2 is a human-centered decision-making process that is supported by Command, Control, Communications, and Intelligence systems (Levis & Athans, 1988). To successfully accomplish C^2, the military commander relies on information about personnel, materiel, and the time available during the period leading up to combat and in combat itself. The history of battle shows that the winner of a battle is not always the side with the most troops; rather, it is the side that can create a timely advantage (through maneuver, deception, or persistence, etc.).

The battle commander employs a C^2 process to develop an overall strategy and an implementation of the strategy (i.e., tactics). The commander manages this task with a command staff and with the aid of available information. That information usually consists of an understanding of both the commander's and the enemy's resources and disposition. Miscommunication and intentional deceptive efforts by the enemy further complicate the task. Key components of the commander's task are to assess the situation, develop courses of action, make decisions, and monitor their implementation.

In what follows in this chapter, we first assess the contributions and shortfalls of the classical approach in supporting the C^2 decision-making process and in providing guidance for training the battle commander. We then describe alternative approaches to the C^2 challenge and show how the NDM approach supports a commander-centered definition of command and control. The concluding section of the chapter reviews the potential implications of this new framework in providing realistic solutions for the enhancements of battle command decision-making performance through decision support and training.

THE CLASSICAL APPROACH

All modern armies have recognized the prime importance of effective command and control (Levis & Athans, 1988). Command and control has been an area of scientific analysis for many years. Early efforts have concentrated on developing a consistent, analytical approach to solving command and control problems (Van Trees, 1989). The classical approach (see Orasanu & Connolly, 1993), using analytical tools, was not designed to represent what commanders wanted to do, or even actually did during the command process. Instead, it tried to guide them to perform tasks that seemed analytically correct.

Training Using the Classical Approach

In the past, U.S. military commanders were often trained to develop three courses of action for a tactical scenario. Then they were to estimate the advantages and disadvantages of each course of action and weigh those evaluations in reaching a decision. It is, perhaps, not surprising that in the modern era scientists, engineers, and mathematicians seeking to aid the battlefield commander have attempted to quantify the decision process. A "Rational Choice" decision aid for the commander might consist of a computational tool that helps to quantify the advantages and disadvantages of alternative courses of action. The final "recommended" decision would then be computed using an algorithm-based technique such as Bayesian analysis or multiattribute utility theory.

Acceptance of the Classical Approach

The military never completely adopted the tools and procedures derived from a classical approach that stressed Rational Choice models. First, command style was always seen as the prerogative of the commander. Each commander's style was different and the freedom to express that style was seen as crucial to the success of the operation (Van Creveld, 1985). In the classical approach there is actually little room for the human as a cognitive processor, who would ordinarily rely on a wealth of experience and knowledge to make decisions. In the Rational Choice model, the commander was largely seen as one who performs the calculations that will lead to optimal decisions. Further, the classical approach minimized the fact that data used for quantitative judgments might be suspect, severely affecting the validity of the resulting calculations. Nonetheless, the classical approach, although honored mostly in theory, has been pervasive in the support systems and training products that resulted from these efforts (Van Trees, 1989). Motivated by the realization that an alternative paradigm was needed, several research teams began the search for alternative approaches to understand C2 decision making.

NEW CONCEPTUALIZATIONS

Various paradigms have been proposed to analyze more effectively the cognitive component in human decision making in military C^2 environments (Kahan, Worley, & Stasz, 1989). Only a few, however, address the issue of command expertise and its effects on the commander's decision style and information needs (Helme & Uhlaner, 1979; Michel & Riedel, 1988; Mutter, Swartz, Psotka, Sneed, & Turner, 1988; Wohl, 1981).

Role of Multiple Alternatives

Military planning doctrine calls for the preparation and evaluation of multiple options in developing the plan. Empirical evaluations of planning effectiveness, however, do not support the advantage of developing multiple distinct options.

Laboratory experiments produced mixed results on this issue. Entin, Needalman, Mikaelian, and Tenney (1988) found little difference in actual performance in an air–land defense scenario between single and multiple option conditions. They noted that the single-option groups appeared to have compensated for the lack of additional explicit options by exhibiting hedging behavior (i.e., the assignment and positioning of forces to be able to execute any of several options, conditional on two or more alternate but implicit hypotheses about enemy position, capabilities, intent, or any combination thereof).

A series of observational and interview studies of decision making in field environments conducted by Klein and his colleagues have also shed doubt on the value of multiple option generation in military decision making (Klein, 1988; Klein, 1989; Klein, Calderwood, & Clinton-Cirocco, 1986). In one experiment (Klein, Wolf, Militello, & Zsambok, 1995), chess players reported on all of the moves that they were considering, even if they could dismiss the moves immediately. The results show that for both high- and low-skill players, the first move considered was often the move that the subject would follow even after 15 minutes of additional study.

These empirical studies represent a combination of systematic observation and real-world settings. A major finding of this body of research is that the experts rarely reported considering more than one option, in contradiction of then-current military doctrine. Instead, experts apparently recognized prototypical situations using feature cues and reached an understanding of the causal dynamics associated with the decision problem. This recognition suggested promising courses of action and generated expectancies. These empirical findings and the emergence of alternative paradigms for C^2 led to a reexamination of the concept.

REDEFINING COMMAND AND CONTROL

Command

In the last few years, especially since the Desert Storm operation, the U.S. armed forces have redefined command and control. This doctrinal shift, as defined in field manuals such as FM100–5 (U.S. Army, 1993) emphasizes the human element. Command is seen as the specific job of the commander, who gives direction to the battle and formulates concepts that drive it. The job of command is thus regarded as an art. Effective commanders are encouraged to visualize the battlefield, assess risk, and anticipate change while leading the battle.

Control

Control is conceptualized as a monitoring function. It is said to rely on the gathering of information that is processed and is then used to describe the battle as it progresses. Control is the "science" of regulating the battle—it describes "how to" do something, rather than "what to" do. Control is oriented to the processing of information, and lends itself to proceduralization and check lists.

With this differentiation between command and control, it can be seen that the classical approach to command and control decision making addresses "control" more than "command." Because it is so closely linked to data, control might be improved through the classical approach to decision making. Analytical tools are designed to effectively process data. Command is neither data-oriented nor proceduralized. It provides the intelligent framework that makes sense of the data and depends heavily on the commander's and staff's knowledge. Commanders must respond quickly to changing battle conditions—not with the application of set responses to specific events, but with experience-based "intuition," creativity, and initiative, as there may be little opportunity during a battle to rely on factors other than experience. Control can be taught through instruction in procedures and rules. Command has fewer rules and because of its infinite variety cannot be easily instructed.

RESEARCH IN NATURALISTIC DECISION MAKING

Characteristics of NDM

The conceptual changes that led to redefinition of command and control is strongly supported by the theoretical dimensions of NDM. The results of NDM research stimulated many changes in the Army's command and control doctrine. Klein, Orasanu, Calderwood, and Zsambok (1993) de-scribed several naturalistic decision making models. Orasanu and Connolly (1993) described the eight key characteristics of naturalistic decision mak-ing settings. Those characteristics are: ill-structured problems, uncertain dynamic environments, shifting or competing goals, action/feedback loops, time stress, high stakes, multiple players, and organizational goals and norms. These characteristics are usually present in battle situations at least as much as in any other human activity. Command and control operations are therefore an almost ideal setting for research to better understand naturalistic decision making processes and for the application of those processes to improve command and control performance.

The Recognition-Primed Decision (RPD) Model

The early work of Klein was performed for the U.S. Army Research Institute, and was motivated by the Army's need to better understand decision making in command and control. Klein (1988) observed that decision makers in difficult situations and under time stress did not appear to use the classical approach to make decisions, even when they were trained in that approach. Instead of seeing this discrepant behavior as an "error," Klein saw it as a reasonable behavior that effectively used the expertise of the decision maker. Klein later termed a specific model of decision making the RPD model. The model emphasizes the importance of situation assessment in expert decision making. Several chapters of this volume discuss the RPD model in detail.

Key Concepts

Perhaps Klein's major accomplishment was the realization that the strategies used by experienced decision makers to make decisions were a legitimate expression of their developed expertise. Early observations were made in standard command and control situations: the behavior of forest fire com-manders, commanders of urban fire brigades, urban emergency rescue teams, and tank platoon commanders.

The RPD model postulates that experienced decision makers generate fewer alternatives for action and that they generate the options sequentially. The decision maker does not directly compare the alternatives but, instead, stops generating alternatives when a satisfactory decision is found. The focus is on understanding the situation and judging its familiarity, not on the generation of options and the direct comparison of the alternatives (perhaps through calculation) to find an acceptable solution. The decision maker is aware of the demands of time and of the possible inaccuracy of the available information.

IMPLICATIONS OF THE NDM APPROACH TO C²

If the classical approaches require the application of analytical and computational tools to make decisions, then what are the implications of the NDM approach?

NDM Strategies as Decision Support Techniques

The NDM approach relies on having the decision maker better understand the conditions under which the decision is made. Decision makers use their previous experience by framing the current situation in terms that are relevant to previous experience and training. When the commander finds a good match between the current situation and past situations, then the course of action becomes clear. A poor match causes an increased effort to acquire additional information. The quality of the match also provides a measure of the commander's confidence in the decision.

NDM approaches to decision making are supportive, rather than replacements for other methods. The classical approach, using the Rational Choice model, is certainly a useful model to use under certain conditions. For example, it is better when there is sufficient information about the situation such that alternative courses of action and their implications are all well understood. Such complete information might provide all the information that is necessary to apply, for example, multiattribute utility theory. Unfortunately, in a battle situation, even in the rare event that such complete information is available, the time demands may be too great to fully apply the methods of Rational Choice.

If NDM techniques rely on the expertise of the decision maker, rather than analytical solutions, then how do we both define and train expertise? The practice of command in battle is not quite like a chess game. Serfaty, MacMillan, Entin, and Entin (chapter 23, this volume) describe a theoretical framework to understand and evaluate expertise in a battle command decision-making environment. They argue that battle command is one of the most complex decision-making environments in which true expertise comes into play. Unlike games like chess or physics problems, battles go on far longer, the forces are not equal, there are external forces acting, and sometimes it is difficult to know whether you won or lost. The definition and measurement of expertise, therefore, are quite difficult. Experience and training, perhaps, are the best measure of expertise. By its nature, the experience of leading troops in battle is now rare. Even those who have done it more times than most have still done it only infrequently.

NDM Strategies as Training Techniques

If modern armies cannot rely on actual war experience to train commanders, they must rely on training. Although training includes both classroom and nonclassroom instruction, because NDM stresses the importance of the special conditions of the battle situation (i.e., time demands, stress, etc.), training that more closely simulates battle conditions is more likely to be effective.

Daniel (1979) demonstrated the importance of individual expertise differences in the use of information in a tactical setting. Daniel arranged a set of tactical battle scenarios that were given to a group of tactically experienced players, who were then asked to decide on a course of action. Players' decisions were scored based on time taken and enemy units destroyed. The results showed that individual differences in the expertise of the players were equally if not more important than quantity of information. Although more information definitely improved the quality of the decisions made by all players, the best players made as good decisions with 20% of ground truth as did the worst players with 80% of ground truth. However, it is not clear whether the performance differences observed in Daniel's study were due to differences in training and experience.

Studies across several expertise domains (Chase & Simon, 1973) indicate that experts organize knowledge about their domain into semantically meaningful units or chunks in long-term memory and that these chunks differ from those formed by novices. Also, expert knowledge in long-term memory is pattern indexed, and this pattern indexing is organized to facilitate achievement of domain-specific goals (e.g., winning the chess game or solving the physics problem). This pattern indexing allows the expert to immediately retrieve the information needed to focus on solving a particular problem.

Training Strategies

Although experience (and the resulting expertise that is developed through experience) is an important factor in making good decisions, several relevant skills can be trained for more effective decision making. Situation awareness can be trained so that the decision maker is more aware of alternative hypotheses and less likely to fixate on a hypothesis. Decision makers can be trained to make fine distinctions that distinguish one situational pattern from another. The recognition of typical patterns is also trainable. Some research indicates that decision makers profit from establishing a "mental model" of the situation they face. The mental model provides a framework on which to hang changing information. Finally, decision makers can also be trained to look further out in time in making their plans.

Building Experience

Even actual combat may not provide good training for commanders. Combat is sufficiently complex that the results of combat may not be easily understood. More and more, both trainers and trainees are using simulation.

Simulation has an immediacy and complexity that can rival combat. Moreover, the control of conditions makes it possible to structure training more effectively. The After-Action Review is also a valuable source of training. Traditional, classroom-based training can also be effective. Understanding past battles and how commanders made decisions may provide valuable insights into the decision-making strategies of experts.

CONCLUSIONS

The naturalistic approach to understanding decision making has made an important contribution to the reformulation of the U.S. armed forces' policy on command and control. NDM considers the time demands of the situation, the lack of adequate information, and the uncertainty that is present. It also recognizes that the uncertainties of war, the time demands put on the commander, and the sheer complexity and dimensionality of the battlefield rarely allow the commanders to make decisions using rational, analytical methods.

The naturalistic framework recognizes that commanders, because of their experience and training, can make important personal contributions to the quality of decisions. Moreover, NDM models are supported by a body of cognitive research on behavior in command and control situations.

Will the NDM paradigm withstand a generalization from the description of the individual decision-making process to the understanding of coordinated decision making in complex organizations? We suggest that the next key challenge for naturalistic C^2 researchers resides in the understanding and modeling of the team cognitive processes that support distributed decision making in large-scale joint operations, wargames, and simulations.

REFERENCES

Chase, W. G., & Simon, H. A. (1973). Perception in chess. *Cognitive Psychology, 4,* 55–81.

Daniel, D. W. (1979). What influences a decision: Some results from a highly controlled defence game. *OMEGA, The International Journal of Management Science, 8,* 409–419.

Entin, E. E., Needalman, A., Mikaelian, D., & Tenney, R. R. (1988). *Experiment II report: The effects of option planning and battle workload on command and control* (TR–388–1). Burlington, MA: ALPHATECH, Inc.

Helme, W. H., & Uhlaner, J. E. (1979). *Relationship between leader knowledge, directive behavior, and performance in administrative, technical, and combat situations.* (Tech. Paper 373). Alexandria, VA: US Army Research Institute.

Joint Pub 1–02. (1994). *Department of Defense dictionary of military and associated terms.* Washington, DC: Department of Defense.

Kahan, J. P., Worley, D. P., & Stasz, C. (1989). Understanding commander's information needs (RAND Tech. Rep. No. R–3761–A), Santa Monica, CA.

Klein, G. A. (1988). Naturalistic models of C^3 decisionmaking. In S. Johnson & A. Levis (Eds.), *Science of command and control: Coping with uncertainty* (pp. 86–92). Washington, DC: AFCEA International Press.

Klein, G. A. (1989). Recognition-primed decisions. In W. B. Rouse (Ed.), *Advances in man–machine system research* (Vol. 5, pp. 47–92). Greenwich, CT: JAI.

Klein, G. A., Calderwood, R., & Clinton-Cirocco, A. (1986). Rapid decision making on the fire ground. *Proceedings of the Human Factors Society 30th Annual Meeting, 1,* 576–580.

Klein, G. A., Orasanu, J., Calderwood, R., & Zsambok, C. E. (Eds.). (1993). *Decision making in action: Models and methods.* Norwood, NJ: Ablex.

Klein, G. A., Wolf, S., Militello, L., & Zsambok, C. (1995). Characteristics of skilled option generation in chess. *Organizational Behavior and Human Decision Processes, 62,* 63–69.

Levis, A. H., & Athans, M. (1988). The quest for a C3 theory: Dreams and reality. In S. Johnson & A. Levis (Eds.), *Science of command and control: Coping with uncertainty* (pp. 4–9). Washington, DC: AFCEA International Press.

Michel, R. R., & Riedel, S. L. (1988). *Effects of expertise and cognitive style on information use in tactical decision making.* (Tech. Rep. 806). Alexandria, VA: U.S. Army Research Institute.

Mutter, S. A., Swartz, M. L., Psotka, J., Sneed, D. C., & Turner, J. O., (1988). Changes in knowledge representation with increased expertise (Tech. Rep. No. 810). Alexandria, VA: U.S. Army Research Institute for the Behavioral and Social Sciences.

Orasanu, J., & Connolly, T. (1993). The reinvention of decision making. In G. A. Klein, J. Orasanu, R. Calderwood, & C. E. Zsambok (Eds.), *Decision making in action: Models and methods* (pp. 3–20). Norwood, NJ: Ablex.

U.S. Army. (1993). *FM 100–5, Operations.* Washington, DC: HQDA.

Van Creveld, M. (1985). *Command in war.* Cambridge, MA: Harvard University Press.

Van Trees, H. L. (1989). C^3 systems research: A decade of progress. In S. Johnson & A. Levis (Eds.), *Science of command and control: Part II. Coping with complexity* (pp. 24–44). Washington, DC: AFCEA International Press.

Wohl, J. G. (1981). Force management requirements for Air Force tactical command and control. *IEEE transactions and systems, man, and cybernetics, SMC11*(9), 618–639.

Chapter 8

Current and Future Applications of Naturalistic Decision Making in Aviation

◆

George L. Kaempf
Klein Associates Inc.

Judith Orasanu
NASA–Ames Research Center

The purposes of the panel discussion that provided the material for this chapter were to address areas in which Naturalistic Decision Making (NDM) methods and models are currently applied in the aviation industry and to identify future applications in which the NDM perspective may contribute to system performance and safety. We chose not to focus on a single dimension of the industry, but to assemble a panel that represented a wide variety of interests and perspectives including: military, commercial, and general aviation; flight-crew performance in fixed and rotary-wing aircraft; air traffic control and flight dispatch; training; and decision aiding. Panel members included: Judith Orasanu (NASA–Ames Research Center), George L. Kaempf (Klein Associates Inc.), Marvin Cohen (Cognitive Technologies, Inc.), Alan Stokes (Florida Institute of Technology), Andrew Lacher (Mitre Corporation), and Carolyn Prince (Naval Air Warfare Center Training Systems Division).

Aviation is an interesting domain for studying decision making. For example, pilots make important decisions frequently. Often, these decisions have a great consequence, but are made with ambiguous information, under great risk, and with very little time. However, the domain attracts considerable attention and has been studied extensively for decades. So, what can we add?

We have known for many years that a critical component of pilot proficiency is the ability to make good decisions and that decision skills could be trained. Decision training has become commonplace in the aviation industry, and these programs have had a positive impact on pilot performance. However, it is well documented that there is room for improvement, and we believe the most profitable avenue for improvement is to revisit the premises underlying our understanding of how pilots make decisions. The early concepts of pilot decision making were based on models that reflected the level of thinking in cognitive psychology some 25 years ago. These models remain the underlying premise of many contemporary decision training programs in aviation (Kaempf & Klein, 1994). Clearly, NDM research provides a more attractive alternative for understanding how pilots make decisions and for designing interventions that will help them make these decisions.

Thus, although the need is great, current applications of NDM research in aviation are rather limited. However, the future does look bright. An increasingly large number of researchers and practitioners recognize the limitations of existing programs and the need for advancement. In addition, researchers have identified a set of decision models and analytic methods that show great promise for identifying decision requirements and for providing a basis for developing effective training and decision supports.

With these factors serving as a backdrop, the panel discussion focused on three general topics: the implications of recognitional decision making, the "what" of decision training interventions, and research needs in the aviation community. This chapter describes highlights of these discussions organized around these three major topics.

IMPLICATIONS OF RECOGNITIONAL DECISION MAKING

The panel devoted considerable discussion to the nature of recognitional strategies observed in aviation settings and the need to support these strategies. Many facets of the aviation domain are procedural in nature. Much thought and many resources have gone into developing checklists and procedures for situations that have occurred or can be expected to occur. Because these checklists prescribe the appropriate courses of action for specified situations, in many cases the task of the decision maker is not to generate and evaluate courses of action. It is to accurately assess the situation. Once the decision maker recognizes the situation, the course of action becomes obvious. The source of information about this action may be either previous experience, a procedure, or a checklist. These findings are consistent with the recognitional decision strategies seen in other domains (e.g., see Kaempf, Klein, Thordsen, & Wolf, 1996).

Flow management in air traffic control (ATC) provides an excellent example. Flow management has both short-term tactical goals and long-term strategic goals. The tactical goals include separation and spacing of aircraft for safety. The strategic goals include maintaining an optimum flow of traffic through the system. Furthermore, the culture as well as the decision-making processes of controllers are driven by large penalties for mistakes. This is a highly procedural domain. Controllers rely heavily on prescribed courses of action derived from their Standard Operating Procedures. Their difficulty arises in determining which antecedent conditions apply.

Studies of expert controllers have produced several interesting findings (e.g., Bayles & Das, 1993; Mundra, 1989). When explaining the nature of a traffic management problem to another controller, experienced controllers spend most of their time explaining the procedures to be used. They describe the decisions made as obvious. However, when working the same problem, the controllers spend 90% of their time processing information rather than focusing on the procedures to be employed. That is, most of their effort is expended building their situation awareness. The actions to be taken become simple, once the situation is identified.

Therefore, decision supports for flow management might aid the controllers in building their situation awareness, not just in selecting or identifying courses of action. A critical component of this assessment may be to know when a situation is unusual and does not have a corresponding explicit procedure. Current studies of flow management examine the information that controllers use and share, the individuals they are sharing information with, and the holes in the information that is available for given situations.

Similar to ATC, many decisions made by flight crews are procedural, call for a recognitional decision strategy, and emphasize the importance of situation assessments. In fact, many of the procedural decisions appear relatively simple. They are frequently taught as simple stimulus–response situations. When situation "x" occurs, then take action "y." Through repetition, pilots learn the antecedent conditions and the appropriate responses. However, research and accident investigations have demonstrated that these "simple" decisions based on simplistic, prescribed procedures can lead to accidents and incidents because more complex decision processes are involved under certain conditions (Besco, 1995). In fact, in training, one often observes a difference between more and less experienced decision makers in how they respond to these "simple" decisions. Less experienced aviators consider these as "no brainers." They assess the situation based on the one or two cues learned in training and perform the response prescribed by the checklist. In contrast, experienced decision makers consider a larger number of cues in building situation assessments, and, under specific circumstances, take actions that appear contrary to those prescribed by checklists.

The Rejected Takeoff provides an excellent example of a flight task that fits this description. As an aircraft accelerates down the runway, the flight crew must make a decision to takeoff or to terminate the takeoff and stop the aircraft. This decision must be made at a specific, predesignated airspeed (V1). This is the highest speed at which the pilot can terminate the takeoff safely. If a malfunction occurs below V1, then the pilot should terminate the takeoff; if a malfunction occurs above V1, then the pilot should continue. When accelerating toward V1, the pilot must examine the health of the airplane carefully and be poised to terminate the takeoff if necessary. Most training programs teach pilots to focus on two sets of stimuli—the aircraft's airspeed and the status of specific aircraft systems. Thus, the student pilot learns to initiate a rejected takeoff if an alert occurs before reaching V1.

However, as pilots gain experience, many learn that a rejected takeoff may not be such a simple decision, and that other factors may come into play. Aircraft weight, runway condition, tire condition, wind conditions, and runway length all affect whether a pilot can reject a takeoff safely. Therefore, an experienced pilot will look beyond the cues of airspeed and system status to include other contextual cues, and may be more conservative with a decision to reject a takeoff when flying a heavy airplane on a short, contaminated runway.

These examples of ATC and flight crew decisions illustrate the importance and predominance of recognitional decisions in aviation and the importance of accurate situation assessments in making those decisions. The panel identified several implications from discussion of these examples. The first is a need to examine the nature of all decisions made, not just those that are reported to be most difficult or ambiguous. Some decisions may be imbedded in accepted procedures, and as a result, their importance is minimized. The Rejected Takeoff decision is only one example of many pilot decisions that appear to be simple procedures, that may become critical under certain conditions, that may receive only trivial attention during training, and that may be more complex than the conditions under which they are learned.

The second implication concerns the relative importance of situation awareness in the decision process. Under conditions of time pressure and ambiguity, decision makers need help to determine what is occurring in the environment around them. This is true for flight crews, air traffic control, and many other decision makers. Therefore, decision aids, interfaces, and training should provide decision makers the tools and skills necessary to accurately and quickly make situation assessments.

Finally, experienced aviators employ checklists and procedures differently when compared to less experienced aviators. Those with less experience often rely on checklists as absolutes. However, experienced aviators learn to consider a broader range of cues and they learn when these cues are relevant. This enables them to employ checklists and procedures as guidelines rather than as absolute prescriptions of courses of action.

TRAINING NEEDS

A second topic for this panel discussion concerned the need for NDM-based training interventions, training that enhances the skills necessary for effective decision making and judgment. A need for such training exists in all areas of aviation. However, military flight training provides several clear examples.

The military has several problem areas that challenge the decision-making skills of pilots while stressing the military's training resources. For example, the military conducts *ab initio* training; that is, it transforms complete novices into minimally proficient and safe pilots. This basic flight training is conducted in formal schools; the graduates are then transferred to operational field units for mission training. In addition, the military has difficulty retaining its most experienced pilots because they are lured away from service by the commercial air carriers. These conditions create a rather large population of aviators who are at an advanced beginner stage of skill development. Therefore, an objective of military flight training is to provide these aviators with decision and judgment skills as quickly as possible.

Despite the recognized need, the military has no formal programs to teach its pilots how to make decisions or to give them the skills needed to make decisions successfully. Some decision training modules are imbedded within Aircrew Coordination Training or Crew Resource Management courses. As part of CRM training, Line-Oriented Flight Training has been an effective means of teaching both technical and decision skills with operational contexts. However, the shortfall of the accepted decision training practices is that they communicate relatively simple, prescriptive decision formulas based on classical decision strategies. For example, the D.E.C.I.D.E. model (Benner, 1975) proposes a six-step process designed to organize the decision maker's thoughts and to prevent overlooking anything that may be important. The decision maker is taught to consider a variety of possible courses of action, evaluate the utility of their possible outcomes, and select the one that would be the most beneficial. These strategies address one type of decision and do not address issues concerning situation awareness.

A second set of issues for the military concerns the nature of the military mission itself. Military pilots must perform a wide array of tasks in addition to getting the aircraft from one point to another. Their primary task is to deliver weapons, troops, and equipment. Flying frequently becomes a secondary task. The military pilot must learn to make decisions and judgments related to mission performance in addition to those decisions related solely to flying the aircraft. These conditions are compounded by the demanding pace of military operations and the "can do" attitudes that are prevalent within the military.

Panelists noted two examples that represent common decision "errors" for advanced beginner pilots. In the first case, pilots recognize the cues that indicate a decision needs to be made, but fail to make the decision. In the second case, advanced beginners fail to recognize the ramifications that a decision regarding one aspect of the mission will have on the remainder of their mission. For example, early in a mission the pilot may experience some equipment malfunction. The pilot accurately assesses the problem and takes corrective action, but does not recognize that the corrective action has changed the remainder of the mission. An example follows.

In one study, army helicopter flight crews conducted a troop insertion mission in a high-fidelity flight simulator. The mission required the crews to cross deep into enemy territory to insert infantry troops into a drop zone within a specific window of time. During the ingress flight, many crews encountered enemy anti-aircraft fire and diverted from the planned course. These diversions enabled the crews to evade enemy contact and continue on the primary mission, but they caused a delay in arriving at the drop zone. The flight crews dealt with the immediate problem posed by enemy anti-aircraft fire, but they failed to recognize the impact that these diversions had on subsequent elements of the mission. Consequently, they did not replan the troop insertion portion of the mission, they arrived at the drop zone too late, and the mission failed.

Decision training needs to be addressed more directly and should be incorporated into the military's training programs that are delivered at the flight training centers as well as in operational field units. Much of the training required to meet this goal may be accomplished without the expenditure of additional training resources. For example, decision training can occur within the context of training already conducted in flight and mission simulators. Simulators provide an excellent venue for teaching in context. That is, students can perform individual tasks within the context of real world environments and missions. Studies of Line-Oriented Flight Training indicate that such context-driven training can enhance situation assessment skills that are necessary for effective recognitional decision making (Robertson & Endsley, 1995). In addition, much of the anecdotal training that occurs currently in operational units is consistent with the NDM approach. Pilots frequently engage in "table talk" sessions in which they discuss specific instances of things that work and things that do not. These opportunities provide an excellent method for transferring the experience and expertise directly from one pilot to another.

However, to optimize the effectiveness of such training, we need to understand the skills that must be developed, the tasks to be performed, and the conditions under which they are performed. A more structured approach to training development using NDM methods could build on this base by identifying the training requirements for judgment and decision making. Incorporating judgment and decision training could lead to more effective training programs for both the institutional and operational settings.

Cognitive Task Analysis (CTA) may provide this structure for the training development process. Cognitive Task Analysis comprises a set of analytic methods for identifying the decision requirements for proficient task performance (Klein & Kaempf, 1994). These decision requirements include the judgments and decisions that must be made, the cues and factors to which the decision maker attends, and the factors that make the decisions difficult. Once they are identified, decision requirements can then be used to guide the development of training objectives that incorporate the cognitive skills necessary for proficient decision making. Cognitive Task Analysis has been used successfully for the development of decision supports (e.g., Miller, Wolf, Thordsen, & Klein, 1992) as well as for training development (e.g., Means & Gott, 1988).

RESEARCH NEEDS

One goal of NDM researchers is to enhance decision making in operational settings. This can be accomplished through the design of better decision supports and interfaces and better training. However, what do we need to know and what do we need to do to develop such interventions? The panel discussed several areas of productive study including research investigating decision processes employed in aviation, identification of classes of decisions, identification of the strategies that are more and less effective for each decision class, and the effects of stress on decision-making processes.

Why is it important to understand the different classes of decisions and the effectiveness of various strategies that decision makers may employ? Put simply, it is essential for designing effective interventions. To be most effective, decision supports and decision training would be tailored to the cognitive processes naturally invoked by the decision maker. Effective interventions will not impose some process or organization that is logical to the designer but foreign to the decision maker. Understanding the strategy in use, the designer can build a system that identifies the pitfalls of the strategy and then serves as an advisor or critic for the decision maker. Such a decision aid might not be smart enough to make all decisions, but it might be smart enough to serve as a "devil's advocate" by presenting information that might contradict the decision maker's tendencies. Thus, it is important to study not only the types of strategies employed, but also the characteristics of the problems and situations that elicit these strategies.

Clearly, all situations requiring a decision are not the same. For example, some situations may have reasonable solutions "ready at hand" as discussed by Dreyfus (chapter 2, this volume). Others may be unanticipated or so improbable that no predetermined solution is available. Decision makers vary in their experiences, skills, and abilities. Consequently, they may vary in the strategies that they have available and that they invoke to solve a particular problem. Decision makers learn through experience how to adapt to a domain. They learn specific ways of doing things and handling difficult situations that are specific to the domain, but could apply to a wide range of problems. It is important to differentiate between the structure of the underlying problem and the way the decision maker responds to the problem.

Investigations of decision classes and strategies have already begun (see Orasanu, 1993; Orasanu & Fischer, chapter 32, this volume). These studies address the needs of commercial air transport pilots focusing on difficult decisions that have critical consequences. They have resulted in a taxonomy of decision classes and the identification of differences between more and less experienced decision makers. Orasanu and her colleagues have identified six types of decisions, including two types of rule-based decisions (Go/No Go and Condition Action) and four types of knowledge-based decisions (Choice, Scheduling, Procedural Management, and Creative Problem Solving). They have identified behaviors that are characteristic of highly effective crews including demonstrating flexibility with a repertoire of strategies, monitoring the environment closely, not overestimating their own capabilities, and planning for contingencies. See Orasanu and Fisher (chapter 32, this volume) for a more complete discussion of these findings.

Another important area of research concerns the effects of stress on decision processes. Stress research has been constrained by early paradigms that examined either physiological responses to stress or stressful stimuli and their effects on measures of traditional information processing. Early studies rarely looked at the effects of stress on decision-making processes. Studies of stress and cognitive processes have relied on a desktop metaphor of cognitive processing examining working memory, attention, and spatial abilities. Consequently, we know very little about the effects of stress on prominent decision strategies that have been discussed in the NDM literature. It is the belief of some researchers that naturalistic strategies are stress-resistant and could represent an accommodation to stressful environments (Klein, 1993; Stokes, Kemper, & Kite, chapter 18, this volume).

Clearly, this area warrants more study and could play a key role in the design of training and decision aids.

SUMMARY

This panel served to bring together a number of experienced researchers and practitioners from divergent areas of the aviation domain. The discussions revealed commonalities in the issues that are crucial to each of the areas.

One such area is the relative importance of situation assessment as a component of the decision-making process and the need to support situation assessments through decision aids and training. A second common area of concern is the need to continue two important lines of research. The first line will identify classes of decisions based on the nature of the underlying problem and the decision strategies that work best for each decision class. The second research line will investigate the effects of stress on how decision makers implement these strategies. The final area of discussion concerned the need for training decision and judgment skills in all areas of the aviation domain.

Finally, it is important to note that the application of NDM research in aviation is in its infancy. A core group of aviation researchers and practitioners are employing an NDM perspective, but the approach has not become commonplace in the industry. As NDM research receives more attention and as NDM methods and models prove more effective, interventions based on NDM principles will follow. These discussions helped each of us address our own concerns and they brought some coherence to a somewhat fractured domain that has a single common goal: safe and efficient flight.

REFERENCES

Bayles, S. J., & Das, B. K. (1993). *Using artificial intelligence to support traffic flow management problem resolution* (Rep. No. MTR93–000245). McLean, VA: Mitre Corporation.

Benner, L. (1975). D.E.C.I.D.E. in the hazardous materials emergencies. *Fire Journal, 69*(4), 13–18.

Besco, R. O. (1995). Improving take-off abort decisions/performances. *Proceedings of the Eighth International Symposium on Aviation Psychology* (pp. 1336–1340). Columbus: The Ohio State University.

Kaempf, G. L., & Klein, G. (1994). Aeronautical decision making: The next generation. In N. Johnston, N. McDonald, & R. Fuller (Eds.), *Aviation psychology in practice* (pp. 223–254). Hants, England: Avebury Technical.

Kaempf, G., Klein, G., Thordsen, M., & Wolf, S. (1996). Decision making in complex command and control environments. *Human Factors, 38*(2), 220–231.

Kaempf, G., Wolf, S., & Miller, T. (1993). Decision making in the AEGIS combat information center. *Proceedings of the Human Factors and Ergonomics Society 37th annual meeting* (pp. 1107–1111). Santa Monica, CA: The Human Factors and Ergonomics Society.

Klein, G. (1993). *Naturalistic decision making: Implications for design.* Dayton, OH: CSERIAC.

Klein, G., & Kaempf, G. (1994, October). *Cognitive task analysis.* Workshop presented at the 37th annual meeting of the Human Factors and Ergonomics Society, San Diego, CA.

Means, B., & Gott, S. (1988). Cognitive task analysis as a basis for tutor development: Articulating abstract knowledge representations. In J. Potska, L. D. Massey, & S. A. Mutter (Eds.), *Intelligent tutoring systems: Lessons learned* (pp. 35–57). Hillsdale, NJ: Lawrence Erlbaum Associates.

Miller, T., Wolf, S., Thordsen, M., & Klein, G. (1992). *A decision-centered approach to storyboarding anti-air warfare interfaces.* Fairborn, OH: Klein Associates Inc.

Mundra, A. D. (1989). *A description of air traffic control in the current terminal airspace environment* (Rep. No. MTR–88W00167). McLean, VA: Mitre Corporation.

Orasanu, J. (1993). Decision-making in the cockpit. In E. Wiener, B. Kanki, & R. Helmreich (Eds.), *Cockpit resource management* (pp. 137–172). New York: Academic Press.

Robertson, M. M., & Endsley, M. R. (1995). The role of Crew Resource Management (CRM) in achieving team situation awareness in aviation settings. In R. Fuller, N. Johnston, & N. McDonald (Eds.), *Human factors in aviation operations* (pp. 281–286). Hants, England: Avebury Aviation.

Chapter 9

Naturalistic Decision Making in Business and Industrial Organizations

◆

Neal Schmitt
Michigan State University

The discussion of decisions in business and industrial contexts at the Second Conference on Naturalistic Decision Making (NDM; June 1994) reaffirmed the characteristics of naturalistic decision making discussed earlier (e.g., Lipshitz, 1993; Orasanau & Connolly, 1993), but it also suggested some other peculiarities of decision making in this context. The purpose of this chapter is to examine the descriptions of decision making in business and industrial organizations provided by the panel (Lee Beach, Scott martin, William Rouse, and Janet Sniezek) on NDM in business and industry.

Over the last several decades, the study of decision making in business and industrial organizations has paralleled the study of decision making in other "realistic" contexts as opposed to purely laboratory contexts. Perhaps the first formal approaches to decision making in which the researchers were interested in applications drew from the Brunswik lens model and Bayes' theorem (Slovic & Lichtenstein, 1971). Early applied Brunswikian research was nearly always directed toward capturing an expert's policy with respect to some decision-making context. The context varied widely, including performance appraisals (Hobson & Gibson, 1983), assessment centers (Schmitt, 1977); and business relocation decisions (Schmitt, Gleason, Pigozzi, & Marcus, 1987). These studies usually focussed on identifying the major cues or information dimensions, and determining how a person combines that information to make a decision or form an evaluation. Research conducted about the same time within the Bayesian paradigm was concerned with why decision makers were generally conservative with respect to the prescriptions of Bayes' theorem (e.g., Edwards, 1968). Partly as an outgrowth of the Bayes approach, researchers relied on multiattribute theory (Keeney & Raiffa, 1976) to prescribe what decision makers ought to decide given the utilities these decision makers associated with different attributes of the decision alternatives.

The policy-capturing research resulted in reasonably accurate (in a regression sense) predictions of decisions, but there were large individual differences in the weights applied to the same information and these differences were not readily explained. In the end, researchers realized there were many ways to arrive at similar predictions about a phenomenon and that little if anything had been discovered about what went on in decision makers' heads (Dawes, 1979).

The disillusionment with policy-capturing approaches to decision making was followed by process-tracing approaches, in which the focus was centered on the process by which the decision maker sought information. There was little attempt to assess whether the decision was accurate or acceptable to the people affected by the decision (see Ford, Schmitt, Schechtman, Hults, & Doherty, 1989). This approach to the study of decision making was used to analyze decision behavior in a wide variety of applied areas including marketing, personnel selection, and performance appraisal. The process-tracing approach did document that decision makers used a variety of "nonlinear" strategies and that these strategies were often the result of the nature of the decision task, particularly the complexity of the decision (see Ford et al., 1989, for a review).

Another stream of basic and applied research in decision making grew out of the cognitive bias literature (see Christensen-Szalanski, 1993, for a review). Again, the range of applied problems in which these biases have been studied has been extremely broad, ranging from auditing problems (Ho & Keller, 1994) to medical decision making (Kassirer & Kopelman, 1989). This research indicates that people use a variety of heuristics in making decisions and that these heuristics sometimes lead to judgmental biases; however, it seems that when expert decision makers are observed in a natural context, they are not subject to the same kinds of cognitive limitations as are students using more abstract or context-deprived decision problems.

DEVELOPMENT OF NATURALISTIC DECISION MAKING

The recognition that experts working in applied contexts did not behave as laboratory decision makers was a major impetus to the development of the area of research now being referred to as Naturalistic Decision Making.

Determining what NDM actually involves is still one of the major foci of researchers in this area. Perhaps the best statements of the characteristics of NDM are contained in a work by Orasanu and Connolly (1993) and a summary of the various models of NDM (Lipshitz, 1993) that have resulted from the study and observation of decision makers, both individuals and groups, in naturalistic settings. These characteristics of NDM are summarized in the introductory chapter to this volume by Zsambok.

NDM IN BUSINESS AND INDUSTRY: SOME OBSERVATIONS

Some participants expressed the notion that NDM researchers should attend more frequently to the outcomes associated with particular decisions and the decision processes they study. This discussion stimulated several remarks about outcomes of decisions in applied contexts. In some situations, an outcome cannot be specified for years, and sometimes there is evidence that decision makers will actually pick an alternative course of action for which they can receive relatively immediate feedback about the consequences of their action, as opposed to a course of action that cannot be evaluated in any specified time frame. Panelist Sniezek pointed out that researchers are also likely to pick short-time criteria in assessing the effectiveness of decisions; she identified this as one more aspect of the criterion problem that plagues many other areas of industrial and organizational psychology.

It also appears that many decisions made in business and industry never get implemented because the parties responsible for implementing the decision are not consulted, the implications of a decision are not clearly specified by those who made the decision, or an organizational bureaucracy or structure is inconsistent with implementation, thereby thwarting it. One is reminded here of frequent attempts to "reinvent" government that seem to have no effect. It also emphasizes that much of decision making in organizations must be collaborative or at least be perceived as collaborative. Another central and related issue is whether the elements of a decision, as traditionally defined, exist in a business organization or whether decisions have important consequences for the organization. Certainly, the notions that there are many involved decision makers, that decisions are often made in steps with feedback at each step, and that organizational norms and goals (culture?) are important reaffirm the early characterization of NDM. However, the recognition that decisions are really only one part of a change process in which many other elements of a people nature are equally or more important, is perhaps new, or at least more common in large work organizations.

Rouse described his approach to aiding decision makers during the discussion. For him, the central issue is how to bring the participants in strategic decision making to make good decisions quickly. He maintained that decision makers sometimes feel they have answers to questions, or can secure answers, but often do not know which questions to ask in formulating plans for strategic decisions. Rouse (1995) proposed a set of questions that he uses as a means of identifying the most important organizational concerns. These questions may all be relevant depending on the situational context in which a given strategy or decision is planned. Once relevant or important questions are identified, he has found that decisions or plans for action are easy to generate and are often consensual. In fact, if decisions or strategies are not consensual, he would maintain that no decision has been reached, because implementation is unlikely.

In considering the capability of doing research on decision making, Rouse had several important observations. First, he has found it impossible to simply observe the decision-making process; organizations want help in making their decisions as a precondition for organizational entry, and are not likely to tolerate a lengthy research and observation phase. Thus the researcher is at least a participant observer in what amounts to a change process. Questions of objectivity and generalizability are obvious, but most participants in the panel agreed that this was likely the price one has to pay for data that actually describe the richness of NDM in business organizations. All decisions are embedded in a situational context in which it is important to know *what* is being decided, by *whom*, and for *what reason*. The fact that these often-assumed aspects of a decision in traditional experimental studies of decision making are simply not identifiable in many natural settings was voiced by all members of the panel and many members of the audience. Often these aspects of a naturalistic decision are very difficult, if not impossible, to ascertain even when one does have first-hand knowledge of how a decision is made.

Another observation about the study of NDM in organizations made by Rouse was the need to employ multiple different methods to gain information. Direct observation and participation in planning or strategy sessions as described earlier is supplemented with corporate histories that generate stories of situation assessment, planning and commitment to projects, and execution and monitoring of project implementations. Interviews are conducted with executives and managers concerning past projects in which they are asked to relate circumstances and tradeoffs that were considered in other recent decisions. Therefore, participant observation, interviews, and archival studies all may be helpful in describing and understanding the decision making in an organization. The result of Rouse's case study research are that data (at least multiple data points) are expensive, accumulate slowly, and are often not likely to be published in traditional outlets. This concern was expressed by several panal participants.

One of the central tenets of NDM is that the process associated with decision making in a natural context is different partly because of the greater level of expertise (relative to college student subjects) possessed by organizational decision makers (Klein, 1993). This assumption was questioned by panelists for several reasons. Expertise is not easily defined in many organizational settings because the outcome of most decisions has multiple causes and effects, most of which any given decision maker is not likely to know or understand. This suggests that expertise may need to be defined as a group-level variable as has been done in some artificial intelligence research. In addition, it is often likely that a particular decision has never previously been addressed by a particular group or individual; hence the proposition that the decision maker has a routinized or well-defined scenario with respect to dealing with the situation can be hard to support. For example, most of the small business leaders studied by Schmitt et al. (1987) had never expanded or moved their business before and had no experience with many of the problems they eventually encountered. If there is any expertise relevant to these new or unique decisions, this expertise is likely to be organizational knowledge (like street knowledge), knowledge of where to go in an organization to obtain the relevant information and resources required to develop an appropriate decision strategy, or general business knowledge.

That decisions in natural settings are often made by groups of individuals and that they often evolve over a long period of time is widely accepted by NDM researchers, but Sniezek described the judge–advisor system of decision making that has heretofore not been widely discussed. She discussed situations when the formal organizational decision maker is advised by a group of subordinates who make recommendations regarding decisions. In this situation, the leader or decision maker may not have all the specialized knowledge or expertise to reach a decision, but must rely on the input of advisors. These advisors will often provide input that is relevant to the decision, but can, if they so desire and are trusted by the judge, make and support their own preferred recommendations. Often, the advisor(s) who is most committed and expert in promoting her or his viewpoint wins the day and in effect makes the decision. In this case, the judge is often more concerned with the reaction various organizational members will have to a particular decision alternative, and will consider implementation issues and the organizational or public relations consequences of a decision rather than the quality of the decision.

A participant in this panel related his observations of decision making among government officials, in which most strategy sessions were spent discussing how to convince the formal decision makers that their approach to a problem is correct, and who to call to martial the informal support that will be crucial in securing the decision they espouse. These relatively low-level staff people have made the decision, but a higher level person may be credited with the decision. The judge–advisor system of decision making again changes the locus of the decision to the group level and the locus of any decision maker expertise becomes the aggregate of the advisor group available.

The notion that decisions are made by groups or by judge–advisor teams suggests that individual differences be considered in studies of NDM. Interestingly, this question was not addressed by any of the chapters in the 1993 volume on NDM. The point was made by Martin, who suggested that cognitive ability has been related to decision making expertise in other studies and ought to be considered in NDM studies. Orasanu and Connolly (1993) also suggested a variety of personality variables like risk taking, creativity, self-reliance, and tolerance of ambiguity and stress that should certainly be related to satisfaction with the organizational decision-making process, if not also with the quality of the decisions that result from this process. There is also some evidence that decision makers in applied situations try to choose from among acceptable, but not necessarily the best, decision alternatives. If so, then motivation to achieve or excel might relate to satisfaction and performance in an organizational decision group or team.

Beach described an interesting intervention in a large banking firm, in which the emphasis was not on how any particular decision was made, but on making sure that the whole host of "decisions" made by various branches of this bank were consistent with the corporate vision. In pursuing this intervention, Beach was guided by the elements of "image theory." The image theory position is that decision makers have three images that guide or constrain the decisions they make: (a) a set of beliefs and values; (b) more specific goals to which the decision maker is striving; (c) specific operational plans for reaching the goals. Decisions involve either adoption or rejection of the potential goals and plans and monitoring progress toward decisions. In the case study described, the corporate vision was to develop a consumer-friendly bank with consistency of quality service. This general climate could not be simply "required" of the 2200 different branches, many of whom were newly acquired. Further, it was recognized that without agreement about the need to develop this climate, setting goals or developing plans to improve customer service would be futile. The intervention, therefore, focused on communicating company philosophy, with the hope that bank managers would develop goals and plans consistent with this philosophy. Over time, the researchers hope to see that measures of climate conform more closely to the corporate image and that individuals within a local site are more homogenous in their perceptions of the local climate. The interesting aspect of this case study of an attempt to provide a "decision aid" is that no specific decision is targeted; rather, only the framework, or image, within which future decisions will be provided is assessed.

CONCLUSIONS

Although most of the discussion reaffirmed the previous descriptions of NDM provided by Orasanu and Connolly (1993) and Lipshitz (1993), some new emphases or directions were suggested as well. The following summary statements evolved from the discussion on NDM in business and industry:

1. The notion of expertise needs to be expanded to include the idea that expertise about decisions may reside in groups, not individuals, and that in some cases, the decision maker(s) do not have the expertise or schematic to guide their decision.

2. Research on NDM is different than traditional research on decision making. It is more likely to be of a case-study nature including aspects of participant observation, interviews (both retrospective and prospective), and archival study.

3. Of the elements of decision making, interventions are likely to focus on problem definition and aspects of implementation and monitoring. Interventions probably will not focus on any given choice, and are likely to be the result of descriptions of previous successful attempts at decision making.

4. There is a need to evaluate decisions in terms of outcomes, but there remains a significant criterion problem. Because of the nature of NDM, it can be difficult to define a single outcome variable and when it can be done, it is often something that cannot be evaluated in the foreseeable future. Because of the inherently dynamic nature of decisions, desired outcomes may also change.

5. Individual differences should be given attention in studies of NDM. Questions concerning who is most effective in decision-making groups and at making decisions in natural contexts should be addressed.

6. The judge–advisor system, how it operates, what its strengths and weaknesses are, and how it compares to other modes of decision making should be described.

REFERENCES

Christensen-Szalanski, J. J. J. (1993). A comment on applying experimental findings of cognitive biases to naturalistic environments. In G. A. Klein, J. Orasanu, R. Calderwood, & C. E. Zsambok (Eds.), *Decision making in action: Models and methods* (pp. 252–261). Norwood, NJ: Ablex.

Dawes, R. M. (1979). The robust beauty of improper linear models in decision making. *American Psychologist, 34,* 571–582.

Edwards, W. (1968). Conservatism in information processing. In B. Kleinmuntz (Ed.), *Formal representation of human judgment* (pp. 17–52). New York: Wiley.

Ford, J. K., Schmitt, N., Schechtman, S. L., Hults, B. M., & Doherty, M. L. (1989). Process tracing methods: Contributions, problems, and neglected research questions. *Organizational Behavior and Human Decision Processes, 43*, 75–117.

Ho, J. L., & Keller, R. (1994). The effect of inference order and experience-related knowledge on diagnostic conjunction probabilities. *Organizational Behavior and Human Decision Processes, 59*, 51–74.

Hobson, C., & Gibson, F. W. (1983). Policy capturing as an approach to understanding and improving performance appraisal: A review of the literature. *Academy of Management Review, 8*, 640–649.

Kassirer, J., & Kopelman, R. (1989). Cognitive errors in diagnosis: Instantiation, classification, and consequences. *The American Journal of Medicine, 86*, 433–441.

Keeney, R. L., & Raiffa, H. (1976). *Decisions with multiple objectives: Preferences and value tradeoffs.* New York: Wiley.

Klein, G. A. (1993). A recognition-primed decision (RPD) model of rapid decision making. In G. A. Klein, J. Orasanu, R. Calderwood, & C. E. Zsambok (Eds.), *Decision making in action: Models and methods* (pp. 138–147). Norwood, NJ: Ablex.

Lipshitz, R. (1993). Converging themes in the study of decision making in realistic settings. In G. A. Klein, J. Orasanu, R. Calderwood, & C. E. Zsambok (Eds.), *Decision making in action: Models and methods* (pp. 103–137). Norwood, NJ: Ablex.

Orasanu, J., & Connolly, T. (1993). The reinvention of decision making. In G. A. Klein, J. Orasanu, R. Calderwood, & C. E. Zsambok (Eds.), *Decision making in action: Models and methods* (pp. 3–20). Norwood, NJ: Ablex.

Rouse, W. (1995). *Enterprises in transition.* Englewood Cliffs, NJ: Prentice-Hall.

Schmitt, N. (1977). Interrater agreement and dimensionality and combination of assessment center judgments. *Journal of Applied Psychology, 62*, 171–176.

Schmitt, N., Gleason, S. E., Pigozzi, B., & Marcus, P. M. (1987). Business climate attitudes and company relocation decisions. *Journal of Applied Psychology, 72*, 622–628.

Slovic, P., & Lichtenstein, S. (1971). Comparison of Bayesian and regression approaches to the study of information processing in judgment. *Organizational Behavior and Human Performance, 6*, 649–744.

Chapter 10

Training Decision Makers for Complex Environments: Implications of the Naturalistic Decision Making Perspective*

♦

Janis A. Canon-Bowers
Naval Air Warfare Center Training Systems Division

Herbert H. Bell
Armstrong Laboratory Aircrew Training Research Division

Perhaps the most significant contribution of the Naturalistic Decision Making (NDM) perspective is that it has forced researchers to reevaluate the assumptions they hold about how people make decisions in "real" environments (Orasanu & Connolly, 1993). A logical extension is to examine the *practical* implications of this theoretical perspective on decision making. In particular, it is of interest to determine how conclusions drawn from the NDM perspective affect the manner in which decision making might be optimized, or at least improved. Therefore, the purpose of this chapter is to examine NDM-related theorizing from the standpoint of what

*The opinions expressed herein are those of the authors, and do not represent the official positions of the orgainzations with which they are affiliated.

it implies for how we train decision makers. It is based largely on the panel discussion devoted to this topic; panelists included the authors, John Schmitt, Hugh Wood, Raanan Lipshitz, and Jon Fallesen.

Before proceeding, we have several important points to make regarding NDM and training, and more specifically, our approach in this chapter. First of all, it became clear from the panel discussion that panelists were not familiar with any empirical investigations of the effectiveness of NDM-generated training interventions. This conclusion was bolstered in an extensive literature search we conducted for the purpose of identifying studies conducted on this topic. Therefore, this chapter is necessarily analytical in nature. Specifically, it focuses on generating propositions for training the cognitive aspects of decision making as suggested by the NDM perspective. Furthermore, we do not attempt to review literature pertaining to this topic (this has been accomplished nicely by Means, Salas, Crandall, & Jacobs, 1993; Orasanu, 1993; and others); instead, we seek to organize and summarize thoughts on this topic.

We also note that the task of generating cognitive training principles consistent with NDM is not as straightforward as it may seem. In fact, the nature of NDM itself provides a significant dilemma when it comes to specifying training interventions because it is in conflict with the assumptions traditionally made about how people make decisions in the real world. According to NDM theories, effective decision making is not characterized by an invariant set of steps or procedures as prescribed by normative theories (Orasanu & Connolly, 1993). Instead, decision making is seen as intertwined with task accomplishment, context-specific, fluid, flexible, and in some respects, "procedure-free" (i.e., lacking prescribed rules as suggested in more classical views of decision making). These are seemingly impossible goals to achieve through traditional training—specifically, how do we train people to be flexible, or to engage in decision making that does not follow a set of predetermined steps?

Therefore, we have to reason about decision-making training, and particularly about what the goals of training might be, with a different point of departure under an NDM formulation as compared with classical decision-making perspectives (Means et al., 1993). In fact, the most fruitful way to characterize NDM-consistent training might be to view it as a mechanism to *support* natural decision-making processes, and as a means to *accelerate* proficiency or the development of expertise. The terms "managed experience," "augmented experience," and "accelerated proficiency" all come to mind as descriptive of the training philosophy that we believe is consistent with NDM theories. However, we also contend that knowledge, skills, and processes that underlie decision making—even naturalistic decision making—can be identified and trained. This is not to say that we can (or should) specify a single set of steps for decision making, only that consistent processes that support expert decision making can be specified as a basis for training.

This position implies that we must be able to describe what expert decision makers *do*—delineate the knowledge, skills, and processes that may underlie effective decision making, and also describe the mechanisms of NDM that we seek to support. To accomplish this here, we first use NDM theories to establish a set of characteristics that describe expert decision makers (based on what are defined as the characteristics of "naturalistic" decisions). We then examine the processes, knowledge, and skills that NDM theories suggest are responsible for expert performance; in other words, *how* does the NDM perspective explain effective decision making performance? Next, we use this information as a basis to generate propositions regarding the manner in which training methods, strategies, and content might be designed in order to improve decision making. Finally, we offer suggestions for future research in this area.

CHARACTERISTICS OF EFFECTIVE DECISION MAKERS

One way to determine the implications of the NDM perspective for training is to address the question of what it suggests are the characteristics an effective decision maker. In order to answer this question, we first examine what we want trainees to be able to do at the conclusion of training. Once established, we can use these attributes as a basis to specify which knowledge, skills, and processes must be trained in order to achieve targeted performance.

As noted elsewhere, characteristics of the naturalistic environment (see Orasanu & Connolly, 1993) contribute to the requirements for effective decision makers. On this basis we hypothesize that effective decision makers can be described as:

1. Flexible—Naturalistic decision making contexts (that are ill-structured, dynamic, complex) require decision makers to be flexible. That is, decision makers must be able to cope with environments that are ambiguous, rapidly changing and complex. The implication of this line of thinking is that expert decision makers have a *repertoire* of decision making strategies that they can draw on in response to particular situational cues.

2. Quick—One of the characteristics of naturalistic decisions is time pressure (Orasanu & Connolly, 1993). This demands that decision makers are able to make rapid decisions, often in the face of severe consequences. It is also perhaps the single most important feature that

mitigatesagainstanalyticaldecisionmaking(i.e.,generatingoptions,assessing options, etc.).

3. Resilient—Several of the characteristics noted by Orasanu and Connolly (1993) are indicative of stress in the decision-making environment. For example, ambiguity, uncertainty, and high stakes (i.e., severe consequences for error), along with other factors such as high workload and adverse physical conditions, are all stressors that can affect decision-making performance (Driskell & Salas, 1991). Therefore, in naturalistic environments, expert decision makers must be trained to operate under these conditions, without suffering degradations in performance (Means et al., 1993).

4. Adaptive—When events unfold rapidly in decision making, or decisions involve multiple goals, "static" models of decision making do not apply (Lipshitz, 1993). Instead, it is clear that a naturalistic view of decision makings suggests that decision makers must engage in a continual process of strategy assessment and modulation. This implies that decision makers must recognize when and how to apply a decision strategy and when to change, or modify that strategy in accordance with problem demands.

5. Risk taking—According to Orasanu (1993), expert decision makers in naturalistic settings have a tendency to use their knowledge more effectively to conduct risk assessment in making decisions. Obviously, when environments are characterized by high stakes, decision makers must- be able to assess the risk associated with various courses of action, and weigh the consequences of error against any potential payoff. This does not imply that decision makers generate multiple options or weigh them against each other; only that successful risk taking is often a crucial part of making decisions in naturalistic settings.

6. Accurate—To say that expert decision makers are accurate is stating the obvious. However, it is probably worth acknowledging that, particularly in light of the fact that many situations provide such high demand on the human decision maker, it is perhaps more surprising when costly incidents and accidents do not occur than when they do.

Now that we have established the characteristics of effective decision makers—essentially delineating a set of criteria that describe the "target" or goals of training—we can turn attention to examining what NDM-related theories have to say about how experts are able to perform as described earlier. The following subsections summarize what we know about the mechanisms of decision making—in particular, how experts are able to perform in the often difficult, challenging, time-compressed naturalistic environment.

MECHANISMS OF NDM AND EXPERTISE

As noted, the value of NDM theories in designing training lies in what they have to offer regarding the knowledge, skills, and processes that underlie expert performance. In addition, the study of expertise has yielded insight that has applicability to the naturalistic environment. Given these perspectives, we now highlight the following mechanisms of expert decision making that have been studied.

Situation-Assessment Skills

Several theorists operating under the NDM umbrella have argued that situation-assessment skills are paramount in effective decision making (e.g., see Klein, 1993; Orasanu & Connolly, 1993). According to Lipshitz (1993), all nine of the NDM-related theories he reviewed included an element of situation assessment, with several arguing that expert decision makers are able to perform situation assessment more quickly and accurately than novices. Overall, it is believed that this superiority in situation-assessment skills accounts for much of the ability of experts to make rapid decisions, and contributes to their decision-making accuracy. Two aspects of situation assessment behavior are:

1. Cue recognition/significance. For some time, researchers have believed that experts recognize decision-making cues differently than novices. For example, it has been concluded that experts recognize cues more quickly and completely than novices, recognize patterns of cues better than novices, and can detect important features of a stimulus more readily than novices (see Druckman & Bjork, 1991; Means et al., 1993). Experts also appear to be better able to frame decision problems so that they can detect the underlying structure of a problem (Orasanu, 1993). These skills all contribute to the decision maker's ability to perform effective situation assessment.

2. Pattern recognition. Several NDM theories place heavy emphasis on the ability of decision makers to perceive and recognize an entire pattern of relevant cues in assessing a decision making situation (e.g., see Klein, 1993). Experts are also believed to be able to "parse" the pattern of cues rapidly, ignoring those that are less relevant. According to Klein, it is this recognitional process that allows experienced decision makers to assess complex situations quickly.

Organized Knowledge Structures

One of the markers of expertise seems to be the organization of knowledge. That is, experts appear not only to know more, but also to organize knowledge more effectively (Druckman & Bjork, 1991). In terms specific to NDM, experience allows experts to build up knowledge *templates* on which they can draw in new decision-making situations. This enables decisions to be made quickly (because the decision maker draws on preexisting memory structures) and contributes to the accuracy of decision making. Details regarding template building follow.

Template Building/Matching.   Several NDM theories suggest that experts may store knowledge in templates, or as reference problems (Noble, 1993). For example, Noble contended that these reference problems contain objective features (i.e., the specific goals to be achieved by the decision maker), action features (i.e., a particular method for accomplishing the objective specified by the objective feature), and environmental features (i.e., criteria for adopting the solution method specified in the reference problem). Although this formulation is more detailed than others within the NDM perspective, many converge on the notion that over time, expert decision makers build well-organized knowledge structures that can be readily accessed and applied in decision-making situations.

Mental Simulation

NDM theories do not support the notion that decision makers generate and compare a series of response options; instead they assume that a recognitional process leads the decision maker to generate a solution from memory. However, when a situation is novel (i.e., a template for solving it does not exist in memory), mental simulation of the potential solution is hypothesized to be a primary mechanism by which a decision maker selects a course of action (Klein, 1993). According to Klein, this process allows the decision maker to determine whether the currently held solution is viable, and leads either to application or adjustment of the solution. It contributes to the accuracy of decision making, and helps decision makers to save time in crucial situations.

Strategy Selection/Modulation

As noted, flexibility is considered by many to be an important marker of expert decision makers. Several NDM theorists suggest that, in contrast to classical theories, no single set of decision-making processes can be delineated that describes the various types of decision making that occur in the natural environment (Lipshitz, 1993). Therefore, skilled decision makers must learn to select decision-making strategies that are best suited to the nature of the decision-making situation. Related to this, the fluid nature of many decision-making situations requires dynamic regulation of decision strategy. That is, decision makers must continually assess and modulate their strategy in accordance with demands of the decision. One way that decision makers do this is via metacognition, a skill that contributes to a decision makers' flexibility, and ability to adapt his or her decision strategy when required.

Metacognition. Expert decision makers appear to be better able to monitor their own processes during decision making (e.g., Chi, Bassock, Lewis, Reimann, & Glaser, 1989; also see Druckman & Bjork, 1991). For example they are superior to novices in understanding their own level of comprehension, and what is necessary to improve their state of knowledge. This "executive" function is crucial in guiding a decision maker, particularly as the problem changes and evolves, because it allows the decision maker to adapt his or her strategy as needed. Orasanu (1990) maintained that effective metacognitive skills also allow decision makers to better manage resources because they have a more accurate picture of their own strengths and weaknesses, and of the nature of the problem.

Reasoning Skills

A host of skills that we refer to loosely as "reasoning skills" can be hypothesized to support NDM functions. In fact, the notion that expert decision makers must be flexible and adaptable suggests that several categories of reasoning skills may be useful in expert decision making. For example, Orasanu (1993) maintained that when problems are ill-defined, decision makers must be able to diagnose situations, which requires them to engage in causal reasoning, hypothesis generation, and hypothesis testing. Orasanu also noted that creative problem solving (i.e., constructing novel solutions to a problem, or applying existing strategies in a new or different way) is required when existing knowledge and procedures do not meet the needs of the current decision. Other reasoning skills include using analogies, critical thinking skills (i.e., testing assumptions, checking facts, seeking consistency among cues), and domain-specific problem-solving skills.

Domain-Specific Problem-Solving Skills. Obviously, experts have richer, more extensive domain knowledge than do novices. However, it is not only the knowledge content that distinguishes experts from novices, but also the manner in which decision strategies are applied to that knowledge (see Hoffman, 1991; Means et al., 1993). Specifically, experts use domain knowledge to determine when and how to apply various decision strategies (Glaser, 1984). This suggests that understanding general problem-solving skills will not improve decision-making performance—it is only in the naturalistic context of a domain that decision-making strategies can aid performance.

IMPLICATIONS OF NDM THEORIES FOR TRAINING

As was stated at the outset of this chapter, we believe that the overriding goal of decision-making training according to NDM theory is to accelerate the natural processes leading to expertise. Now that we have examined those processes more closely, and briefly described relevant assumptions from research on expertise, it is possible to begin generating hypotheses about how NDM-consistent decision-making training might be designed. In addressing this issue we distinguish two types of propositions for training. The first involves NDM-derived suggestions for *what* should be trained—that is, the content of training. The second concerns implications from NDM theories regarding *how* decision makers should be trained—that is, the context, methods, strategies, and media employed in training. These categories of implications are presented in detail in the following sections.

Implications for What to Train Decision Makers

In considering what to train decision makers, NDM theories imply strongly that context-specific domain knowledge is a crucial aspect of expert decision making. In addition, there are several cognitive *processes* and *skills* that appear to be required for effective decision making (which are generated directly from the list of characteristics and mechanisms delineated previously) that should be trained. These include: (a) *metacognitive skills* (including the ability to select decision-making strategies, to modulate strategies as problems unfold, to engage in effective resource management, and to self-assess and adjust as necessary); (b) *reasoning skills* (including analogical reasoning, causal reasoning, creative problem solving, and critical thinking); (c) *domain-specific problem-solving skills* (although training "generic" problem solving strategies does not appear to be fruitful, training decision makers in problem solving specific to the domain may be more successful); (d) *mental simulation skills* (including the ability to know when to apply mental simulation as a means to evaluate a potential solution); (e) *risk assessment skills* (i.e., accurately assessing the risk associated with various courses of action); and (f) *situation assessment skills* (including the ability to make rapid, accurate assessments of the decision situation by improving pattern recognition skills and learning to assess the significance of such cues).

In addition, *knowledge organization*—fostering the organization of knowledge to support NDM—is an important training goal. Specifically, the decision maker must be exposed to the "typical" situations, cues, cue patterns, and responses that characterize a domain. This process is crucial as a means to foster development of necessary decision-making templates in memory.

Implications of NDM Research for How to Train Decision Makers

NDM theories provide a number of implications for how decision-making training should be conducted. We are particularly interested in proposing training strategies that we believe can help accelerate the acquisition of proficiency or the achievement of expertise—that is, to aid decision makers in learning and organizing domain knowledge in a manner that supports complex decision making.

Simulations. Several researchers have noted the importance of simulation in training NDM skills (e.g., see Means et al., 1993). Simulation is a valuable tool for training NDM skills because it can accelerate proficiency by exposing decision makers to the kinds of situations they are likely to confront in the real world. Moreover, simulation can be controlled—the characteristics of decision problems, situational cues and cue patterns, and decision outcomes can be provided as a means to aid in development of situation awareness, pattern recognition, and template building. For this reason, scenario/exercise design becomes a crucial aspect of simulation-based training. In fact, the effectiveness of this type of training will depend largely on the extent to which scenarios capture and display important cues (and the relationship among them) along with associated responses, so that necessary templates can be developed. Simulations are also an effective means to train reasoning skills, metacognitive skills, and risk-assessment skills.

Guided Practice and Feedback. Several studies have shown that allowing trainees to practice on a task without feedback may produce suboptimal decision-making performance (Means et al., 1993). In the current context, we contend that it is crucial to provide feedback as a means to reinforce important "cue → strategy" associations. In this manner, guided practice (i.e., practice that incorporates measures of performance and specific feedback) can be thought of as managed experience—providing decision makers with examples that enable them to characterize cue patterns, build templates, and associate effective responses to cues and cue patterns.

Embedded Training. NDM theories generally suggest that decision making cannot be removed meaningfully from the context in which it occurs. This implies that training decision makers in the environment in which they operate is recommended if possible. One way to accomplish this is through embedded or organic training. In such cases, training is incorporated into the operational system or equipment so that training takes place on the job (with appropriate safeguards). The success of such systems depends on their ability to support the development of expertise via performance measurement, specific feedback, and exposure to important decision cues. Therefore, scenarios and exercises should be developed that allow decision makers to build templates that represent a broad base of problems they are likely to encounter in their actual decision environment.

Cognitive Apprenticeships. Recently, another on-the-job training technique that has gained attention is cognitive apprenticeship (see Druckman & Bjork, 1991). Like traditional apprenticeships, which train skilled physical performance, cognitive apprenticeships allow the trainee to operate in his or her actual context, and require the trainee to work closely with an expert. Briefly, the overriding goal of this type of training is first for the expert to demonstrate effective performance, and then gradually to lead the student through a series of constructive activities that allow him or her to deepen, clarify, integrate, and synthesize knowledge. As such, it may be a useful means to train all of the skills noted earlier.

Multi-Media Presentation Formats. Investigation into the manner in which knowledge presentation formats affect knowledge organization and template building is needed. Initial evidence suggests that employing graphics and animation may aid in mental model development (e.g., White, 1984) and hence, may also have a positive impact on NDM processes. The impact of textual, graphic, animated, video, and audio presentation of knowledge on decision-making performance has the potential to provide important guidelines for training system designers.

CONCLUSIONS AND FUTURE RESEARCH

In this chapter we have attempted to provide ideas generated from an NDM perspective for training decision makers. At the risk of bemoaning the point, *our strong recommendation is that empirical assessments of NDM-generated training be conducted.* Until such studies are completed, we will only be able to generate theoretically derived propositions, as was done here. In closing, we offer a number of specific research issues that we believe require further consideration as a basis to design training:

1. On what basis do decision makers perceive similarity in situations? That is, what triggers the activation of a template in memory, or causes a

decision maker to seek additional information (or engage in mental simulation) when he or she does not perceive the template to match sufficiently?

2. In what manner might NDM-consistent training be incorporated into more traditional training environments? For example, given that decision making is not easily removed from its context, how might NDM training be integrated into other aspects of task training in classroom situations, simulations, on the job, or any combination thereof.

3. In knowledge-rich environments—that is, those that require a decision maker to hold and access a large volume of knowledge in making a decision—how should expert knowledge be organized so that it fosters access of necessary information in decision making? That is, how can we initially present knowledge in a manner that fosters development of templates that are accessible, flexible, complete, and useful?

4. What does the NDM perspective suggest for the manner in which we evaluate decision making? For example, how do we know when someone becomes an expert? What are the descriptors or criteria associated with expertise? What do we mean by proficiency? What are appropriate assessment strategies for decisions made under conditions that are dynamic, rapidly changing, and fluid?

In summary, NDM theories offer a rich and exciting basis on which to specify training that will improve decision-making performance. As such, it has the potential to contribute both to the science and practice of training (Cannon-Bowers, Tannenbaum, Salas, & Converse, 1991).

REFERENCES

Cannon-Bowers, J. A., Tannenbaum, S. I., Salas, E., & Converse, S.A. (1991). Toward integration of training theory and technique. *Human Factors, 33,* 281–292.

Chi, M. T. H., Bassock, M., Lewis, M., Reimann, X., & Glaser, R. (1989). Self-explanations: How students study and use examples in learning to solve problems. *Cognitive Science, 13,* 145–182.

Driskell, J. E., & Salas, E. (1991). Overcoming the effects of stress on military performance: Human factors design, training and selection strategies. In R. Gal & A. D. Mangelsdorff (Eds.), *Handbook of military psychology* (pp. 183–193). London: Wiley.

Druckman, D., & Bjork, R. A. (1991). Modeling expertise. In D. Druckman & R. A. Bjork (Eds.), *In the mind's eye: enhancing human performance* (pp. 57–79). Washington, DC: National Academy Press.

Glaser, R. (1984). Education and thinking: The role of knowledge. *American Psychologist, 39,* 93–104.

Hoffman, R. (1991) *The psychology of expertise: Cognitive research and empirical AI.* New York: Springer-Verlag.

Klein, G. A. (1993). A recognition-primed decision (RPD) model of rapid decision making. In G. A. Klein, J. Orasanu, R. Calderwood, & C. E. Zsambok (Eds.), *Decision making in action: Models and methods* (pp. 138–147). Norwood, NJ: Ablex.

Lipshitz, R. (1993). Converging themes in the study of decision making in realistic settings. In G. A. Klein, J. Orasanu, R. Calderwood, & C. E. Zsambok (Eds.), *Decision making in action: Models and methods* (pp. 103–137). Norwood, NJ: Ablex.

Means, B., Salas, E., Crandall, B., & Jacobs, O. (1993). Training decision makers for the real world. In G. A. Klein, J. Orasanu, R. Calderwood, & C. E. Zsambok (Eds.), *Decision making in action: Models and methods* (pp. 306–326). Norwood, NJ: Ablex.

Noble, D. (1993). A model to support development of situation assessment aids. In G. A. Klein, J. Orasanu, R. Calderwood, & C. E. Zsambok (Eds.), *Decision making in action: Models and methods* (pp. 287–305). Norwood, NJ: Ablex.

Orasanu, J. (1990, July). *Shared mental models and crew decision making.* Paper presented at the 12th annual conference of the Cognitive Science Society, Cambridge, MA.

Orasanu, J. (1993). Decision-making in the cockpit. In E. L. Weiner, B. G. Kanki, & R. L. Helmreich (Eds.), *Cockpit resource management* (pp. 137–168). San Diego: Academic Press.

Orasanu, J., & Connolly, T. (1993). The reinvention of decision making. In G. A. Klein, J. Orasanu, R. Calderwood, & C. E. Zsambok (Eds.), *Decision making in action: Models and methods* (pp. 3–20). Norwood, NJ: Ablex.

White, B. (1984). Designing computer games to help physics students understand Newton's law of motion. *Cognition and Instruction, 1,* 69–108.

Chapter 11

Naturalistic Decision Making Research and Improving Team Decision Making

◆

Caroline E. Zsambok
Klein Associates Inc.

Naturalistic Decision Making (NDM) researchers have looked at how teams make decisions, and they have begun applying these findings to help teams improve. This chapter addresses four questions: What have we learned about team decision making (TDM) since the NDM conference 5 years ago? What are the major problems NDM researchers face when conducting research on teams? What problems are encountered when applying research findings to help teams improve and how can they be overcome? What issues are ripe for future TDM research?

This chapter captures central themes, not commonly found in literature about teams, from a panel discussion of these questions at the 1994 NDM conference. Panelists included David Gaba, an anesthesiologist at Stanford University, who represents a consumer's view. Gaba has adapted Crew Resource Management that Helmreich developed for aircraft cockpit crews (Helmreich & Foushee, 1993) to the training of anesthesiologists working in surgical teams. The other four panelists conduct team research: Bob McIntyre, Eduardo Salas, Daniel Serfaty, and Caroline Zsambok.

QUESTION 1: WHAT HAVE WE LEARNED?

We Know What Good Teams Do

We are producing solid data to support conclusions about what good decision-making teams do. Different researchers use different terms to describe these behaviors, and they place them at different levels of specificity in their models. However, many researchers agree that good teams monitor their performance and self-correct; offer feedback; maintain awareness of roles and functions and take action consistent with that knowledge; adapt to changes in the task or the team; communicate effectively; converge on a shared understanding of their situation and course of action; anticipate each other's actions or needs; and coordinate their actions (McIntyre & Dickinson, 1992; Orasanu, 1993; Salas, Dickinson, Converse, & Tannenbaum, 1993; Serfaty, Entin, & Deckert, 1994; Zsambok, 1993).

It seems obvious that good teams would engage in these behaviors. In fact, organizational development consultants have been conducting team training in our nation's businesses and industries for two decades that addresses these sorts of team behaviors.

So what's new here? What's new is that we now have data to validate that good teams do these things. As two panelists put it: "The gurus of teamwork were operating at an intuitive level, and some of them were doing good work. But they didn't agree on the components of teamwork" and, "It may seem unglamorous, but researchers can now say with confidence what good teams do. Without that we couldn't do anything else."

Yet, although we believe our ideas about the nature of these components are similar, we lack a common vocabulary. A significant contribution to the field would be an effort to reconcile semantic differences, to make explicit any differences of substance that remain, and to recommend a research agenda that would test their validity (Cannon-Bowers, Tannenbaum, Salas, & Volpe, 1995).

We Can Measure Team Processes

The major reason for our confidence in knowing what good teams do is that we have developed measures of team processes and outcomes that capture moment-to-moment aspects of TDM. Depending on the need, we can state team behaviors in very detailed domain-specific terms (usually for checklists used by researchers or observer–controllers). And we can state them as domain-independent, midlevel behaviors that either represent the aggregate of lower level behaviors or that represent directly measurable midlevel behaviors. We have gone from concepts about TDM to measures of TDM to data-based adjustments of the concepts. Virtually all of the researchers cited so far have been engaged for some time in research programs dedicated to this process.

Our Ideas Are More Concrete

At our last conference, not only were we less confident about what good teams do, but we did not talk much about *how* they do it. We now have more concrete hypotheses about TDM processes, which are helping us to focus on what to look for, what to measure.

For example, one concept that has spawned fruitful research about how decision-making teams operate is "shared mental models" (Cannon-Bowers, Salas, & Converse, 1993; Orasanu, 1990). Other researchers have decomposed this concept further, including Serfaty. He distinguishes between the situational mental model (the common blackboard that everyone is reading from, containing knowledge of the task, the mission, and the situation) and the mutual mental model (knowledge of interaction between oneself and other team members that is relevant to team tasks). Serfaty found that teams who develop more overlap across their members in what is on the blackboard perform better than teams who develop less overlap, provided they have the teamwork skills to act on that shared knowledge (Serfaty, Entin, & Volpe, 1993).

The concept of situational mental models is strikingly similar to the concept of team situation assessment described by Zsambok and her colleagues in their model of Advanced Team Decision Making (Zsambok, 1993). Through observations, they found that one of the characteristics of advanced teams is their ability to form realistic situation assessments that are understood by all members of the team. Specifically, in planning teams, some of the *hows* of situation assessment involve seeking diverging opinions, querying members about gaps and ambiguities in information or analyses, and then converging on (and updating) a commonly held assessment.

We Can Treat Teams as Cognitive Entities

From the consumer's side, Gaba believes that progress has been made by placing more emphasis on the cognitive, as opposed to the social aspects of team decision making. Pilots receiving the early Crew Resource Management training were overheard to call it "Charm School." There was resistance among them and Gaba's anesthesiologists to training based primarily on social elements of teams. Recently, however, cognitive elements have been emphasized in both CRM pilot training and in the training Gaba gave to his anesthesiologists working in teams (Gaba, Howard, & Fish, 1994; Howard, Gaba, Fish, Yang, & Sarnquist, 1992). Gaba reported that resistance is down.

Likewise, Zsambok found that organizations receiving Advanced Team Decision Making training are particularly attracted to the model's components of team conceptual level and team self-monitoring. They are interested in moving beyond seeing teams as social entities to seeing them as thinking entities, and they are seeking training to improve their teams' cognitive and metacognitive abilities.

We began with the behavioral side—discovering what good teams do. We are now trying to understand the cognitive side. We are learning how teams process and synthesize information, use resources, and generate courses of action. Our models are beginning to emphasize cognition and metacognition in teams.

We Are Developing True Team Tasks

We are beginning to see more research and training tasks that are not just an extension of individual decision making. These tasks cannot be completed successfully unless individuals perform interdependently, as a team. One class of such tasks are contained in some high-fidelity simulators. For example, SIMNET, at Ft. Knox, simulates tank platoon movements on the battlefield, and requires a variety of team coordination skills in order to achieve battle success. Another class is high-fidelity wargame exercises. If teams do not use collective TDM skills (as opposed to compiling individual decisions), they cannot succeed in these exercises. The existence of these tasks and technologies means that we can collect outcome and process data directly related to teams, not just individuals.

We Know That Training Makes a Difference

Training TDM skills affects team outcome performance even in the absence of training to improve task skill proficiency. The panel was adamant about the importance of this. Individual studies support this conclusion; a contribution to the literature would be a comprehensive review of them.

Among recent examples, Salas (Fowlkes, Lane, Salas, Franz, & Oser, 1994) has shown that training experienced helicopter crews helps them improve on a set of skills critical to good TDM. They include assertiveness, leadership, communication, flexibility, situation awareness, and mission analysis. These findings are consistent with and extend the work of other TDM researchers such as Helmreich, Serfaty, Cannon-Bowers, and Orasanu.

Likewise, Serfaty (Serfaty, Entin, Sovereign, Kemple, & Green, 1994) recently trained ad hoc Navy teams to perform better in stressful scenarios by teaching them how to improve certain coordination skills. Preliminary results show improvement in team performance after intervention. Similarly, Zsambok (1993) found that Air Force teams participating in standard training for logistics planning enhanced their performance if also trained on 10 behaviors related to team self-monitoring, team identity, and team cognition.

These researchers used different training approaches in their studies. Entin and Serfaty (1994), and Prince and Salas (1993) used variations on a behavioral modeling approach (i.e., instruction, demonstration, practice, feedback). Zsambok, Klein, Kyne, and Klinger (1993) taught teams how to self-assess on TDM behaviors, and how to use those assessments as inputs to developing strategies for improved TDM. These are just two examples of training methods that worked in particular cases. TDM researchers are at an exploratory stage in understanding approaches to team training. We need much more data about questions like: What methods are best for teaching team behaviors, and under what conditions? How much team training and practice is required? How and when should feedback be provided? Who should provide it? How long can the effects of various training methods be expected to last, and under what conditions? (See Swezey & Salas, 1992, for a review.)

QUESTION 2: WHAT ARE THE MAJOR RESEARCH PROBLEMS?

The second major question for this panel was to identify problems we face when trying to conduct research on teams. Instead of producing a laundry list, panelists devoted their attention to a fundamental problem raised by one of the members: "In team research, who should we be studying? Who are we claiming is making the decision—an individual, a collection of individuals, or the team as a single entity?"

McIntyre believes that *individuals* make decisions and take actions, not teams as single entities. Individuals make many microdecisions that affect the team's performance. Sometimes individuals feed information to commanders or team leaders who then make major decisions for the team. In either case, it is individuals on teams who should be the unit of analysis, not teams as a whole (McIntyre & Salas, 1995). McIntyre fears we may be reifying a concept by calling this process *team* decision making.

McIntyre also is concerned about using the term *team decision making* to refer to the host of activities that are related to team performance, such as identifying a problem, assessing the situation, generating and evaluating a course of action, monitoring for feedback, and adjusting plans. By calling all of this team decision making, instead of teamwork or team performance, he is concerned that we lose focus on research surrounding the moment of choice—the "decision."

By raising these concerns, McIntyre offered us an opportunity to extend a discussion begun at the 1989 NDM conference about decision making in general to teams in particular. At the 1989 conference, there was considerable debate about classical versus naturalistic decision research (see Zsambok, chapter 1, this volume, for a summary). Among the conclusions was that decision makers operating in real-world settings often use processes and strategies very different from those used to choose from an array of options, which is a focus of classical decision research. NDM researchers found they could not maintain a narrow focus on the "choice point" if they were to understand how experienced people make decisions while performing their jobs.

This theme was represented in reactions from other panelists to McIntyre's concerns: "If there is one contribution that NDM research has made, it's the extension of the definition of decision making. It's not just the moment of choice. It's the whole decision making process including the choice point" and, "For teams, each component of the decision making process is distributed across many individuals, so the team *is* a relevant and necessary unit of analysis." Audience opinions about these issues were mixed, and probably are representative of the larger NDM research community. We are unresolved about how team decision making should be defined, although majority sentiment favors a liberal definition both in terms of *who* is making the decision and *which processes* constitute decision making.

QUESTION 3: WHAT ARE SOME PROBLEMS IN APPLYING RESEARCH TO TEAMS?

Generalizability

Do we need a grand theory of teams that identifies team characteristics and maps them to conditions under which particular elements of teamwork, taskwork, and team training are relevant? This question arose in response to a problem raised by a consumer of TDM research. Gaba found no guidance in the literature to help him select which theories or training approaches were likely to generalize to his domain. As it happened, he discovered he could adapt CRM theory and training originally developed for command and control teams to anesthesiologists on surgical teams. However, on surgical teams, responsibility for the patient is shared equally by the surgeon and the anesthesiologist; there is no official team leader, and no formal command and control structure. What factors concerning his teams, tasks, and environment allow training from one domain to be useful in another, even in the face of such seemingly major differences?

Classification systems of team, task, and environmental factors (and their relationships) would help us develop a grand theory of teams. Some classification systems of teams do exist, but they are based on limited varieties of teams. Currently, McIntyre and his colleagues are collecting data about many types of teams to determine what defines their relevant differences. They are developing a quantifiable classification system for teams and team leadership. We look forward to publication of this work, as one more step towards the grand theory of teams.

Applied researchers, however, must move forward without a master theory. Training applications *are* being exchanged across domains. For example, Serfaty is translating measures of communication developed for military tactical teams to helicopter cockpit crews and trauma care teams in operating rooms. Zsambok is translating team decision training applications from strategic planning teams to emergency response teams in nuclear power plants. Other applications are ongoing, and this is a rich area of progress.

Expense

Because of our uncertainty about generalizability of training applications, there is a time-consuming, expensive analysis that often precedes applications of TDM research (i.e., training interventions) transported from one domain to another. Therefore, intervention analyses are needed to understand elements like the nature of the team and its tasks, the context in which the team operates, and the team's learning history with prior team training systems. Also, there is often a need to adapt old or develop new materials when we apply training from one setting to another. We cannot assume that exercises, simulations, or feedback procedures that were successful in one organizational setting will be appropriate or even feasible in another.

Finally, there is the expense of gathering and analyzing data about the effectiveness of modified applications. This is critical to maintaining a cohesive research program. The problem is this: At some point, a TDM researcher develops a set of findings and a successful training intervention based on those findings. In applying the intervention to more and more organizations, and in tailoring it to meet their needs, there is a drift away from the intervention that at one time was tightly coupled to the TDM research findings and theory. If the researcher does not continue to collect data about key elements of each new setting, and analyze it in terms of training modifications and their effectiveness, she or he eventually has only weak theory or data behind the TDM interventions. Nor can she or he make serious claims about generalizability. However, funding for rigorous back-end analysis of an intervention is difficult to obtain from users.

QUESTION 4: WHAT ARE THE KEY FUTURE RESEARCH TOPICS AND THEMES?

Listed in previous sections are targets of active research likely to continue into the future. Also mentioned was the need to develop common vocabulary about components of TDM and to answer key questions about approaches to team training. Other critical areas for future research are discussed in the following.

Team Structure

One of the new frontiers of TDM research concerns effects of team structure on team performance. A few dimensions of team structure have been identified, such as the structures of information (who knows what), expertise, resource ownership, authority (e.g., hierarchical versus heterarchical), responsibility (e.g., geographical versus functional), and communication (Pete, Pattipati, & Kleinman, 1993). Decision-making teams contain a certain overlap across team members on some or all of these dimensions. For example, information is rarely either totally partitioned or centralized across members.

With growing numbers of distributed teams, it is becoming increasingly more critical to answer a host of questions about team structure, such as: Can teams change their structure in real time to adapt on the fly to task demands? What are the costs and benefits of training teams to reconfigure different structural components? How does task structure map onto team structure?

Long-Term Team Training Effects

NDM researchers typically are not funded to collect long-term effectiveness data, or to monitor the effects of extraneous variables (e.g., organizational norms and goals; incentive systems) on trained team behaviors. Although our TDM theories and hypotheses are more solid today, we need to collect longitudinal data so we are in a position to produce cost/benefit analyses to potential users, and to advocate for organizational support requirements to maximize team training effects.

Even with funding, longitudinal research on teams is a particularly difficult challenge, because team membership changes over time. Studying teams is exponentially more complex than studying individuals. When the target of study—the team itself—is subject to change, the challenge is even greater. For example, how are training effects on the host of processes distributed across team members to be interpreted when one new member comes on the team? Or two new members? Or, when trained members leave the team?

Methodological Bridges

TDM researchers are using a variety of approaches to study teams. They vary from normative—such as mathematical models tested by experiments (Kleinman, Luh, Pattipati & Serfaty, 1992)—to structuralist—such as simulation models (Perdu, Zaidi, Sadigh, Lehner, & Levis, 1994)—to naturalistic–structuralist—such as interaction models and theories tested in realistic experiments (Serfaty, Entin, & Deckert, 1993), to purely naturalistic–descriptive models—such as models developed and tested in the field (Zsambok, 1993). We need to bridge these approaches, perhaps through an experimental database. Models and hypotheses resulting from these approaches must be tested empirically more often.

CONCLUSION

NDM researchers have only just begun the process of understanding and aiding team decision making. We have seen real progress since the last conference in the form of good data about what decision making teams do; nonetheless, much work lies ahead. We need a common vocabulary of TDM processes and behaviors. We need to understand more about interactions among task, team, and environmental factors. We need to continue our investigation into the cognitive aspects of TDM. And, we need much greater understanding about team training approaches: their critical elements, their impact on team product or output, their cost compared to benefit, and the impact of posttreatment factors on longevity.

REFERENCES

Cannon-Bowers, J., Tannenbaum, S., Salas, E., & Volpe, C. E. (1995). Defining competencies for establishing team training requirements. In R. Guzzo & E. Salas (Eds.), *Team effectiveness and decision making in organizations* (pp. 353–380). San Francisco: Jossey-Bass.

Cannon-Bowers, J., Salas, E., & Converse, S. (1993). Shared mental models in expert team decision making. In N. J. Castellan (Ed.), *Individual and group decision making: Current issues* (pp. 221–246). Hillsdale, NJ: Lawrence Erlbaum Associates.

Entin, E. E., & Serfaty, D. (1994). *Team adaptation and coordination training* (Tech. Rep. 648). Burlington, MA: ALPHATECH, Inc.

Fowlkes, J., Lane, N., Salas, E., Franz, T., & Oser, R. (1994). Improving the measurement of team performance: The TARGETS methodology. *Military Psychology, 6*(1), 47–61.

Gaba, D., Howard, S., & Fish, K. (1994). *Crisis management in anesthesiology.* New York: Churchill-Livingstone.

Helmreich, R. L., & Foushee, H. C. (1993). Why resource management? In E. L. Wiener, G. B. Kanki, & R. L. Helmreich (Eds.), *Cockpit resource management* (pp. 3–45). San Diego, CA: Academic Press.

Howard, S., Gaba, D., Fish, K., Yang, G., & Sarnquist, F. (1992). Anesthesia crisis resource management training: Teaching anesthesiologists to handle critical incidents. *Aviation, Space, and Environmental Medicine, 63,* 763–770.

Kleinman, D. L., Luh, P. B., Pattipati, K. R., & Serfaty, D. (1992). Mathematical models of team performance: A distributed decision-making approach. In R. Swezey & E. Salas (Eds.), *Teams: Their training and performance* (pp. 177–217). Norwood, NJ: Ablex.

McIntyre, R. M., & Dickinson, T. L. (1992). *Systematic assessment of teamwork processes in tactical environments.* Prepared under contract N61339–91–C–0145 for the Naval Training Systems Center, Orlando, FL.

McIntyre, R., & Salas, E. (1995). Measuring and managing for team performance: Emerging principles from complex environments. In R. Guzzo & E. Salas (Eds.), *Team effectiveness and decision making in organizations* (pp. 149–203). San Francisco: Jossey-Bass.

Orasanu, J. M. (1990). *Shared mental models and crew decision making* (Tech. Rep. No. 46). Princeton, NJ: Princeton University, Cognitive Sciences Lab.

Orasanu, J. M. (1993). Decision-making in the cockpit. In E. L. Wiener, B. G. Kanki, & R. L. Helmreich (Eds.), *Cockpit resource management* (pp. 137–172). San Diego, CA: Academic Press.

Perdu, D., Zaidi, A., Sadigh, A., Lehner, P., & Levis, A. (1994). On a methodology for team design using influence diagrams. In *Proceedings of the 1993 Symposium on Command and Control Research* (pp.77–84). McLean, VA: SAIC.

Pete, A., Pattipati, K. R., & Kleinman, D. L. (1993). The effect of team structure on team decision performance. In *Proceedings of the 1993 Symposium on Command and Control Research* (pp. 27–36). McLean, VA: SAIC.

Prince, C., & Salas, E. (1993). Training and research for teamwork in the military aircrew. In E. L. Wiener, G. B. Kanki, & R. L. Helmreich (Eds.), *Cockpit resource management* (pp. 337–366). Orlando, FL: Academic Press.

Salas, E., Dickinson, T. L., Converse, S. A., & Tannenbaum, S. I. (1993). Toward an understanding of team performance and training. In R. Swezey & E. Salas (Eds.), *Teams: Their training and performance* (pp. 3–29). Norwood, NJ: Ablex.

Serfaty, D., Entin, E. E., & Deckert, J. C. (1994). Implicit coordination in command teams. In S. E. Johnson & A. H. Levis (Eds.), *Science of command and control: Coping with change* (pp. 87–94). Fairfax, VA: AFCEA International Press.

Serfaty, D., Entin, E. E., Sovereign, M., Kemple, W., & Green, L. (1994). Team adaptation and coordination training. In *Proceedings of the 1994 JDL Command and Control Research Symposium* (pp. 279–281). McLean,VA: SAIC.

Serfaty, D., Entin, E. E., & Deckert, J. C. (1993). *Team adaptation to stress in decision making and coordination with implications for CIC teams training* (Tech. Rep. 564). Burlington, MA: ALPHATECH, Inc.

Serfaty, D., Entin, E. E., & Volpe C. (1993). Adaptation to stress in team decision-making and coordination. In *Proceedings of the Human Factors and Ergonomics Society 37th annual meeting* (pp. 1228–1232).

Swezey, R. W., & Salas, E. (1992). Guidelines for use in team-training development. In R. W. Swezey & E. Salas (Eds.), *Teams: Their training and performance* (pp. 219–245). Norwood, NJ: Ablex.

Zsambok, C. E. (1993). Advanced team decision making in C^2 settings. In *Proceedings of the 1993 Symposium on Command and Control Research* (pp. 45–52). McLean, VA: SAIC.

Zsambok, C. E., Klein, G., Kyne, M., & Klinger, D. (1993). How teams excel: A model of advanced team decision making. In *Proceedings of the 1993 NSPI conference* (pp. 135–145). Bloomington, IN: NSPI Publications.

Chapter 12

Analyzing Decision Making in Process Control: Multidisciplinary Approaches to Understanding and Aiding Human Performance in Complex Tasks

♦

Emilie M. Roth
Westinghouse Science and Technology Center

Process control has long served as a focus for research on decision making in complex dynamic settings (e.g., Bainbridge, 1979; DeKeyser, 1992; Leplat, 1981; Rasmussen, 1986). One reason is that process control work environments, such as chemical plants, nuclear power plants, and offshore oil and gas installations, exemplify the dimensions of complexity identified with naturalistic decision making (NDM) settings (Orasanu & Connolly, 1993). The environments are inherently dynamic, with high demands for skilled performance and team work; high uncertainty because of the complex processes and impoverished state information displays; high stakes because of the potentially devastating consequences of accidents; and high organizational and sociopolitical influences.

Process control faces many of the same issues confronted by other NDM environments (Lee & Morgan, 1994; Orasanu, 1993; Sarter & Woods, 1995). With the increased introduction of digital technology and automation, the operator's role is shifting from manual control to supervisory control. Common issues include: how to develop new displays and decision aids to improve decision making; how to deal with increased automation; how to support distributed decision making; how to train cognitive skills, team skills, and the ability to make decisions under high-stress conditions. In many cases, process control has been in the vanguard of tackling these issues.

This chapter summarizes research issues raised during the panel on process control at the 1994 NDM conference. Panel members included: Kevin Bennett (Wright State University); Lia Di Bello (Laboratory for Cognitive Studies of Activity, City University of New York); Rhona Flin (Offshore Management Center, The Robert Gordon University); Emilie Roth (Westinghouse Science & Technology Center); and Penelope Sanderson (University of Illinois). The first part of the chapter describes a sampling of research topics that have grown out of issues confronting process control. A commitment to NDM studies as a basis for identifying actual demands faced by practitioners serves as a common thread. Part two focuses on methodological issues associated with NDM studies.

MULTIPLE DISCIPLINES AND RESEARCH PERSPECTIVES

One of the exciting aspects of process control research is that it has a long tradition that draws on multiple disciplines and research perspectives. Because the consequences of human error in process control environments are typically high, process control has traditionally opened its doors to behavioral researchers from a variety of disciplines, each contributing different tools and research perspectives to a common set of problems: how to understand and aid human performance in complex work environments. Much of the early work drew on the European work analysis tradition that was naturally sympathetic to field studies and naturalistic observation techniques (e.g., Leplat, 1981). Other strands of research grew out of engineering disciplines (e.g., Rasmussen, 1986), industrial and experimental psychology (e.g., Bennett, Toms, & Woods, 1993; Sanderson, Haskell & Flach, 1992; Vicente, Christoffersen, & Pereklita, 1995; Woods, 1991); and more recently, developmental psychology (e.g., Di Bello, chapter 16, this volume); and sociology and anthropology traditions (e.g., Heath & Luff, 1992). The members of the panel on process control reflected this diversity of research perspectives. One of the aims of this chapter is to illustrate how NDM research has benefited from the intermixing of these different disciplines and traditions.

IMPROVING OPERATOR PERFORMANCE:
SELECTED TRENDS

As an applied field, our main focus is on finding ways to improve human performance. The process control domain has provided rich opportunities for developing and testing new approaches to aiding performance. This section focuses on two current topics of research: Representational Aiding and Cognitive Skills Training. In both cases there is an emphasis on uncovering actual demands confronting practitioners as a foundation for defining requirements for improved performance, and a commitment to NDM studies.

Representational Aiding

Traditionally, the user interfaces provided in the process control industry have been very impoverished. In many cases a "single sensor–single parameter display" approach was used so that the burden of collecting and integrating plant parameter data to derive meaningful state information was left to the operator. With the advent of advanced digital technology there is a growing trend to develop displays that integrate plant parameter data for the operator.

An area of research that has emerged to support the need for guidance on integrated display design has been variously termed *representational aiding* (Bennett et al., 1993; Woods, 1991) and *ecological interface design* (Vicente et al., 1995). The basic premise of this approach is to develop graphic representations of the domain that enable critical task-relevant information to be perceptually apparent. The goal is to minimize the need for deliberate cognitive processing and problem solving for situation awareness, and to capitalize on perceptual processes instead. In this respect, the goals of representational aiding research are consistent with models of NDM that argue that much of expert decision making in dynamic emergencies is recognition-primed (Klein & Calderwood, 1991).

At the same time, a primary premise of representational aiding research is that complex situations unanticipated by system designers are bound to arise, and that preplanned response strategies can never cover all contingencies (Roth, Bennett, & Woods, 1987; Roth, chapter 17, this volume). A primary goal of representational aiding research is to design for adaptability to unforeseen circumstances. The goal is to develop *external representations* of domain goals and constraints that facilitate the kinds of knowledge-based problem solving and planning that is required to handle unanticipated situations (Rasmussen, 1986).

Within this research area there has been a particular emphasis on understanding the factors that contribute to effective integral or object displays (Bennett et al., 1993; Sanderson et al., 1992). A main finding in this literature is that object displays do not always lead to improved human performance. The results indicate the importance of understanding the domain constraints and critical data relationships present in the domain, and crafting object displays that map these critical data relationship into perceptually salient properties of the display. The findings underline a main thesis of the representational aiding approach, which is the importance of analyzing the cognitive demands of the domain as a prerequisite for display design.

Another recent trend in representational aiding has been to examine requirements for group decision making, and ways technology can be used to enhance *group situation awareness* and *computer-supported cooperative work*. These studies have relied on naturalistic observation of group work settings as a tool for uncovering the actual demands imposed by a domain and the strategies domain practitioners have developed for dealing with them.

One example is a study that was recently conducted examining multimember interaction in London Underground Line Control Rooms, with the aim of understanding crew member activities that enabled shared situation awareness (Heath & Luff, 1992). They found that the "openness" of the physical work environment supported shared views, enabling workers to maintain awareness of each other's activities and build on them without need for explicit communication. The results of the study led the authors to argue against proposed control room changes that would have replaced a large fixed line diagram that provided a common representation simultaneously visible to all crew members, with individual "private" graphic displays that would have disrupted mutual monitoring of each other's activities, and created new communication demands. Similar analyses of the requirements for group decision making are being conducted in the nuclear power plant industry (Mumaw & Roth, 1994). The objective of this work is to define requirements for planned wall-panel information systems intended to provide a shared representation of plant state to facilitate group situation awareness and problem solving.

The experience in representational aiding makes salient the importance of a thorough analysis of the cognitive demands imposed by the domain and how domain practitioners respond, as a prerequisite for effective display and decision-aid design. Naturalistic decision making studies contribute to this understanding by uncovering the actual demands practitioners face and how they deal with them.

Training Cognitive Skills and Decision Making

Another active area of research has centered around training problem-solving and decision-making skills. As in the case of representational aiding, a common element of the training research has been a commitment to analyze the cognitive demands of the domain, and a reliance on observational field studies as an important source for understanding these demands.

One thread of research has examined training requirements for decision making under dynamic, uncertain, and high-risk conditions. The focus has been on analysis of the cognitive and team interaction demands imposed by emergency situations, and the knowledge and skills that need to be developed to enable people to adequately handle the situations. For example, Flin and Slaven (1994) examined the characteristics of crises in offshore oil and gas installations, and the decision-making demands that crises place on the on-scene commander. They argued that examination of how people actually perform tasks, as opposed to how they are "supposed" to perform them, leads to different approaches to the design of training programs. In the case of offshore oil and gas installations managers, they found that Klein's recognition-primed decision-making theory (Klein, chapter 27, this volume), and the cockpit resource management training methods developed by the airline industry, provided a better foundation for training decision making under emergency conditions than previous training approaches that focused on formal decision-analysis methods.

Mumaw, Swatzler, Roth, & Thomas (1994) performed similar analyses to identify the cognitive skills and decision-making training requirements for handling complex nuclear power plant accidents. Their report provides an overview of training techniques, targeted at different aspects of cognitive and team-interaction skills required for decision making under dynamic, high-risk conditions. Their approach contrasts with traditional approaches to emergency-response training in process control, which focus on providing practice in responding to anticipated emergencies, and familiarizing operators with the use of administrative and emergency procedures for handling these events.

Another thread of research in the training area focuses on the learning process itself. Much of the available guidance on training stems from examination of traditional classroom learning situations, yet a great deal of job-related knowledge and skills are developed through more informal means. Recently there has been an interest in studying more informal kinds of "situated" learning, such as learning from coworkers and apprenticeship training, that take place in work settings.

One of the main research methodologies employed in this literature is naturalistic field observations. The results of these studies are leading to new approaches to training that are better suited to the workplace context. For example, Di Bello and her colleagues (chapter 6, this volume) are studying the introduction in manufacturing plants of a new computer-based decision-support system for material requirements planning called MRP, and the process by which individuals with different educational and occupational backgrounds learn to use it. They found that classroom instruction was ineffective for developing the kind of flexible mastery needed to use MRP. On-the-job activity proved to be critical for developing the necessary skills. One of the most important contributors to on-the-job learning was exposure to situations where the goals to be achieved were clear, but the solution path for achieving them was not well specified and had to be worked out by the individual through what Di Bello and her colleagues have called *constructive activity*. This result held across a variety of backgrounds. Di Bello and her colleagues have argued that "constructive" activities result in more effective learning because they tap into people's cognitive entry points. When faced with situations requiring "constructive" activity, people attempt to bring their current way of thinking to bear on the problem. This enables them to notice deficiencies, and make refinements in accordance with the new concepts being introduced. The results are being used to develop new approaches to workplace training that are designed to address the training needs of individuals from a variety of occupational and learning backgrounds, by tapping into their own individual cognitive entry points.

NATURALISTIC DECISION-MAKING RESEARCH: SELECTED METHODOLOGICAL ISSUES

One of the salient points to be drawn from the review of current research trends in process control is the central role that naturalistic field studies play in defining the requirements for improved training, displays, and support systems. There has been a long tradition of drawing on naturalistic field studies in process control (e.g., Bainbridge, 1979). This tradition has enjoyed renewed vigor and legitimacy with the rise in interest in naturalistic decision making studies in other domains and the introduction of ethnographic methods to the analysis of work environments (e.g., Jordan, 1996).

Although NDM research reflects a commitment to empirical investigation, this can range from retrospective analyses of critical incidents (Flin & Slaven, 1994; Klein & Calderwood, 1991); to field studies of "naturally occurring" ongoing work practice (Di Bello, chapter 16, this volume; Heath & Luff, 1992; Jordan, 1996); to field experiments (Roth et al., 1987; Roth, Mumaw, & Lewis, 1994; Sarter & Woods, 1995).

One of the exciting aspects of process control research is that it has attracted researchers from multiple disciplines and research perspectives—industrial engineering, experimental psychology, developmental psychology, forensic psychology, cognitive engineering, anthropology, and sociology—each contributing research methods and traditions. This section discusses some methodological issues currently confronting NDM research and some of the approaches that have arisen from the different research traditions.

Ethnographic Methods and Cognitive Apprenticeship: Entering the World of Domain Practioners

One of the challenges to NDM studies is that the researcher typically begins with limited knowledge of the domain of practice that he or she is studying. This lack of expertise in the domain makes it difficult for the researcher to understand the environment from the domain practitioner's point of view.

Ethnographic methods, which have grown out of the anthropology tradition, provide one approach to addressing this problem.

Ethnographic methods typically involve a combination of observation and "in-situ" question asking, carried out while participating in the ongoing activities of the work group (Jordan & Henderson, 1995). As example, Di Bello (chapter 16, this volume) used this approach to study work practices of electricians maintaining old New York subways. She and her coworkers literally put on overalls and workboots and crawled under the trains with the mechanics, acting as apprentices. Through this process, and through interviews conducted while acting as apprentices (e.g., asking questions like "I notice you did something that I would never know to do from looking at the procedure manual. What were you thinking? What are you paying attention to?") an understanding emerges of what the perceptual world is like for the domain practitioners, what they pay attention to, and what kind of information they need. Di Bello and her colleagues refer to this knowledge acquisition methodology as *cognitive apprenticeship* and are currently codifying the skills required to be an effective cognitive apprentice.

Strategies for Parsing and Interpreting Observational Data

NDM research entails a change in orientation from traditional behavioral decision-making research that relies on established methods of experimental design and statistical analysis. NDM research relies on limited sample field observations and participant-centered ways of studying work. This requires a change in analytic framework and scientific foundation (Jordan & Henderson, 1995; Sanderson & Fisher, 1994).

One issue regards the basis for analyzing observational data and drawing generalizations from limited, and in some cases single, instances. Field studies can result in masses of observational data. How does one begin the analysis? How do we parse the ongoing flow of activity into analysis units? How do we determine what is interesting? There are many different models and potential explanations that one can come up with. How do we find a model that best captures the observed phenomena?

Researchers have confronted these issues from a variety of perspectives. At one end of the spectrum are researchers from the ethnographic research tradition who consciously attempt to approach the analysis of observational data with minimal preconceptions (Jordan & Henderson, 1995). Toward the other end of the spectrum are researchers from the cognitive engineering tradition who begin with a formal representation of the underlying domain as a guide to interpretation and analysis of observational data (Rasmussen, 1986; Roth & Woods, 1988, 1989).

A variety of heuristics have been employed to support the parsing and interpretation of observational data. Most of the time these are utilized without explicit mention of the analytic framework employed or explicit discussion of its epistemological underpinnings. Jordan (1996; Jordan & Henderson, 1995), who coined the term *analytic framework*, provided the most explicit and comprehensive discussion of analytic frameworks available for parsing observational data. In most cases, the heuristics she describes rely on abstract structures that do not require knowledge of the domain to parse the flow of participant activity into meaningful units of analysis. Examples of analytic frameworks she covered include: (a) examining the structure of events (e.g., beginnings and endings); (b) examining the temporal organization of activity (e.g., periodicity); (c) examining "participatory" structures (e.g., turn taking); (d) looking at the spatial organization of activity (e.g., whose turf is it? who has more ready access to relevant resources? who is in a best position to overview scene? who is in the best position to maintain eye contact with everyone else?); and (e) looking for instances of "trouble and repair" (i.e., paying particular attention when the usual stream of activity is disrupted in some way).

Whereas the methods described by Jordan do not depend on detailed knowledge of the domain, several other researchers have developed analytic frameworks that are based on formal analyses of the domain. Examples include the analysis of operator performance in a complex process control task that was conducted by Roth and Woods (1988). In their study they began with a formal function-based analysis of the control task itself, the control requirements, and the information available to operators. This function-based cognitive task analysis provided the framework for interpreting operator performance. The framework revealed the paucity of process state information available that made the control task difficult for any agent—man or machine. It made salient that the control strategies observed to be used by expert operators, while enabling them to compensate for the poor process state displays, remained suboptimal; and pointed to new representational aids to increase level of performance.

A related analytic framework begins with a "canonical" representation of how operators "are supposed to" perform according to some authoritative source (e.g., the perception of the system designers, managers, or regulators), looking for discrepancies in behavior from this prescriptive norm, and trying to understand and document why they arise. Examples of this strategy include a study that was conducted examining use of an expert system to support an electronics trouble-shooting task (Roth et al., 1987), and a study examining the use of emergency operating procedures in handling complex simulated nuclear power plant emergencies (Roth et al., 1994).

The various analysis frameworks described earlier are all valuable to include in our analysis toolkit. More effort is needed to formalize and systematize the methods available for analyzing NDM research along the lines that Sanderson and Fisher (1994) and Jordan and Henderson (1995) have begun.

Automated Tools to Support Analysis of Observational Data

Analysis of observational data is time and resource intensive. Jordan and Henderson (1995) estimated that it takes 10 to 20 hours to analyze an hour of videotaped observational data. In our experience it has taken an average of 8 hours to analyze an hour of videotape. Either estimate is substantial. There is an urgent need to develop more streamlined and automated analysis techniques if NDM research is to continue to flourish and expand. Fortunately, there are growing efforts to develop computer-based "exploratory sequential data analysis" methods (Sanderson et al., 1994).

SUMMARY AND CONCLUSIONS

Process control has a long tradition of NDM research. This chapter introduced a sample of current research topics, a common theme being the importance of analyzing the cognitive demands imposed by the domain and how practitioners respond to these demands, as a prerequisite for design of effective displays, decision aids, and training programs. The second part of the chapter focused on methodological issues associated with NDM studies, a salient theme being the multidisciplinary perspectives that have contributed to NDM research methods.

Naturalistic decision making research necessitates a shift from traditional methods of experimental design and statistical analysis to analyses of field observations and participant-centered ways of studying work. This shift requires a change in the scientific basis for the research we conduct and the analysis methods we employ. Work is beginning on formalizing analysis methods and their epistemological underpinnings, but more effort is needed in this direction.

REFERENCES

Bainbridge, L. (1979). Verbal reports as evidence of the process operator's knowledge. *International Journal of Man–Machine Studies, 11,* 411–436.

Bennett, K. B., Toms, M. L., & Woods, D. D. (1993). Emergent features and graphical elements: Designing more effective configural displays. *Human Factors, 35,* 71–97.

DeKeyser, V. (1992). Why field studies? In M. Helander & M. Nagamachi (Eds.), *Design for manufacturability: A systems approach to concurrent engineering and ergonomics* (pp. 305–316). Washington, DC: Taylor & Francis.

Flin, R., & Slaven, G. (1994). *The selection and training of offshore installation managers for crisis management* (Rep. No. OTH 92374 Offshore Safety Division, HSE). London: HSE Books.

Heath C. C., & Luff, P. (1992). Collaboration and control: Crisis management and multimedia technology in London Underground Line Control Rooms. *Computer Supported Cooperative Work Journal, 1,* 69–94.

Jordan, B. (1996). Ethnographic workplace studies and computer supported cooperative work. In D. Shapiro, M. Tauber, & R. Traunmüller (Eds.), *The design of computer-supported cooperative work and groupware systems* (pp. 17–42). Amsterdam, The Netherlands: North Holland/Elsevier Science.

Jordan, B., & Henderson, A. (1995). Interaction analysis: Foundations and practice. *The Journal of the Learning Sciences, 4,* 39–103.

Klein, G. A., & Calderwood, R. (1991). Decision models: Some lessons from the field. *IEEE Transactions on Systems, Man, and Cybernetics, 21,* 1018–1026.

Lee, J. D., & Morgan, J. (1994). Identifying clumsy automation at the macro level: Development of a tool to estimate ship staffing requirements. In *Proceedings of the Human Factors and Ergonomics Society 38th annual meeting* (pp. 878–882). Santa Monica, CA: Human Factors and Ergonomics Society.

Leplat, J. (1981). Task analysis and activity analysis in situations of field diagnosis. In J. Rasmussen & W. B. Rouse (Eds.) *Human detection and diagnosis of system failures* (pp. 287–300). New York: Plenum.

Mumaw, R. J., & Roth, E. M. (1994). Design of group overview displays to support operator performance. *Transactions of the 1994 annual meeting of the American Nuclear Society* (p. 115). La Grange Park, IL: American Nuclear Society.

Mumaw, R. J., Swatzler, D., Roth, E. M., & Thomas, W. A. (1994) *Cognitive skill training for nuclear power plant operational decision making.* Washington, DC: U. S. Nuclear Regulatory Commission. (NUREG/CR–6126)

Orasanu, J., & Connolly, T. (1993). The reinvention of decision making. In G. Klein, J. Orasanu, R. Calderwood, & C. Zsambok (Eds.), *Decision making in action: Models and methods* (pp. 3–20). Norwood, NJ: Ablex.

Orasanu, J. (1993). Decision-making in the cockpit. In E. L. Wiener, B. G. Kanki, & R. L. Helmreich (Eds.), *Cockpit resource management* (pp. 137–172). San Diego: Academic Press.

Rasmussen, J. (1986). *Information processing and human machine interaction: An approach to cognitive engineering.* New York: North-Holland.

Roth, E. M., Bennett, K., & Woods, D. D. (1987). Human interaction with an "intelligent" machine. *International Journal of Man–Machine Studies, 27,* 479–525.

Roth, E. M., Mumaw, R. J., & Lewis, P. M. (1994). *An empirical investigation of operator performance in cognitively demanding simulated emergencies.* Washington DC: U. S. Nuclear Regulatory Commission. (NUREG/CR–6208)

Roth, E. M., & Woods, D. D. (1988). Aiding human performance: I. Cognitive analysis. *Le Travail Humain, 51*(1), 39–64.

Roth, E. M., & Woods, D. D. (1989). Cognitive task analysis: An approach to knowledge acquisition for intelligent system design. In G. Guida & C. Tasso (Eds.), *Topics in expert system design* (pp. 233–264). New York: Elsevier.

Sanderson, P. M., & Fisher, C. (1994). Exploratory sequential data analysis: Foundations. *Human–Computer Interaction, 9,* 251–317.

Sanderson, P. M., Haskell, I., & Flach, J. M. (1992). The complex role of perceptual organization in visual display design theory. *Ergonomics, 35,* 1199–1219.

Sanderson, P. M., Scott, J. J. P., Johnston, T., Mainzer, J., Watanabe, L. M., & James, J. M. (1994). MacSHAPA and the enterprise of Exploratory Sequential Data Analysis (ESDA). *International Journal of Human–Computer Studies, 41,* 633–681.

Sarter, N. B. & Woods, D. D. (1995). How in the world did we get into that mode? Mode error and awareness in supervisory control. *Human Factors, 37,* 5–19.

Vicente, K. J., Christoffersen, K., & Pereklita, A. (1995). Supporting operator problem solving through ecological interface design. *IEEE Transactions on Systems, Man, and Cybernetics, SMC–25,* 529–545.

Woods, D. D. (1991). Representation aiding: A ten-year retrospective. In *Proceedings of the 1991 IEEE International Conference on Systems, Man, and Cybernetics* (pp. 1173–1176). Charlottesville, VA: IEEE.

Chapter 13

Cognitive Task Analysis

◆

Sallie E. Gordon
Richard T. Gill
University of Idaho

In naturalistic decision making, analysts are frequently interested in studying how people in a particular field make decisions in complex, dynamic, real-time environments. Cognitive task analysis (CTA) is often used for this purpose. Cognitive task analysis addresses a central and longstanding difficulty in a diverse number of fields: the need to capture the knowledge and processing used by experts in performing their jobs. In this chapter, we first consider the emergence of CTA, and attempt to define what most people mean by the term. Then we provide a generic description of CTA, and very briefly describe several representative CTA methods. Finally, we conclude with a brief review of issues that emerged during the panel discussion—issues that are central questions as analysts and researchers work to further develop cognitive task analysis methods.

THE NEED FOR CTA

Historically, when analysts wished to study job performance, they would perform a *task analysis*. This task analysis typically resulted in a list of tasks and subtasks, as well as the input, actions, output, conditions for performance, and proficiency requirements (Ryder, Redding, & Beckshi, 1987). However, because we live in an era of technological complexity, many jobs are beginning to involve complex cognitive processing, rather than simply a collection of overt or observable behaviors. Because of this shift, the traditional *behavioral* task analysis methods are often inadequate for describing how people do their jobs. Analysts have begun looking for methods that can capture the richness of the knowledge base and cognitive processing embodied in expert performance of a task (e.g., Gordon & Gill, 1992; Gott, 1989; Hoffman, 1987; Klein, Calderwood, & MacGregor, 1989; Means & Gott, 1988; Rasmussen, 1986; Redding, 1990; Rouse, 1984; Ryder & Redding, 1993; Zaff, McNeese, & Snyder, D. E., 1993). Such methods are generically termed *cognitive task analysis*.

Defining CTA

CTA does not have a single, well-accepted definition. However, we can list the types of information that researchers try to identify when performing a CTA. Analysts speak with experts, journeymen, novices, or any combination therein and analyze their behavior, in order to determine some subset of the following, as they relate to a person's job:

- Concepts and principles, their interrelationships with each other, and their relationship to the task(s).
- Goals and goal structures (including methods of achieving the goals, and initiatiing conditions or "triggers" for goals and methods).
- Cognitive skills, rules, strategies, and plans.
- Perceptual learning, pattern recognition, and implicit or tacit knowledge.
- Mental models (how experts represent and run models of the system).
- Problem models (how experts represent a problem and work within the problem space).
- How novices move through all of the above in various stages to become expert.
- Difficulties in acquiring domain knowledge and skills.
- Instructional procedures useful for moving a person from novice to expert.

Most analysts do not try to determine all of these factors for a given case. However, they do try to develop a fairly complete theory or *model* of how experts in a field perform their tasks (Ford & Bradshaw, 1993). This model can then be used to develop system interfaces, intelligent support or decision aids, training programs, and other artifacts.

Deciding to Perform a CTA

CTA often is very time consuming and labor intensive, which means it is also expensive. Because of this, analysts frequently wonder when a CTA is appropriate or necessary. Those who perform CTA frequently respond with a reference to the task, environment, or both; an activity will require this type of analysis if it involves (a) *complex, ill-structured tasks* that are difficult to learn, (b) *complex, dynamic, uncertain, and real-time environments,* and perhaps (c) *multitasking* where the person must perform many tasks simultaneously.

Another approach is to simply evaluate the cognitive complexity of the task. That is, CTA is appropriate if performance of a task requires: (a) the use of a large and complex conceptual knowledge base; (b) the use of complex goal/action structures dependent on a variety of triggering conditions, or (c) complex perceptual learning or pattern recognition (DeVries & Gordon, 1994b). DeVries and Gordon (1994a) developed a software tool that helps an analyst estimate the cognitive complexity of a task as well as its component subtasks. This allows the analyst to determine which subtasks would benefit most from CTA.

It is important to point out that an analyst does not just blindly perform both an in-depth behavioral task analysis *and* a CTA. The extent of each should depend on the nature of the task or subtask in the particular domain being analyzed. Many times, a variety of task analysis methods will be combined, where behavioral and cognitive components are studied to the extent appropriate for both the domain and the final goal of the project (Gordon, 1994).

METHODS OF CTA

Because it attempts to identify how (primarily) experts perform a cognitive task, CTA is similar to, and draws heavily from, knowledge engineering methods that originated in artificial intelligence. As with those methods, performing CTA is challenging because experts have large bodies of knowledge they have accumulated through experience. Capturing an expert's body of knowledge is difficult, first because it is large and complex, but also because it involves perceptual/cognitive skills that are frequently hard for the expert to verbalize, especially without actually performing the task in a realistic environment (Gordon, 1992; Hammond, 1987). For this reason, we will see that many if not most methods rely on some type of analysis in situated, task-specific conditions with experts actually solving problems or making decisions.

Generic Methods

CTA methodologies differ along several dimensions. These include the type of knowledge *representation* or formalism used to encode the information obtained from the analysis, the *methods* used for eliciting or identifying expert (or novice) knowledge, and the *type of tasks and materials* used in the analysis. Typical representation formats include lists and outlines, matrices and cross-tabulation tables, networks, flowcharts, and problem spaces with alternative correct solution paths. Some specific representational formats, such as COGNET, include combinations of these.

Elicitation methods vary widely, but the most commonly used ones include unstructured or structured interviews, group interviews, having experts draw maps or pictures of some sort, observation of task performance (with videotaping or some other type of data collection methodology), concurrent verbal think aloud protocols (e.g., an expert performs the task and "thinks out loud" during the process), retrospective verbal protocols (an expert reviews previous task performance with the analyst and provides comments or answers questions), and analysis of previous "critical incidents." Many analysts suggest the use of more than one because they get at different types of knowledge, or are biased in different ways (Gordon, 1992; Hoffman, Shadbolt, Burton, & Klein, in press).

Finally, the tasks and task materials can vary in a number of different ways. First, they may be simulated at levels of fidelity, or may be performed in the actual working environment. They may be generated by the analysts, or generated by one expert to be solved by another expert. Tasks may be chosen so as to be familiar, atypical or tough, contrived, or constrained (Hoffman et al., in press). Contrived tasks are deliberately changed in some way by the analyst, usually so as to present difficulties for the expert. Constrained tasks are ones where normal, familiar tasks are constrained somehow, such as time or information constraints. Several sources, including Hoffman et al., provide a review of knowledge elicitation methods.

Examples of Specific Methods

There is a great diversity in the methodologies used for CTA. This is undoubtedly perplexing to analysts outside of the field. If we are all trying to identify the same thing, why would we have such widely varying techniques or methods? To answer this question, one need only inspect publications reviewing methods of *behavioral* task analysis, a process much more straightforward because behavior is directly observable. Kirwan and Ainsworth (1992) recently published a review of task analysis techniques. There were no fewer than 41 methods discussed in that volume. It is certainly not surprising that a variety of methods for CTA are emerging, and we should expect the numbers to increase. The following sections provide a very brief overview of several methods that have been used to study complex decision-making tasks.

Concept Mapping and Expert Design Storyboarding.   Researchers at Armstrong Laboratory are using two methods where experts are directly involved in development of the CTA knowledge base (McNeese, Zaff, Citera, Brown, & Whitaker, in press; Zaff et al., 1993). The first, concept mapping, is performed by an analyst working individually with an expert. Working together in unstructured interviews, they draw concept maps of the knowledge required for task performance. The concept maps, which consist of unstructured graphs containing concept nodes interrelated by labeled links, are drawn on white boards. The maps provide a communication mechanism and memory aid to support the elicitation of knowledge from the expert. The graphs are eventually entered onto a computer and reviewed by both analyst and expert. Ultimately, graphs from multiple experts can be combined to identify invariance or commonalities.

Once the concept maps have been finalized, experts are asked to draw sequences of sketches showing appropriate interfaces for the task under analysis. That is, experts are asked to storyboard an interface design. The designs generated by experts provide a medium for expressing perceptual or other nonverbal types of knowledge.

COGNET. COGNET (COGnitive NETworks of tasks) is a set of design tools that allow analysts to model human–computer interaction in complex, real-world, decision-making tasks where people must share attention among multiple tasks (Ryder & Redding, 1993; Seamster, Redding, Cannon, Ryder, & Purcell, 1993; Zachary, Ryder, Ross, & Weiland, 1992). The COGNET model is a specific analysis methodology based on a very formal and precise representational formalism. Essentially, it models behavior as a set of tasks that people do, that compete for behavior. To accomplish this, the knowledge representation has several components: a mental model or problem representation shared by multiple competing tasks in the form of a generalized multipanel blackboard structure; a hierarchical notational system for task procedural knowledge; and a rule-based system for dealing with perceptual information and its introduction into the problem representation (blackboard contents). The COGNET model is computationally runable, and allows for consideration of situation effects, experience and knowledge of the operator, and attention switching among tasks.

In developing COGNET models of an actual domain, Zachary and colleagues have predominantly relied on knowledge elicitation methods such as "detailed timelines of human operator–computer interactions for a number of complete problems" (Zachary et al., 1992, p. 387). These analyses, based on many experts, are augmented with verbal protocol data generated by experts reviewing problem replays. Other applications using COGNET have used a combination of methods including structured interviews, critical-incident interviews, paired paper problem solving, simulated performance modeling, and structured problem solving (Seamster et al., 1993).

Conceptual Graph Analysis. Conceptual graph analysis embodies both a representational formalism and a specific set of knowledge acquisition methods (Gordon & Gill, 1992; Gordon, Schmierer, & Gill, 1993; Moore & Gordon, 1988). The representational formalism is termed *conceptual graph structures* (Graesser & Clark, 1985), which are semantic networks with a very specific syntax (unlike concept maps, which do not have any graph syntax). In conceptual graph structures, nodes may be simple concepts, but can also be more complex events, states, goals, or actions. They are linked by labeled arcs, where the labels must be one of a "legal" set. Conceptual graphs can be used to represent definitional knowledge, causal or system knowledge, goal/action structures with their associated triggering conditions, and spatial representations.

The knowledge-acquisition methods are all performed to support development of conceptual graph structures representing the domain knowledge (much like analysis for COGNET). The methods, in the order usually performed, include: (a) document analysis, unstructured interviews to obtain information to initiate a graph, or both; (b) structured interviews using probe questions such as "Why do you x," "When do you x," "What happens as a consequence of doing x," etc.; and (c) recording multiple experts performing a wide variety of task scenarios (both familiar and "tough") in real or simulated environments, with a postperformance review by analyst and expert working together. Note that, unlike concept mapping, the experts are not asked to draw the graphs, but speak freely about the tasks; this information is translated into graphical form by the analyst.

PARI.   This method has been used to perform cognitive task analysis for tasks that are complex and ill-structured, in particular, diagnosis of hydraulic, electronic, or mechanical failures (Gott, 1989; Gott, Porkorny, Kane, Alley, & Dibble, in press). The analysis focuses on having experts first generate problems with which they are familiar. Each of these experts is then paired with another expert who then proceeds to solve the diagnostic problem. This process is repeated with multiple problems and a relatively large number of experts. As the experts work, an analyst records the information. This analyst also asks probe questions at appropriate points such as "What would be the first thing you do?" or "Why are you going to do that?" Experts are also asked to draw pictures or diagrams that provide insight into the problem mental model. Finally, experts and analysts spend time looking over the analyses to identify aspects such as precursors or triggering conditions, actions, results, interpretations, and so forth.

Several types of knowledge are identified in this type of analysis. One is knowledge of the system, or declarative knowledge. A second consists of the various procedures required to accomplish goals (in this case, usually actions required to further the diagnostic process). Finally, the analysis uncovers strategic knowledge, which is abstract top-level knowledge about how to decide when to perform various procedures.

Critical Incidents, Tough Cases, and Constrained/Contrived Cases. Several analysts use methods where the focus is on special kinds of problem solving or decision making cases (Hoffman, 1987; Hoffman et al., in press; Kaempf, Wolf, Thordsen, & Klein, 1992; Klein, 1987; Klein et al., 1989). The rationale is that having an expert perform normal, familiar tasks will not be particularly insightful. To truly capture the person's expertise, one must ask about the near misses and difficult, tough, or unusual cases. Klein and colleagues have captured expertise by having experts recall critical incidents from their past (e.g., Klein et al., 1989). Using preplanned probe questions, the analyst works with the expert to identify various types of information such as decision points, as well as triggering conditions that led to certain decisions. This information can be represented in flow diagrams, lists, or decision tables.

Comparative Aspects

We can see that the methods for performing CTA described earlier differ along a number of dimensions. Some of these include:

1. Although some methods use generic discussion of the task through interviews or concept mapping, most use verbalization or performance in a very realistic, task-specific, and context-specific environment (either actual or simulated). In different words, one must ask the expert to do the job (or perhaps at least remember instances of doing the job).

2. Methods vary on how, and by whom, the scenarios or test cases are generated. Some researchers have experts recall past experiences, some have experts generate scenarios, and some rely on the analyst for generating scenarios with the desired characteristics.

3. All methods involve at least one analyst and one expert, who work closely together in a sort of "apprenticeship." PARI uses pairs of experts working with an analyst.

4. The methods vary in terms of the type of cases analyzed. Some, such as conceptual graph analysis, specify a very wide variety of scenarios, both familiar and "tough," whereas others specify only that some should be "tough" or "critical incidences."

5. The methods vary in the type of representational formalisms used to embody the expert knowledge. COGNET represents one end of the continuum, with a representation that is so precise that it has built-in computational capabilities. Having a formal representational syntax has the advantages of being easier to develop into final material, and being more reliable across experts and analysts. It is also easier to combine information from multiple experts, and it guides what types of information to look for.

However, less precise representational formats, such as concept maps or decision tables, have the advantages that they are easier for the analyst to learn initially, and during interviews, the expert can be directly involved in building the knowledge base.

The previous contrasts focus on some differences among the methods; there are also commonalities important to successful CTA. These are (a) the use of many experts; (b) the use of many tasks, many different scenarios revolving around the same task, or both; (c) performance of actual problem-solving or decision-making tasks in a realistic environment; and (d) the advantages, and perhaps the need, for the analyst to assume an apprenticeship role in the task analysis.

ISSUES FOR THE FUTURE

We have seen that CTA methods differ widely. Are the differences important? If one were to use one method as opposed to another, would it impact the final system design and its success? Cognitive task analysts are sometimes asked to defend their methods and the invariance of the results that would come from different methods. We find this somewhat odd, because in several decades of traditional behavioral task analysis, no one has asked analysts to verify that different methods result in the same data. In fact, behavioral task analysis methods explicitly end up with different types of data (Kirwan & Ainsworth, 1992). A better focus is to identify what *types* of information are likely to be yielded by the different methodologies, both qualitatively and quantitatively. A system designer might then evaluate the *type* of information required for a particular project, and the *type* of information that predominates in the task, and then choose the CTA methods most appropriate.

There are other important questions concerning the validity and reliability of the various CTA methods. How do we know that we are getting "good data?" There are many versions of this question; for example, how do we know if the method is yielding unbiased information or complete information? Different analysts might apply the same method and obtain different results. We do not know if these differences are actually critical for system design. On a related note, if knowledge engineering means building models, should the analyst be a person who is good at building and analyzing models? This is probably a very different set of skills than that possessed by the typical training analyst or interface designer.

On the question of validity (i.e., have we adequately captured domain expertise?), there are certain ways that we can know if we have built a good model of expert performance on a particular task. The major one is the same method we use to validate theories in science. One first builds the model and then derives predictions for future problem solving or decision making (i.e., on new scenarios or test cases). Presumably, if the method is valid, the model is valid, and will be able to adequately predict expert performance (this cannot be performed if the knowledge representation has no predictive power). A second way that we know the method has been successful is to look at the success of the artifact it has supported. Unfortunately, the success of artifacts is dependent on the quality and completeness of the cognitive task analysis and on the design solution itself.

One of the difficulties facing cognitive task analysts is the lack of publications, seminars, or courses outlining *how* to perform the procedures. Many of us have experienced the frustration of telling people it is a skill that simply takes time and essentially an apprenticeship to learn. As CTA methodologies mature, it would be desirable to identify some relatively simple methods that can be taught to training and system analysts.

Finally, there is a general need for guidance on how to translate the results of CTA into a final system, whether it is a training program, human–computer interface, decision support system, and so on. The process of going from the knowledge base to a specific design solution is probably more responsible for differences in the resultant artifacts than the cognitive task analysis per se. Questions such as this relating to the application of the obtained knowledge base will become increasingly important.

REFERENCES

DeVries, M. J., & Gordon, S. E. (1994a). COG–C: A tool for estimating cognitive complexity and the need for cognitive task analysis. In *Proceedings of the Human Factors and Ergonomics Society 38th annual meeting* (pp. 943–944). Santa Monica, CA: Human Factors Society.

DeVries, M. J., & Gordon, S. E. (1994b). Estimating cognitive complexity and the need for cognitive task analysis. In *Proceedings of the Human Factors and Ergonomics Society 38th annual meeting* (pp. 1023–1027). Santa Monica, CA: Human Factors Society.

Ford, K. M., & Bradshaw, J. M. (Eds.). (1993). *Knowledge acquisition as modeling.* New York: Wiley.

Gordon, S. E. (1992). Implications of cognitive theory for knowledge acquisition. In R. Hoffman (Ed.), *The psychology of expertise: Cognitive research and empirical AI* (pp. 99–120). New York: Springer-Verlag.

Gordon, S. E. (1994). *Systematic training program design: Maximizing effectiveness and minimizing liability.* Englewood Cliffs, NJ: Prentice-Hall.

Gordon, S. E., & Gill, R. T. (1992). Knowledge acquisition with question probes and conceptual graph structures. In T. Lauer, E. Peacock, & A. Graesser (Eds.), *Questions and information systems* (pp. 29–46). Hillsdale, NJ: Lawrence Erlbaum Associates.

Gordon, S. E., Schmierer, K. A., & Gill, R. T. (1993). Conceptual graph analysis: Knowledge acquisition for instructional system design. *Human Factors, 35,* (3) 459–481.

Gott, S. P. (1989). Apprenticeship instruction for real-world tasks: The coordination of procedures, mental models, and strategies. In E. Z. Rothkopf (Ed.), *Review of research in education.* Washington, DC: American Educational Research Association.

Gott, S. P., Pokorny, R., Kane, R., Alley, W. E., & Dibble, E. (in press). *Development and evaluation of Sherlock II: An avionics intelligent tutoring system.* Technical Report. Brooks Air Force Base, AZ: Armstrong Laboratory.

Graesser, A. C., & Clark, L. F. (1985). *Structures and procedures of implicit knowledge.* Norwood, NJ: Ablex.

Hammond, K. R. (1987). *Reducing conflict among experts*. (Tech. Rep. No. AAMRL–TR–87–015). Wright Patterson AFB, OH: Armstrong Aerospace Medical Research Laboratory.

Hoffman, R. R. (1987). The problem of extracting the knowledge of experts from the perspective of experimental psychology. *The AI Magazine, 8*, 53–66.

Hoffman, R. R., Shadbolt, N. R., Burton, A. M., & Klein, G. A. (in press). Eliciting knowledge from experts: A methodological analysis. *Organizational Behavior and Human Decision Processes.*

Kaempf, G. L., Wolf, S., Thordsen, M. L., & Klein, G. (1992). *Decision making in the AEGIS combat information center* (Tech. Rep. under contract N66001–90–C–6023). San Diego, CA: Naval Command, Control and Ocean Surveillance Center.

Kirwan, B., & Ainsworth, L. K. (Eds.). (1992). *A guide to task analysis*. London: Taylor & Francis.

Klein, G. A. (1987). Applications of analogical reasoning. *Metaphor and Symbolic Activity, 2*, 201–218.

Klein, G. A., Calderwood, R., & MacGregor, D. (1989). Critical decision method for eliciting knowledge. *IEEE Transactions on Systems, Man, and Cybernetics, 19*, 462–472.

McNeese, M. D., Zaff, B. S., Citera, M., Brown, C. E., & Whitaker, R. (in press). AKADAM: Eliciting user knowledge to support participatory ergonomics. *International Journal of Industrial Ergonomics.*

Means, B., & Gott, S. P. (1988). Cognitive task analysis as a basis for tutor development: Articulating abstract knowledge representations. In J. Psotka, L. D. Massey, & S. A. Mutter (Eds.), *Intelligent tutoring systems: Lessons learned* (pp. 35–57). Hillsdale, NJ: Lawrence Erlbaum Associates.

Moore, J., & Gordon, S. E. (1988). Conceptual graphs as instructional tools. In *Proceedings of the Human Factors Society 32nd annual meeting* (pp. 1289–1293). Santa Monica, CA: Human Factors Society.

Rasmussen, J. (1986). *Information processing and human–machine interaction: An approach to cognitive engineering*. New York: North-Holland.

Redding, R. E. (1990). Taking cognitive task analysis into the field: Bridging the gap from research to application. In *Proceedings of the Human Factors Society 34th annual meeting* (pp. 1304–1308). Santa Monica, CA: Human Factors Society.

Rouse, W. (1984). Models of natural intelligence for fault diagnosis tasks: Implications for training and aiding of maintenance personnel. In *Artificial intelligence in maintenance: Proceedings of the Joint Services Workshop* (Rpt. No. AFHRL–TR–84–25; pp. 193–212). Lowry AFB, CO: Air Force Human Resources Laboratory.

Ryder, J. M., & Redding, R. E. (1993). Integrating cognitive task analysis into instructional systems development. *Educational Technology Research & Development, 41*(2) 75–96.

Ryder, J. M., Redding, R. E., & Beckshi, P. F. (1987). Training development for complex cognitive tasks. In *Proceedings of the Human Factors Society 31st annual meeting* (pp. 1261–1265). Santa Monica, CA: Human Factors Society.

Seamster, T. L., Redding, R. E., Cannon, J. R., Ryder, J. M., & Purcell, J. A. (1993). Cognitive task analysis of expertise in air traffic control. *International Journal of Aviation Psychology, 3*(4), 257–283.

Zachary, W., Ryder, J., Ross, L., & Weiland, M. Z. (1992). Intelligent computer–human interaction in real-time, multi-tasking process control and monitoring systems. In M. Helander & M. Nagamachi (Eds.), *Human factors in design for manufacturability* (pp. 377–401). New York: Taylor & Francis.

Zaff, B. S., McNeese, M. D., & Snyder, D. E. (1993). Capturing multiple perspectives: A user-centered approach to knowledge and design acquisition. *Knowledge Acquisition, 5*, 79–116.

Chapter 14

Key Issues for Naturalistic Decision Making Researchers in System Design

◆

Thomas E. Miller
Klein Associates Inc.

David D. Woods
The Ohio State University

System design is an area where Naturalistic Decision Making (NDM) research is beginning to have an impact. In order to design good systems, NDM research must address all the steps involved in transitioning the cognitive requirements of a decision maker into a support system that aids the decision making. After the system is built, the final step is to test the effectiveness of the design, which is commonly conducted in a usability laboratory. This chapter summarizes the design and usability testing issues raised during a panel discussion at the 1994 NDM conference. The participants on the panel were all experienced system designers who use an NDM perspective in designing and evaluating systems. The panel members were: Joseph Dumas (American Institutes for Research), Thomas Miller (Klein Associates Inc.), Gunilla Sundstrom (GTE Laboratory), Kim Vicente (University of Toronto) and David Woods (The Ohio State University).

The first part of this chapter raises issues associated with using NDM research in system design and with taking a cognitive systems perspective. The second part deals with difficulties associated with usability testing and the evaluation of new products.

DEMANDS, ARTIFACTS AND PRACTIONERS: INVESTIGATING NDM

When approaching system design from an NDM perspective, there are interrelated influences or components that need to be acknowledged:

- In complex settings, cognitive systems are frequently distributed over multiple agents, some of which are machines. This complex system of cognitive agents is interrelated to the extent that changing one aspect of the system will influence other aspects of the system. For instance, adding new automation changes the human's role and creates new tasks such as supervisory control, monitoring, deciding when to intervene, and the need to coordinate activity (Billings, in press).
- The cognitive tools available for performing a task influence the kind of cognitive activities, strategies, and tools that operators employ. For example, cumbersome online help systems that are difficult to use encourage users to adopt other strategies for solving problems, such as using trial and error or asking other users.
- Operators shape the available tools based on their interests, constraints and task demands. For example, with computer software that is very complex to use, practioners develop and stick with stereotypical routes or methods to avoid getting lost in large networks of displays, complex menu structures, or complex sets of alternative methods.

This mutual shaping means that anything a designer does to improve a process changes that process, sometimes in unexpected ways.

If we accept this cognitive systems perspective, we cannot just study cognitive strategies (e.g., mental simulation) in isolation. We must recognize that they are influenced by other properties of the cognitive system, including the artifacts that operators use (Bennett, Toms, & Woods, 1993; Vicente, 1995a; Woods, 1991; Woods, 1994). The tools we introduce in a setting (e.g., automation, displays, organizational changes) transform the nature of the task. This has been described as "writing on water"; just as you cannot write on water without influencing the water surface, the act of performing a task in an environment influences the nature of the environment (Schmidt, 1990). The issue that this raises for designing complex systems is that we are put in the situation of studying an existing system in order to design artifacts that will change the system we just studied.

How are studies of a current system relevant? The artifact will transform the setting, causing different kinds of cognitive activities to be engaged, making different kinds of strategies viable, and making other kinds of strategies error prone or difficult to carry out. There is potentially no end to this task–artifact cycle (Carroll, Kellogg, & Rosson, 1991). As soon as an artifact is changed, operators will adapt to it, changing the nature of the task. New systems, or redesigns of old systems, create new possibilities and inevitably change operator behavior, making it difficult to know when to stop redesigning.

There are implications of this task–artifact cycle that NDM researchers should keep in mind. What we observe in naturalistic settings is necessarily constrained by the current set of possibilities in the setting:

- Some of the constraints are related to the nature of the domain and the tasks that have to be done.
- Some of the constraints are due to the design of the current artifacts.
- Some of the constraints are due to the individuals in the setting.

Discovering and understanding these constraints and their relationships is a challenge to NDM researchers. Understanding these relations allows the investigator to model error and expertise in the domain in question. Modeling the sources of error and expertise provides requirements for the design of new artifacts and informs the search for concepts about how to aid practitioner cognition and performance.

LINKING NDM STUDIES TO DESIGN

Wiener (1989) provided an example where investigations of the interaction of practitioners and technology in the field have led to new design directions. Wiener uncovered a kind of breakdown on the flightdecks of highly automated commercial transport aircraft—automation surprises. In these cases, the flightcrew instructs and manages a set of automated systems that fly the aircraft. The crew expected the automation and the aircraft to behave in one way but were surprised when they suddenly realized that the aircraft was behaving differently. For example, they might have expected the automation to begin the descent portion of the flight but were surprised when they realized that the aircraft had continued to fly at the cruise altitude well beyond the planned top of descent point. In general, situations arose where crews were asking themselves questions such as, "What is it [an automated system] doing?" and "What will it do next?" In some cases the abnormal behavior was not detected until the situation had become quite dangerous, even contributing to incidents or accidents.

Sarter and Woods conducted a series of NDM studies (e.g., Sarter & Woods, 1995) that provided converging evidence on the factors contributing to these automation surprises. They showed that the surprises were based on a kind of breakdown in the coordination between crew and automated systems—mode awareness. The complexity of, and interactions across, automated systems led to situations where the crew's perception of what the automation would do and what it was actually doing were different. Two factors led to crews being surprised by actions taken (or not taken) by the flight computers: gaps in flightcrews' understanding of how the automated systems worked and low observability of automated system behavior. For example, in various studies crews generally did not detect the miscommunication with the automation from displays about the automation's state. Instead, they picked up that a miscommunication had occurred only when the aircraft's behavior became sufficiently abnormal.

The studies of mode awareness also uncovered factors that influenced when crews are most vulnerable to this new form of mode error. As the automation increased in its capabilities, situations arose where an automated system could change modes of control of the aircraft without explicit instruction and immediately preceding instructions from the crew. These indirect mode changes challenge the pilot's ability to track mode transitions. In the words of one pilot, "Changes can always sneak in unless you stare at it."

The result of this and other related research is a set of functional requirements for new directions in the design of flightdeck displays in order to address the problems associated with automation surprises and mode awareness. One such requirement is the need for displays that provide improved feedback about the new approaches to inform crews about automation activities. These displays can include:

- Transition-oriented—provide better feedback about events, targets and transitions.
- Future-oriented—the current displays generally capture only current configuration; the goal is to highlight operationally significant sequences and reveal what should happen next.
- Pattern-based—pilots should be able to scan at a glance and pick up possible unexpected or abnormal conditions about automation activities rather than having to read and integrate each individual piece of data to make an overall assessment.

The result of these studies was not a specific design change or a specification for system developers to follow. Rather, the cycle of data collection and conceptual development guided further investigation to set forth a direction for innovation and provided a set of goals to assist designers.

Moving from studies of practitioners interacting with technology in a field of practice to the design of new artifacts involved more than simply characterizing the problem of mode awareness and indirect mode transitions. The requirements for new aiding concepts is also driven by our accumulating research base on how artifacts shape cognition: concepts and examples about how human cognition and performance is affected by fundamental properties of cognitive tools (e.g., Vicente, 1995b; Woods, 1995). As Winograd (1987) has said, there is an increasingly pressing need to develop "a theoretical base for creating meaningful artifacts and for understanding their *use and effects*" (p.10, italics added). In the case of mode awareness, Sarter and Woods (1995) used results derived from other research about what kinds of displays can aid human performance. This research revealed the role of event-oriented, pattern-based displays in aiding human supervisory control and it guided Sarter and Woods in identifying implications of their studies relevant to design.

NDM researchers can look at the interaction of demands, tools and strategies across agents in the current world. The difficulty is that the introduction of new artifacts can change the field of practice in significant ways. Practioner strategies may be tied very closely to the existing collection of artifacts. For example, a common belief in the system design world is that an interface design should be compatible with an expert operator's mental model. A problem with this belief is that if a current artifact makes only some cognitive strategies possible, then even the expert operators in the current setting may not have an accurate or complete mental model of the process. Thus, a new system is based on an incomplete model. Supporting an operator's existing mental model through system design may result in supporting a model that is incomplete or inaccurate. We need design approaches and tools that will help us discover more complete models on which to base system design. We need to recognize the need for other sources, such as engineering and designer models of the systems.

Another aspect about a cognitive system to consider has to do with the strategies that operators choose to use. Artifacts make it easy to adopt certain cognitive strategies in a given setting, and they make it more difficult for operators to adopt other strategies (Miller, Militello, & Heaton, 1996; Sundstrom, 1993; Zhang & Norman, 1994). For example, strategies that are easy to employ are not necessarily the best strategies and strategies that are hard to employ are not necessarily bad strategies. The implication for NDM studies is that it is important to look at what good strategies *can* be used but that might not be used because the current set of possibilities is highly constrained by the current system. A different interface may help operators perform a task in a different way that would lead to more productivity, but finding these solutions is challenging.

INTEGRATING INFORMATION GENERATED
BY NDM STUDIES

Developers of systems work under resource and time pressure and this commonly results in a short-term, narrow focus. Often their goal is to refine a prototype toward a final product. Studies of the current world seem irrelevant or a luxury they cannot afford. For example, a product developer may ask for assistance with a product. A prototype has been developed, but the designers are worried about it; they are not sure it will do what they want it to do and there is only a month left before it is too late to do anything. What can we do in a month?

One reality of the design process that we must face is that we are usually dealing with short-term problems—there has never been an unlimited resource problem. Projects vary in how limited the resources are, but it is too common to find seat-of-the-pants designs because "the resources are too limited." If we are ever going to get anywhere, we need to take some of the resources and invest them in the future by conducting basic research in parallel with the applied efforts. By forming a foundation of basic research, the next time we confront the same issue in some other setting we will have learned something from prior work. Not enough of this is happening.

NDM can help because it starts to provide the concept base so that every project can have both immediate and long-term benefits. For example, NDM offers models of situation assessment and decision making as starting points. The payoff is local because there is some immediate payoff for the project at hand. However, this work also can add to the theoretical base by documenting new strategies. The NDM approach offers one way to talk about what that long-term concept base is so that the next time we get a 1-month deadline, we have more knowledge and experience to do a better job, rather than generating another short-term solution. That is, the larger view provided by NDM studies is not necessarily incompatible with the short-term focus of developers.

One place to start is in developing tools for people outside the NDM community to help them transition user requirements into system design. We use concepts to analyze user informational requirements and if we can build tools that translate those requirements into design requirements, this would be a bridge to the engineering side of design work. Commercially available tools are not often designed to address the analysis phase of the software engineer's task. For example, in an object-oriented approach, where do the objects come from? There are currently no tools for software engineers to determine what an appropriate set of objects would be for a particular problem. NDM can offer Cognitive Task Analysis (CTA) tools. There is currently work being done to develop a set of simplified Applied CTA (ACTA) tools for non-NDM and noncognitive systems practitioners (Crandall, Klein, Militello, & Wolf, 1994).

A final issue is about who has power over the system design. If a system is designed from the "outside in rather than from the inside out," then the holder of the user interface has the power over the system. This is a problem for practitioners, especially if we do not use common software terminology. There is often a power struggle between designers of the interface and software engineers over who owns the system design. However, as companies start to use terms like "ergonomic" and "user friendly" in their marketing approaches to convince consumers that their products are better, designers of the interface will have more of a say over the design of the systems.

USING NDM RESEARCH IN USABILITY TESTING AND EVALUATION

Usability testing to identify design problems and hidden functionality becomes critical in the development process of products. A trend in the development of complex products over the past few years is that the functionality of products is becoming more and more hidden in software. Before, the functionality of a product was on the surface. Functionality used to be controlled by switches or dials, but is now buried in the product's software and can only be reached by knowing the right pathway through the menus and prompts (Dumas & Redish, 1993). Some of the issues facing usability testing and how NDM research can be used to address these issues are discussed in the following.

There can be wide variations in how usability testing is conducted, but every usability test shares five characteristics (Dumas & Redish, 1993):

- The goal is to improve the usability of a product, such as the ease of navigation through the menus or the acceptance of the interface by both novice and experienced users.
- The participants represent real users. This group must have the same range of experience and abilities as the target population.
- The participants do real tasks that they will do with the product on their jobs, which means that the users' jobs and tasks must be understood.
- The tests require observations and records of the participants and the comments they make.
- Steps following observation are to analyze the data, diagnose the problems, and recommend changes to fix those problems.

Usability testing provides an intermediate area between studying people in naturalistic environments and studying them in highly controlled experiments. Testing in a usability laboratory brings new products (or prototypes) together with representatives of the user population to perform realistic tasks with a product. The goal of this testing is to help move a product further along in the development phase by uncovering problems. Although usability testing may not uncover all of the problems in a software's design, it allows for sampling and trying out parts of a product.

Having the product designers present while users interact with the product is an important aspect of usability testing. Getting designers to realize the complexity of performing tasks from a user perspective can be a major challenge for people working in the usability field. From the designer's perspective, they are under pressure and competition to get products out and to develop products faster. Under these conditions, it can be easy to overlook the complexities that are being introduced by technology, or to overlook the complexities of the tasks being performed by the operators. Witnessing a user struggle with a product gets the attention of designers and has the potential to help change attitudes and ultimately change the design process.

There are two kinds of problems typically found in software designs: local problems and global problems. Local problems are found on a single screen or a page. Global problems are found throughout the product and show up repeatedly, with participant after participant and task after task. Because usability laboratories provide feedback on many participants, most of the global problems can be revealed by watching 5 to 10 people doing a set of tasks in a usability laboratory (Dumas & Redish, 1993).

A potential problem in a usability laboratory is that a laboratory cannot recreate all the complexity that is in the real environment. When a simulated setting is used, such as a usability laboratory, we must determine what aspects of the real environment are the most important to capture in the simulation. What is important about the nature of the task and how the product will be used to perform the task? When doing usability testing, it is necessary to first go into the environment where the devices are being used (or will be used) and to become familiar with the task and domain, thus accomplishing the front-end work that the NDM approach advocates prior to designing a product.

NDM researchers believe that usability laboratories should have a broad perspective that is part of the ongoing design process. However, usability laboratories are commonly asked to take a narrow focus that tests products feature by feature to identify obvious glitches before the consumer finds them. User involvement typically ends up having too narrow a focus and occurs too late in the process to result in significant changes or input (Dumas & Redish, 1993). NDM research needs to be involved from the beginning of the development process.

Another barrier that NDM researchers face is that they are often unable to be involved in the initial product development phase because most projects revolve around updating a product that already exists. Additionally, there is a fundamental problem of designers and users not talking. The design process of many companies isolates designers/developers in the back room and gives control of user input to the marketing staff.

How do we get involved in the front end of technology development and start to define what would be useful in the actual field of practice? The NDM panel agreed that this is an area where NDM research can make a major contribution. Although the NDM approach can be used to build better simulations for usability testing, its real strength is in the larger view. This larger view allows us to study domains and to model what is going on in a way that allows us to generate innovative ways to use technology in complex settings that often involve multiple cognitive agents. The NDM approach can help generate new ways of using technology, rather than being at the mercy of what has already been designed.

REFERENCES

Bennett, K. B., Toms, M. L., & Woods, D. D. (1993). Emergent features and graphical elements: Designing more effective configural displays. *Human Factors, 35,* 71–97.

Billings, C. E. (in press). *Aviation automation: The search for a human-centered approach.* Mahwah, NJ: Lawrence Erlbaum Associates.

Carroll, J. M., Kellogg, W. A., & Rosson, M. B. (1991). The task–artifact cycle. In J. M. Carroll (Ed.), *Designing interaction: Psychology at the human–computer interface* (pp. 74–102). New York: Cambridge University Press.

Crandall, B., Klein, G., Militello, L. G., & Wolf, S. P. (1994). Tools for applied cognitive task analysis (Contract Summary Rep. N66001–94–C–7008). Fairborn, OH: Klein Associates Inc.

Dumas, J. S., & Redish, J. C. (1993). *A practical guide to usability testing.* Norwood, NJ: Ablex.

Miller, T. E., Militello, L. G., & Heaton, J. K. (1996). Evaluating air campaign plan quality in operational settings. In A. Tate (Ed.), *Advanced planning technology* (pp. 195–200) Menlo Park, CA: The AAAI Press.

Sarter, N. B., & Woods, D. D. (1995). How in the world did we ever get into that mode? Mode error and awareness in supervisory control. *Human Factors, 37*(1), 5–19.

Schmidt, K. (1990). *Analysis of cooperative work: A conceptual framework* (Risoe–M–2890). Roskilde, Denmark: Risoe National Laboratory, Cognitive Systems Group.

Sundstrom, G. A. (1993). User modelling for graphical design in complex dynamic environments: Concepts and prototype implementation. *International Journal of Man–Machine Studies, 38,* 567–586.

Vicente, K. J. (1995a). A few implications of an ecological approach to human factors. In J. Flach, P. Hancock, J. Caird, & K. Vicente (Eds.), *Global perspectives on the ecology of human–machine systems* (pp. 54–67). Hillsdale, NJ: Lawrence Erlbaum Associates.

Vicente, K. J. (1995b). Supporting operator problem solving through ecological interface design. *IEEE Transactions on Systems, Man and Cybernetics,* SMC–25, 529–545.

Wiener, E. L. (1989). *Human factors of advanced technology ("glass cockpit") transport aircraft* (Tech. Rep. No. 177528). Moffett Field, CA: NASA Ames Research Center.

Winograd, T. (1987). *Three responses to situation theory* (Tech. Rep. No. CSLI–87–106). Stanford, CA: Center for the Study of Language and Information, Stanford University.

Woods, D. D. (1991, November). *Representation aiding: A ten-year retrospective.* Paper presented at the IEEE International Conference of Systems, Man, and Cybernetics, Charlottesville, VA.

Woods, D. D. (1994, August). *Observations from studying cognitive systems in context.* Paper presented at the sixteenth annual conference of the Cognitive Science Society, Atlanta, GA.

Woods, D. D. (1995). Towards a theoretical base for representation design in the computer medium: Ecological perception and aiding human cognition. In J. Flach, P. Hancock, J. Caird, & K. Vicente (Eds.), *Global perspectives on the ecology of human–machine systems* (pp. 157–188). Hillsdale, NJ: Lawrence Erlbaum Associates.

Zhang, J., & Norman, D. A. (1994). Representations in distributed cognitive tasks. *Cognitive Science, 18,* 87–122.

Chapter 15

Naturalistic Decision Making Perspectives on Decision Errors

◆

Raanan Lipshitz
University of Haifa

In his critique of Naturalistic Decision Making (NDM), Doherty (1993) asserted that "naturalistic decision making is simply silent on what constitutes an error"(p. 380). Later Doherty (personal communication, May 1995) clarified his assertion by raising three questions: What constitutes an error according to the NDM paradigm? Can decision errors conceived this way be detected online without the benefit of hindsight? What is the positive contribution of NDM besides criticism of the treatment of error within the "heuristics and biases" tradition? Although I disagreed with the blanket assertion that NDM is "silent on error" (e.g., Reason, 1990; Woods, Johannessen, Cook, & Sarter, 1994), I thought the questions that Doherty raised were worthy of serious attention. I therefore presented them to a panel of six NDM researchers: Marvin Cohen, Richard Cook, Randall Mumaw, James Shanteau, Alan Stokes, and myself. Based on the panel's discussion and my own reading of the literature, I will try to answer each of Doherty's three questions.

WHAT IS A DECISION ERROR?

Between 70% and 90% of accidents such as Bhopal, Chernobyl, and less publicized accidents are attributed to "human error" (Cook & Woods, 1994). Because errors of operators in complex work environments such as

operating rooms and nuclear power stations may have disastrous consequences, much work has been done to make such systems less prone to
"human error." Most of the work on decision errors in the NDM paradigm
is a by-product of this effort. A working definition consistent with much of
this work is that decision errors are deviations form some standard decision
process that increases the likelihood of bad outcomes. This definition helps
to draw the similarities and differences between the treatment of error in
the rather loosely related collection of models labeled NDM (Klein,
Orasanu, Calderwood, & Zsambok, 1993) and the equally loosely related
set of models collectively labeled Behavioral Decision Theory (BDT;
Hogarth, 1987). Similar to BDT, some NDM work on decision error (to
which I refer as the cognitive perspective) suggests a probabilistic relationship between decision errors and bad outcomes: Decision errors are likely
to produce bad outcomes, but some bad outcomes are produced by perfectly
sound decisions. As I shortly point out, the cognitive perspective and BDT
differ in the cognitive processes they use as standards for identifying decision
errors. Two additional NDM perspectives are more radically different from
BDT's treatment of decision errors. The interpretive perspective suggests
that decision errors are "produced," in the final analysis, by analysts and
researchers. Going a step further, the adaptive/systemic perspective suggests
that tracing bad outcomes to decision errors of individual decision makers
produces erroneous conclusions. I now discuss these three NDM perspectives in some detail.

The Cognitive Perspective

According to this perspective bad outcomes can be traced to faulty cognitive
processes in complex causal chains that consist of (a) a bad outcome; (b)
an inappropriate action, or substandard performance of an appropriate
action; (c) a fault in one of the elements of the decision-making process
(situation analysis, action selection, action planning, and implementation);
(d) a breakdown in the cognitive mechanisms that control action; and (e)
situational factors such as time stress or a task structure that overload or
mislead the cognitive system (Rasmussen, 1993).

A central proposition of the cognitive perspective is that decision errors
have different causes at different levels of cognitive action control (Rasmussen, 1993). At the skill-based level, action is controlled automatically
by a mental representation of the task situation. Error at this level is caused
by inadequate precision of sensorimotor control of action or by capture of
control by overlearned inappropriate representation, as, for example, when
a person, who decides to pick up something on the way to work, forgets to
take a turn and mindlessly drives straight to work. At the rule-based level,
action is controlled by stored rules or procedures. Error at this level is caused
by incorrect activation of action owing to underspecified rules, inaccurate

recall of rules and omission of acts that are isolated from well-rehearsed action sequences (e.g., forgetting to switch on a system after performing a test operation), or both. Finally, at the knowledge-based level, action is controlled by explicitly formulated goals and situation analysis. Error mechanisms at this level are not as well specified as in the two lower levels of action control. In general, errors at this level are caused by inappropriate (e.g., truncated) information search, or by inadequate processes of reasoning (e.g., failing to consider potential side effects or nonlinear causal relations). For examples of knowledge-based errors see Dörner (1987) and Zapf, Maier, Rappensperger, and Irmer (1994).

Reason (1990) elaborated the situational/cognitive perspective by adding to Rasmussen's three-way classification of skill-based, rule-based, and knowledge-based action a three-way classification of error: *mistakes* (errors in planning), *slips* (errors in execution), and *lapses* (errors in storage). Skill based slips and lapses, and rule- and knowledge-based mistakes are the flip side of the *attentional* and *schematic* modes of the cognitive control of effective action. The attentional mode is used consciously to set goals and to design, implement, monitor, and, if necessary, modify action. The schematic mode is used out-of-attention and consists of a "vast community" of schemata, highly specialized information processors or knowledge packages. The advantages and disadvantages of the attentional and schematic mode are complementary. The attentional mode is relatively effective in novel situations but requires effort that can be sustained for only brief periods. The schematic mode requires little effort. Its effectiveness, however, is limited to familiar situations.

Rule-based action is mostly controlled by the schematic mode. Rule-based mistakes typically have a *strong-but-wrong* form: the activation of a readily available but irrelevant or inefficient action routine. Most of these mistakes are by-products of two heuristics that are used in equivocal situations. The first heuristic is *similarity matching* (judging the similarity between the triggering conditions and stored attributes of appropriate action). The second is *frequency gambling* (choosing among partially matched options in terms of the frequency of their past use). Similarity matching and frequency gambling are efficient and generally effective heuristics. Unfortunately, they also produce a tendency to rely on familiar cues and well-tried solutions when risk taking is called for.

According to Reason (1990), knowledge-based mistakes have two basic causes: incomplete information and bounded rationality. Rasmussen (1993) and Reason suggested that knowledge-based mistakes do not have a typical form, such as the *strong-but-wrong* form that is typical of skill-based slips and rule-based mistakes. Reason treats knowledge-based mistakes by listing biases and fallacies compiled from the literature on deductive and inductive inference. This treatment is not particular to NDM research and suffers from the basic weakness of all such lists, be they in BDT (e.g., Hogarth, 1987) or in NDM (e.g., Endsley, 1994), which are invariably ad hoc and arbitrary (Wallsten, 1983). An alternative approach to understanding knowledge-based mistakes is through the study of "buggy knowledge," that is, deficiencies in the construction and use of mental models that drive decision making (Cook & Woods, 1994; Lipshitz & Ben Shaul, chapter 28, this volume).

The Interpretive Perspective

Decision errors attract great attention because they are presumed to increase the likelihood of bad outcomes. But how does one establish that a certain bad outcome can be attributed to a certain action which, thus, becomes a decision error? The standard procedure (e.g., in accident analysis) is to construct a causal chain of events from the bad outcome and trace backward until its "root cause" (e.g., a decision error) is encountered somewhere "upstream." Thus, in practice, decision errors are "human acts which are judged by somebody to deviate from some kind of reference act" (Rasmussen, 1993, p.1), or "an attribution that (a) the human performance immediately preceding [a bad outcome] was unambiguously flawed, and, (b) the human performance led directly to the bad outcome" (Cook & Woods, 1994).

Granted that decision errors are deviations from some standard decision process that increase the likelihood of bad outcomes, the interpretive perspective points out that identifying particular decisions or cognitive operations ("the X heuristic") as decision errors is very problematic. One particularly worrisome problem is that there are no objective a priori rules for determining the level of detail at which to conduct an analysis or where to stop the construction of a causal chain that links bad outcomes to bad decisions. A second, equally worrisome problem is the absence of objective rules for determining the appropriate conceptual framework for interpreting a particular causal chain or decision process. Thus, what may seem to be an error using one conceptual framework may make perfect sense using another framework (Lipshitz, 1993). Another problem concerns differences in the amount of information available to decision makers and decision analysts, be they experimenters, observers, or retrospective evaluators. Outcome knowledge—never available to the former and nearly always available to the latter—has particularly pernicious effects on the evaluation of the quality of decision processes and the appropriateness of actions (Lipshitz & Barak, in press). As a consequence of these problems, identification of decision errors is controlled by pragmatic considerations regarding the scope of analysis and the aims, knowledge, and preconceptions of the analyst (Rasmussen, 1993; see Klein, 1989, for a vivid demonstration).

The Adaptive/Systemic Perspective

According to the cognitive perspective, decision errors are amenable to objective observation and analysis. The *what* of a decision error is independent from *how, where, when,* and by *whom* it is observed. According to the interpretive approach, these issues are inextricably intertwined. Not only are decision errors context-dependent subjective entities, but tracing bad outcomes to decision errors invites infinite regress: One can always trace decision errors further back from the sharp to the blunt ends of the decision process, from the decision errors of the machine operator to the decision errors of the maintenance crew, the plant management, the machine's designer, and so forth (Cook & Woods, 1994). The adaptive/systemic approach avoids this regress by reframing the meaning of decision errors. Instead of conceiving bad outcomes as products of decision errors that should be tracked to their source (a decision maker) and avoided, the adaptive/systemic perspective conceives of bad outcomes as products of a state of lack of system–person fit, in which decision errors are inevitable and potentially adaptive.

Successful adaptation involves expending less effort to produce more and better output. In terms of the skill–rule–knowledge framework, this means moving downward in the level of cognitive control. As decision makers become more proficient they shift from knowledge-based action to rule-based action, and then to smooth, effortless, and effective performance at the skill-based level. Unfortunately, retaining effectiveness and efficiency while progressing this way is domain and context dependent. Knowledge, rules, and mental models that guide effective action in one environment may spell disaster in another. Thus, tracing bad outcomes to bad decisions assumes stable conditions. In unstable conditions, "correct" models cannot be specified a priori, so the only way to test one's decisions is to judge the response from the environment. Under these conditions, the correct adaptive response is to deliberately break the standards that define "good decisions" because adherence to them no longer increases the likelihood of good outcomes.

In conclusion, the upshot of the cognitive approach is that causes of decision errors can be traced back to identifiable failures in cognitive mechanisms of action control. The upshot of the interpretive and adaptive/systemic approaches is that what may seem to be an error from one standpoint or context may not be so from different standpoints or contexts. Bad outcomes are thus caused not by the failure of individuals but by the failure of systems, and the results of tracing bad outcomes to their sources are always contingent on the context and method of analysis.

DETECTING DECISION ERRORS ONLINE

Detecting decision errors online (without outcome knowledge) is desirable both to preempt bad outcomes and to avoid outcome bias (Hawkins &

Hastie, 1990). Without disputing this argument, BDT and NDM present radically different approaches to the detection of error. Doherty (personal communication, July 1994) summarized the BDT approach while criticizing that of NDM:

> When an investigator [in BDT] asserts that a subject commits a conjunction fallacy, you and I know exactly what the assertion means: the subject behaved in such a way that does not correspond to the dictates of a formal model. . . . I see no attempt to commit NDM to anything like a behavioral criterion for saying that an error had been made . . . we all agree that an undesirable outcome is not a sufficient criterion.

The NDM approach, as presented by Zapf et al. (1994), confirms Doherty's assertion that BDT and NDM differ in their attitude toward outcome knowledge. The BDT approach is that decision errors should be identified independently of outcome information. The NDM approach is that decision errors cannot be identified without this information: "If we want to know whether or not an error has occurred we have to know the user's goal. . . . As action is feedback driven . . . error detection depends on knowledge about the action outcome and the user's goal" (p. 501).

Two explanations are possible for the difference between BDT's and NDM's attitude towards outcome knowledge. The first explanation concerns the feasibility of identification by process. The second concerns the ambiguity of the distinction between a process and an outcome.

The Feasibility of Identification by Process

In the quote presented earlier, Doherty assumed identifying the nature of deviations between observed and standard decision-making processes is not problematic. This may be the case with laboratory tasks such as those typically used by Judgment researchers. In naturalistic settings, unambiguous detection of "biased heuristics" is difficult, if not impossible, because it is not clear which normative standard applies, and because these standards are inevitably underspecified. The first problem is illustrated by Anderson (1990), who noted that research on probability matching shows that subjects are extremely sensitive to base rates (thus exonerating them from the Base Rate Fallacy); nevertheless, probability matching is considered an error because it is a suboptimal prediction strategy.

The underspecification problem was illustrated by Klein (1989). Responding to the claim that the commander of USS Vincennes exhibited confirmation bias (because he erroneously interpreted available information as indicating an Iranian attack), Klein noted that had the fate of USS Vincennes repeated that of USS Stark, "behavior decision theorists would have claimed that this was an example of a typical bias they have been warning against: the failure to use bases rates" (p. 2). Thus, it is far from certain that identifying specific behaviors as errors without outcome knowledge is feasible. My favorite illustration for this claim is the "simplification bias" (Hogarth, 1987), the underrepresentation of the "true" complexity of a problem. The implicit assumption underlying this bias is that problems are objective entities whose complexity can be measured similar to, say, measuring the height of the Empire State Building. This, of course, is false. Consistent with the interpretive perspective, identifying the simplification bias depends on consensus opinion, or, more convincingly, on outcome knowledge: Simple representations that fail are "simplistic." Those that succeed are "elegant." Similar, albeit less precise arguments can be presented with respect to all the biased heuristics that constitute decision errors in BDT. Consistent with the adaptive/systemic perspective, bad outcomes are produced not by deviating from a generally applicable normative model, but from using decision strategies that are often effective in situations for which they are not designed (Funder, 1987; Lipshitz, in press).

Decision Processes vs. Decision Outcomes

The decision processes typically studied in BDT research consist of a single decision (or judgment) and a single outcome. In contrast, the decision processes typically studied in NDM consist of a series of decisions or a sequence of intermediate outcomes. Under these conditions the distinction between process and outcome is blurred, and the injunction against detecting error on the basis of outcome information applies only to the "final outcome." In fact, once we distinguish between intermediate and final outcomes, the difference between Doherty (1993) and Zapf et al. (1994) becomes more apparent than real. This, however, still leaves open a modified version of Doherty's question: What criteria does NDM offer to identify online errors in the *cognitive* aspects of the decision process?

The online identification of cognitive errors is performed in BDT by comparing observed processes to standards obtained from Utility Theory and Bayesian probability. In NDM these standards are replaced by ones obtained either by cognitive task analysis (identifying the cognitive processes that are required to perform a task), or by developing descriptive models of expert decision making. Both methods have been used to develop training programs in operational decision making (Mumaw, Swatzler, Roth, & Thomas, 1994).

THE POSITIVE CONTRIBUTION OF NDM

The feeling that all NDM research has to offer is criticism of BDT can be attributed to the origins of the new paradigm. Clearly, NDM research was originally motivated by dissatisfaction with BDT and the desire for a clean break with its underlying Rational Choice model (Klein et al., 1993). This motivation, however, was least evident in the research programs that produced the three perspectives on decision errors reviewed here, which were motivated by the need for safe, operator-friendly work environments. Most of this work proceeded independently of BDT and, if anything, was receptive rather than critical of it (e.g., Reason's 1990 treatment of knowledge-based mistakes).

The "added value" of NDM research besides criticism of BDT is a more sophisticated treatment of decision errors. Although it sometimes muddies the water and presents a very complex analysis of decision making, that complexity is real.

Whereas BDT offers a single cognitive perspective grounded in Decision Theory, NDM offers an alternative cognitive perspective and an interpretive and an adaptive/systemic perspective with no parallel in BDT. The added value of these perspectives is both conceptual and pragmatic.

The BDT treatment of decision errors presumes a cognitive process in which alternatives are compared in terms of the attractiveness, and sometimes likelihood, of their outcomes. As this conception is grounded in Decision Theory, it leads to a straightforward treatment of what decision errors are, and how to correct them. Basically, decision errors are deviations from the prescriptions of Decision Theory, and the way to correct them is to train individual decision makers to apply the Theory in everyday situations (informally if necessary), and to sensitize them to deviation producing heuristics (Fischhoff, 1982).

The valuable contribution of BDT, which is not disputed by NDM, is that BDT research has convincingly shown that unaided choice behavior deviates consistently from the Rational Choice model. What is at issue between the two paradigms is whether these deviations qualify as decision errors. The NDM research argument is that Rational Choice models are neither relevant to how decisions are actually made, nor applicable in naturalistic settings. Neither deviations from Rational Choice nor bringing cognitive processes closer to this model matter, therefore, to the likelihood of bad outcomes, except in artificial laboratory settings.

As a consequence of this viewpoint, the cognitive perspective of NDM research differs from that of BDT in two respects. First, its standards for identifying decision errors pertain to situation assessment, mental models, and sequential option generation/evaluation rather than concurrent choice. Second, its corrective training methods are domain specific and embedded in situations of practice (Lipshitz & Ben Shaul, chapter 28, this volume; Mumaw et al., 1994).

The situational aspects of decision errors, which are wholly missing in BDT, are further emphasized in NDM's systemic/adaptive perspective. According to this perspective, bad outcomes are not produced by bad decisions, period. Rather, they are produced by the interaction between decisions and incompatible system states. Conceiving decision errors this way has profound implications. First, under certain conditions it makes no sense to attribute bad outcomes to bad decisions, that is, to trace bad outcomes to faulty cognitive processes. Is it reasonable, for example, to trace outcomes to "bad decisions" when latent error tendencies, which are innocuous individually, appear together with lethal consequences. Similarly, is it reasonable to attribute bad outcomes to inappropriate action that is strongly suggested by cues in the situation ("garden path" errors, Cook & Woods, 1994). The second implication, which I have already discussed, is that when a change in standards is called for, decision errors are adaptive and hence should be made, rather than avoided. The third implication is pragmatic. Whereas the cognitive perspective entails correcting decision errors by training, this perspective points to a variety of design countermeasures such as improving decision makers' ability to detect error and the ability of the system to recover from error (Cook & Woods, 1994; Rasmussen, 1993).

Finally, a contribution of the interpretive perspective is conceptual. Basically, it brings the discussion of decision errors from the sophistication of Newtonian physics, where effects are traced to causes independently of the measuring method, to that of quantum physics and fractal geometry where *what* is observed is inseparable from *how* it is observed. What are the long-term pragmatic implications of this transition? I do not know. Nevertheless, my own gut feeling is that this may be the most significant contribution of NDM to the understanding of decision errors.

REFERENCES

Anderson, J. R. (1990). *The adaptive character of thought*. Hillsdale, NJ: Lawrence Erlbaum Associates.

Cook, R. I., & Woods, D. D. (1994). Operating at the sharp end: The complexity of human error. In M. S. Bogner (Ed.), *Human error in medicine*. Hillsdale, NJ: Lawrence Erlbaum Associates.

Doherty, M. E. (1993). A laboratory scientist's view of naturalistic decision making. In G.A. Klein, J. Orasanu, R. Calderwood, & C. Zsambok (Eds.), *Decision making in action: Models and methods* (pp. 362–388). Norwood, NJ: Ablex.

Dörner, D. (1987). On the difficulties people have in dealing with complexity. In J. Rasmussen, K. Duncan, & J. Leplat (Eds.), *New technology and human error* (pp. 97–109). Chichester, England: Wiley

Endsley, M. R. (1994, March). *A taxonomy of situation awareness errors*. Paper presented at the 21st conference of the Western European Association for Aviation Psychology, Dublin, Ireland.

Fischhoff, B. (1982). Debiasing. In D. Kahneman, P. A. Slovic, & A. Tversky (Eds.), *Judgment under uncertainty: Heuristics and biases*. New York: Cambridge University Press.

Funder, D. C. (1987). Errors and mistakes: Evaluating the accuracy of social judgment. *Psychological Bulletin, 101*, 75–90.

Hawkins, S. A., & Hastie, R. (1990). Hindsight: Biased judgmemts of past events. *Psychological Bulletin, 107*, 311–327.

Hogarth, R. M. (1987). *The psychology of judgement and choice* (2nd ed.). San Francisco, CA: Jossey-Bass.

Klein, G. A. (1989). Do decision biases explain too much? *Human Factors Society Bulletin, 22*(5), 1–3.

Klein, G. A., Orasanu, J., Calderwood, R., & Zsambok C. (Eds.). (1993). *Decision making in action: Models and methods*. Norwood, NJ: Ablex.

Lipshitz, R. (1993). Decision making in three modes. *Journal for the Theory of Social Behavior, 24*, 47–66.

Lipshitz, R. (in press). The road to "Desert Storm": Escalation of commitment and the rational vs. single-option paradigms in the study of decision making. *Organization Studies.*

Lipshitz, R., & Barak, D. (in press). Hindsight wisdom: Outcome knowledge and the evaluation of decisions. *Acta Psychologica.*

Mumaw, R. J., Swatzler, D., Roth, E. M., & Thomas, W. A. (1994). Cognitive skill training for nuclear power plant operational decision making (NUREG/CR–6126). Pittsburgh, PA: Westinghouse Electric Corp.

Rasmussen, J. (1993, March). *Perspectives on the concept of human error.* Paper presented at the Conference on Human Performance and Anesthesia Technology, New Orleans, LA.

Reason, J. (1990). *Human error.* Cambridge, England: Cambridge University Press.

Wallsten, T. S. (1983). Processes and models to describe choice and inference behavior. In Wallsten, T. (Ed.), *Cognitive processes in choice & decision behavior,* (pp. 215–237). Hillsdale, NJ: Lawrence Erlbaum Associates.

Woods, D. D., Johannessen, L. J., Cook, R. I., & Sarter, N. (1994). *Behind human error: Cognitive system, computers and hindsight.* Wright-Patterson AFB, OH: Crew Systems Ergonomics Information Analysis System.

Zapf, D., Maier, G. W., Rappensperger, G., & Irmer, C. (1994). Error detection, task characteristics, and some consequences for software design. *Applied Psychology: International Review, 43,* 499–520.

Part III

Research Reports

Chapter 16

Exploring the Relationship Between Activity and Expertise: Paradigm Shifts and Decision Defaults Among Workers Learning Material Requirements Planning

◆

Lia Di Bello
CUNY Graduate School and University Center
and
University of California, San Diego

With increasing advances in technology, the normal divisions between "manual" and "intellectual" labor are collapsing; as more industries move toward mediating and controlling work using computerized tools, a greater number of workers at all levels are being compelled to conceptualize work and judge situations on a very different level of abstraction than before. In addition, because of the nature of the processes being controlled, a background in the details of a specific industry is often proving a better prerequisite for effective technology use than, for example, a background in computer systems or computer mediated management. However, many efforts to implement advanced technologies fail because these systems are difficult for many people to learn, regardless of background. The study described in this chapter concerns three levels of workers in a large remanufacturing facility learning the logic of MRP (Material Requirements Planning) systems. As becomes clear, the important issues may not have to do with identifying who should or can learn these systems, but rather how learning occurs.

This specific study is part of larger program of research being conducted at CUNY's Laboratory for Cognitive Studies of Activity. The focus of our work concerns the cognitive impact of the introduction of technology into the workplace. Specifically, we are interested in exploring how workers' ways of thinking and understanding are affected by changes in the nature of work and workplace organization. Many of our questions have been addressed under a number of headings, such as "novice–expert shift" (e.g., Chi, Glaser & Farr, 1988) "situated cognition" (e.g., Rogoff & Lave, 1984), or "naturalistic decision making" (e.g., Orasanu & Connolly, 1993) and our work has been influenced by the methods and theoretical models from all of these various approaches. However, because the focus of our inquiry concerns the *development* of different ways of thinking in different domains, the research has been most influenced by the theories and methods of developmental psychology and particularly the developmental theories of Vygotsky (1987).

Some of this early research focused on identifying the factors associated with learning MRP. Material Requirements Planning was selected as a domain because it represents a class of technology that is widely known to require users who understand its underlying principles (in the sense described by Dreyfus & Dreyfus, 1986, or Polanyi, 1986) and it has a high failure rate because it is so difficult to learn. However, we do not consider our findings from these studies to be applicable only to MRP learning, but rather to the broader issue of learning complex technologies.

In one study of workers using MRP in two different factories (Scribner, Di Bello, Kindred, & Zazanis, 1992)—one with a successful implementation and one with an unsuccessful implementation — classroom instruction was shown to be an ineffective strategy for developing the kind of flexible mastery needed to effectively use MRP. However, some individuals do manage to master these systems. On-the-job activity proved to be critical, with particular kinds of activity responsible for the difference. An analysis of day-to-day job activity among workers revealed two distinct patterns of activity, "constructive" and "procedural." Briefly, constructive activities are those that have clearly defined goals and poorly defined means. The employee is compelled to develop in an iterative fashion a procedure, form, tool, or artifact that accomplishes the goal. In contrast, procedural activities are those that have clearly specified means and order of execution, whereas understanding the ultimate goal becomes secondary. Constructive activities are associated with an in-depth understanding of MRP's underlying logic, and procedural activities are not, even if the employees perform essentially the same functions and even execute the same kinds of actions most of the time. However, this study also showed that opportunities for constructive activities are usually fortuitous and ill structured. For example, they were most frequently encountered when an employee was forced to invent means to ends due to the sudden unavailability of documented procedures or knowledgeable persons (e.g., due to firings or sudden resignations).

In the study described here, the relationship between "constructive" and "procedural" activities is further explored in an ongoing study of skilled manufacturing workers (individuals with expertise in the skilled trades) who are learning MRP at a very large public rail transport remanufacturing facility. This study differs in a number of significant ways from our previous work and from other research on employees using complex computer systems. First, these workers were introduced to MRP in a 2-day workshop that we designed in order to engage workers with constructive and procedural activities in a more controlled way. We used a hands-on simulation or "game" that permitted participants to invent procedures for running a factory with MRP logic, but with little actual risk. Second, in direct challenge to the notion that MRP and such systems require "prerequisite" general formal education or computer experience, the training format did not differ for employees in different kinds of jobs or with different educational backgrounds.

WHAT IS MRP?

MRP is a family of computer-based systems that integrates information from all aspects of a company's operations and uses it to make decisions (recommendations) regulating production and inventory. MRP has also been characterized as a theory of manufacturing. It instantiates certain key economic concepts such as *zero inventory* and *just-in-time* production and is based on principles of manufacturing (for example, formulas regulating how future orders are forecast) developed over the last several decades (Hendrick & Moore, 1985; Timms & Pohlen, 1970). Its objects and procedures are generically defined and the system is content-free until implemented in a particular plant. Its power as a predictor is contingent on the data used (the content on which the logic operates) and the extent to which its assumptions match the way things are actually made in a given setting.

Employees working with the system must translate the company's anticipated demand into a form that the MRP system can "understand." This is done via a Master Production Schedule (MPS), which the system then interprets as a set of long range, abstract production goals for the company's finished goods. With the information the system has on "what" a particular finished good is (e.g., what parts go into it, what operations are involved, how long it takes to make each of its component parts and assemble it finally), it makes recommendations for every action leading up to the company's preset goals. This includes deciding on start dates and quantities for production orders and determining the most efficient pattern of purchasing. There are three "deeper" principles that organize the logic of MRP, and mastery of MRP seems clearly related to grasping these organizing ideas. This is not surprising, in that numerous studies in the "expertise" literature point to the importance of underlying concepts (e.g., Chi, Glaser, & Rees, 1982). A full exposition of these principles is beyond the scope of this chapter, but in order to make the sense of the results reported here, it is necessary to briefly describe them.

First, MRP conceives of all parts, assemblies, and finished goods as *hierarchically arranged items*, residing on "levels" that map onto how a given item is manufactured. Second, rather than using a linear, chronological representation of time, the timing of events is calculated beginning with a future date and moving back to the present. This is referred to as *phased time*. Third, quantities are not absolute, but relative to time and item. When making inquiries about how many of a given part are in inventory, the system calculates a *virtual* or *relative quantity*, based on a number of time-sensitive factors. Although these principles seem quite simple on the surface, they organize the data and operations of MRP in ways that strike many people as counterintuitive. In essence, it is MRP's counterintuitive structure that can make it so difficult to use.

METHOD

Subjects

Subjects came from three job categories of the transportation facility's Overhaul Division: Air Brake Maintainers (unionized mechanics), Supervisors and Managers (salaried and nonunion), and Analysts (a general office position for those involved in planning and special projects). All participants were drawn from a pool of employees who volunteered to participate in the MRP workshops. All expected to have their jobs affected by the planned MRP implementation and for many, our workshops would be their only form of introductory training.

Workshops

The workshops used with subjects were developed with the help of an MRP expert and others familiar enough with MRP logic to recognize activities and behavior that most richly represented the logic of MRP systems. For 2 days, the participants were required to run a miniature manufacturing and re-manufacturing facility using simple materials in the form of a "game" that simulated actual production, planning, ordering, and budgeting. The participants were required to produce three models of an origami "starship" with both common and unique parts. They purchased materials from a "vendor," sold finished ships to a "customer," and managed a "stockroom" made of foam-board compartments. They had a specific shipping schedule to meet, with a specific budget. A variety of standardized forms as well as blank paper were made available (for the asking) to participants as they set up and organized their "factory." They were explicitly instructed to keep inventory low and to maximize profits. Each "week" the game was 20 minutes long.

During the first morning of the first day, participants were allowed to run their factory according to any organizational scheme they wished. This invariably failed to achieve MRP goals (low inventory, increasing cash flow) and by the end of the first morning, most teams of participants went bankrupt, were unable to deliver, or began arguing with such intensity that the game had to be stopped. When this occurred, the game was paused and participants filled out a number of forms that helped them examine their patterns of decision making, evaluate their assets and losses, and reconstruct what had happened. These results were then compared by the participants with an MRP ideal of purchasing, production, and cash flow. After this phase, the participants were facilitated in the construction of a manual MRP system and introduced through various activities to its logic and overall functioning. Once they had a fully built and implemented "system," they played the game again with the same customer orders and budget, but with very different results.

Testing the Impact of "Constructive" and "Procedural" Activities

In order to examine the different effects of constructive and procedural activities, the workers were engaged with the three main principles of MRP via either "constructive" or "procedural" activities, but not both. The workshop included equal numbers of activities that would "constructively" engage participants with MRP's examples of Hierarchical Items and Phased Time. Considerable "procedural" activities and practice drill exposed the subjects to principles of virtual quantity. In general, the amount of time spent on each concept was equal across activities.

Pretesting Using Performance Probes

Prior to participating in the workshops, each participant filled out a form with information on previous work history, formal education, computer experience, and MRP experience. Subsequently, subjects were interviewed about their notions of manufacturing, and asked to do a number of performance tasks that elicited their strategies in response to a manufacturing task. These knowledge elicitation "probes" had been developed in previous research on MRP and were thoroughly tested using known MRP experts. Generally, these tasks were developed to tap into workers' understanding of key MRP and manufacturing concepts and were constructed to invite a variety of strategies.

This chapter mainly presents the results of one kind of performance task, the "scheduling" problems. Of the four kinds of performance tasks in the battery, the "scheduling" problems in their various forms were designed to invite the most comprehensive MRP, traditional manufacturing strategies, or both, and were the best measures of overall competency. (For a complete description of all probes, see Scribner et al., 1992; Scribner, Sachs, Di Bello, & Kindred, 1991). Briefly, these tasks required the subject to plan all aspects of purchasing and producing a specific quantity of some finished product by a specific due date. All the information MRP required to perform this function was provided to the subject.

Other probes were designed to explore individuals' grasp of very specific aspects of MRP in isolation from the system as a whole. The results from these probes are reported briefly in order to explicate some of the general findings. Importantly, all probes were designed to elicit either MRP *or* traditional manufacturing-based strategy, *or both*, so that it could be observed if an individual were using one or the other, or both.

Posttesting

Posttesting was conducted about 2 months after the workshops. It was expected that by then any rote learning that had taken place would be forgotten, and that MRP strategies exhibited by the subjects would reflect what they had actually come to understand about MRP. The "scheduling" postperformance probes were structurally similar to the pretest but were applied to more kinds of objects (i.e., the "abstract unknown," the "starship" from the workshop, and a familiar kind of button-top pen). These tasks again elicited strategies that would reveal understanding of MRP and traditional manufacturing principles.

Scoring Pretests and Posttests

The strategy elicited on each performance probe was scored for the presence, partial presence, or complete absence of 12 different behaviors that have been shown in previous work to be associated with in-depth understanding of MRP principles. The same protocols were also scored for typical traditional manufacturing strategies. Thus, each task protocol was given an "MRP total" and "Manufacturing total." These numbers were then divided by the total number of possible behaviors for each domain, giving a *proportional* score for each domain (MRP or Traditional manufacturing).

RESULTS

In general, traditional manufacturing strategies were replaced with MRP strategies after the workshops. The pre- and postworkshop differences were found to be statistically significant for all groups ($F(1,30) = 6.79 \, p. < .001$). In fact, the postworkshop scores were found to be somewhat higher than expected and comparable to those seen in plants where the system had been fully functioning for at least 2 years (Scribner et al., 1992). Table 16.1 presents a summary of the preworkshop and postworkshop means among groups.

As can be seen in Table 16.1, among the three job categories, Supervisors (SUPs) began with the most initial MRP knowledge, as a group, but this was largely attributable to two of the four individuals. After the workshops, Analysts showed higher aggregate scores for MRP knowledge. An analysis of variance comparing the Air Brake Maintainers (ABMs) and Analysts indicates that these differences were not significant. (Because of the small number of Supervisors, they were excluded from statistical analyses.)

Interestingly—although not statistically significant—the postworkshop difference between the two groups was due largely to one version of this probe. ABMs as a group seemed to have difficulty with "post-unknown abstract," which was a far simpler scheduling task (with 3 parts as opposed to the starship's 13 parts) but was abstract rather than concrete or "contexted." When scores for this probe were not used in the calculation of averages, group differences almost disappeared, with ABM's averaging .71 and Analysts averaging .75. However, an analysis of variance comparing the Analysts' and ABMs' performance on the "abstract unknown" (alone) after the workshops did show significant differences between the two groups, $F(1,30) = 6.09 \, p < .02$.

A few individuals showed an interesting deviation from the trend toward replacement. These subjects attempted to produce scheduling solutions that integrated MRP with traditional manufacturing strategies. A closer look at their scheduling strategies revealed a rather sophisticated attempt to integrate plant and labor capacity considerations into their scheduling solutions.

TABLE 16.1

Average Scores Between Domains for Each Occupational Group, Showing Replacement of Traditional Manufacturing Strategies With MRP Strategies

	Preworkshop		Postworkshop	
	Manufacturing	MRP	Manufacturing	MRP
SUPs	.52	.56	.42	.69
ABMs	.80	.13	.25	.67
Analysts	.52	.42	.06	.75

For example, although these individuals showed the "backwards time" scheduling strategy inherent to an MRP strategy, they attempted to spread out requirements evenly to accommodate plant or labor capacity. Capacity is one real-world aspect of scheduling that MRP cannot consider. In previous work, this kind of integration is an indication of rather sophisticated understanding of MRP concepts and the system's relationship to actual production. Such an understanding usually takes considerable time to develop. Not surprisingly, the postworkshop interviews revealed that these four individuals were among seven who had begun to use the system intensively (after our training) at least on a weekly basis in a pilot project to schedule the work in their division of the Pneumatic shop (compressors). Observations indicate that these activities were largely "constructive" as they involved setting up a new system without preset procedures. Supervisor 3, on the other hand, showed considerable MRP knowledge before the workshops and no manufacturing strategies (on the scheduling task). His subsequent experience with the system is not known.

Supervisor 3's profile suggests that replacing traditional manufacturing strategies with MRP strategies might precede integrating the two approaches. He moved from an MRP-dominated approach to a more integrated one. This may have been what occurred with ABMs 5, 6, and 7, and Supervisor 2 in the period between interviews, but this cannot be known for certain. However, the profiles of these four do suggest that certain kinds of on-the-job activity after training might effect integration. This is further suggested by the complete absence of integration among the Analysts, who have no opportunity for seeing how schedules are implemented on the floor.

Uneven Change in the Three Organizing Principles

As predicted, the learning did not occur evenly over areas covered by the workshop. As indicated, the workshops were constructed to provide extensive practice with the notion of "Relative Quantity," but subjects were given specific MRP calculation procedures and were not encouraged to invent or explore MRP algorithms for obtaining Virtual Quantities.

Table 16.2 shows relative performance in the three conceptual areas and the probe data scored. These data are percentages of subjects, taken from several probes, as noted. As can be seen, the fewest number of subjects achieved a truly flexible notion of Virtual Quantity. In addition, all subjects who came to understand Virtual Quantity also developed a thorough understanding of Phased Time and Hierarchical Item.

Table 16.2

Differences in Overall Performance on Three Core Concepts

	Relative Quantity	Phased Time	Hierarchical Item
Pretest	[a]0%	[b]12%	[d]15%
Posttest	[a]6%	[b]48%	[d]22%
Posttest "Starship" tasks	[a]NA	[c]74%	[e]79%

Note. Data from "tree," "scheduling," and "card sort" probes. Values Represent percentages of subjects with strategies showing full grasp of the principle involved. [a]Tree Virtual Quantity probe. [b]Phased time analysis from "scheduling abstract unknown" probe. [c]Phased time analysis from "scheduling starship" probe. [d]Hierarchical item analysis from "unknown object card sort" probe. [e]Hierarchical item analysis from "starship card sort" probe.

DISCUSSION

It is clear from the post test results that 2 months after the "constructive" activity training, workers performed at a level of mastery that is normally seen after 18 to 24 months in workers who had managed to learn to use the system effectively (in other sites) after no training, or classroom-based training. More important, workers exhibited these strategies spontaneously and appeared to have replaced old ways of doing the same problems with new approaches. In other words, knowledge was reorganized rather than added. Many subjects reported that they realized they had done it differently before but could not even replicate or remember their previous approach.

Their performance on the floor supported these results. Shop floor personnel who had system access made use of the system to manage their own work areas, instructed management personnel in MRP, and took upon themselves data entry and system upkeep.

The question remains—how did constructive activities affect this transformation? Although tapes from the actual workshops are still being analyzed along with other materials generated by participants, a great deal is suggested by preliminary analyses.

It appears that "constructive" activities permitted individuals to reorganize the implicit mental models driving the decision process. As indicated, during the first morning of the workshops, the teams of participants could "run" their factory in any way they chose. Without exception, under this kind of pressure, all groups "defaulted" to the strategies they normally used at work. Comparing the performance in the workshops (as indicated by the cash and material flow patterns as well as the participants' notes and inventory control records) with work history indicates a strong correspondence between current work practices and decisions under pressure. For example, mechanics almost always delivered a high-quality product on time and thereby managed to stay financially above water by ensuring income. However, they often did so by buying finished assemblies (as opposed to raw materials) at great cost, by accelerating production (which has the real-world equivalent in "overtime"), and by overbuying all materials. Thus, they sacrificed profits.

In contrast, Analysts and Supervisors were often *not* successful at meeting customer orders but they were very hesitant to overspend; often they could not meet orders because they had waited too long to decide to purchase the necessary materials. It is important to note that individuals resorted to their default strategies even when *explicitly instructed* to operate differently and when they were aware that the goal of the game was to make the most profit and buy the least material.

As indicated, during the workshops' first pause, participants evaluated their strategies and decisions using various tools. During the discussions that followed the self-evaluation, participants began to reflect on these default strategies and examined how a different approach might have accomplished their goals. It was only at this point that participants became aware of the assumptions guiding their decisions under pressure and the attendant results. We have come to call this the "deconstruction" phase and the beginning of "reorganization." During the MRP-oriented "constructive" activities that followed in next day and a half, participants were introduced to MRP concepts as they continued a pattern of inventing a solution using their intuitive understanding of manufacturing, noticing its mismatch with their goals, reordering what they had done, and comparing it again to their goals. At the end of this process, all participants arrived at the same solution but, it is important to note, they had all gotten there via different paths and had begun the process through sometimes radically different entries.

It seems that constructive activities work—and work for a wide variety of people—because they are both multimodal and self-modal. They permit a variety of entries to knowledge and they permit the individual's entry to be in terms of his or her existing way of thinking in that domain. Rather than overlaying existing knowledge, constructive activities permit the reorganization of knowledge that occurs when learners are compelled to bring their current way of thinking to bear on the problem, notice its mismatch with desired outcomes, and make refinements in accordance with the tools they are discovering (in this case MRP).

To further illustrate, let us return to some of differences found among the workers. Although there were no significant differences between the ABMs and Analysts on *aggregate* performance, differences in their patterns of response on different probes indicate they were "coming to" the same notions through different means. For example, on the "scheduling" tasks, almost all of the variance in average aggregate MRP scores was due to performance on the "abstract unknown" probe. Although this task was much simpler than the same probe using the starship or pen (3 parts as opposed to 13 parts with various lead times), Mechanics exhibited the most sophisticated strategies with the more complex, but identifiable objects.

The paradox of the more complex (but more concrete) being simpler for one class of subjects suggests interesting possibilities. It seems as though this group was not oriented to approaching problem solving when the context was presented abstractly, although they were capable of thinking abstractly and symbolically about contexted objects. Conversely, the Analysts were able to take better advantage of the simplicity of the "abstract unknown" and were more likely to make calculation errors on the starship task. This result is particularly interesting given that abstract problem solving is sometimes invoked as more sophisticated than "concrete" or contexted problem solving.

The Mechanics' ability to operate with the complex but concrete makes sense given their jobs. They are required to work with highly complicated assemblies having many small but essential parts. In order to perform well on the job they must pay attention to small details, and the impact of small assembly errors. It may be that the requirements of their jobs have oriented them to grasp general relationships or concepts through the specifics, while at the same time being more able to handle a large amount of detail in parallel. The difference between the ABMs and the Analysts supports the more general notion that job-specific ways of thinking may influence ways of learning, and that activities or learning situations that permit multiple ways to attend to information and concepts may erase normally observed group differences in overall ability to learn.

Another illustration is provided by the effects of the "procedural" activities. Despite considerable practice with MRP's methods of calculating virtual quantities, all but two participants (6% on Table 16.2) defaulted to their previous methods of calculating quantities during the postprobe interviews. In other words, no "replacement" took place for most when it came to quantity strategies. Clearly, hearing about concepts and even practicing formulas does not necessarily engender real change or in-depth understanding.

Although this chapter presents only a small portion of a much larger study—with much data still being analyzed—it makes a case for rethinking training for adults who already have considerable skills. These results suggest that the "default" modes or intuitive understanding that are developed with experience may act as both an impediment and bridge to learning new technologies in one's domain of expertise. Constructive activity-based instruction may offer a way to make an impediment into a bridge without sacrificing current experience and expertise.

REFERENCES

Chi, M., Glaser, R., & Farr, M. J. (Eds.). (1988). *The nature of expertise*. Hillsdale, NJ: Lawrence Erlbaum Associates.

Chi, M., Glaser, R., & Rees, E. (1982). Expertise in problem solving. In R. J. Sternberg (Ed.), *Advances in the psychology of human intelligence* (Vol. 1, pp. 7–75). Hillsdale, NJ: Lawrence Erlbaum Associates.

Dreyfus, H. L., & Dreyfus, S. E. (1986). *Mind over machine: The power of human intuition and expertise in the era of the computer.* New York: The Free Press.

Hendrick, T. E., & Moore, F. G. (1985). *Production/operations management* (9th ed.). Homewood, IL: Irwin.

Orasanu, J., & Connolly, T. (1993). The reinvention of decision making. In G. A. Klein, J. Orasanu, R. Caderood, & C. Zsambok (Eds.), *Decision making in action: Models and methods* (pp. 3–20). Norwood, NJ: Ablex.

Polanyi, M. (1986). *The tacit dimension.* Garden City, NY: Doubleday.

Rogoff, B., & Lave, J. (Eds.). (1984). *Everyday cognition: Its development in social context.* Cambridge, MA: Harvard University Press.

Scribner, S., Di Bello, L., Kindred, J., & Zazanis, E. (1992). *Coordinating knowledge systems: A case study.* Unpublished manuscript, Laboratory for Cognitive Studies of Activity, City University Graduate School, New York.

Scribner, S., Sachs, P., Di Bello, L., & Kindred, J. (1991). *Knowledge acquisition at work* (Tech. Paper No. 22). National Center on Education and Employment.

Timms, H. L., & Pohlen, M. F. (1970). *The production function in business: Decision systems for production and operations management* (3rd ed.). Homewood, IL: Irwin.

Vygotsky, L. S. (1987). Thinking and speech. In R. Rieber & A. S. Carton (Eds.), *The collected works of L. S. Vygotsky; Volume 1: Problems of general psychology* (pp. 39–288). New York: Plenum.

Chapter 17

Analysis of Decision Making in Nuclear Power Plant Emergencies: An Investigation of Aided Decision Making

◆

Emilie M. Roth
Westinghouse Science & Technology Center

One of the hallmarks of naturalistic decision making is that in naturalistic environments the decision makers are seldom left completely to their own devices. Often aids are provided to support the decision-making process. These aids can range from policies and guidelines that constrain the decision space to highly prescriptive approaches that attempt to direct all aspects of decision making. One aiding approach that has evolved has been to provide preplanned diagnostic and response strategies that are intended to be followed "verbatim." Examples of this approach include checklists, paper-based procedures, and conventional expert systems that are used to guide performance in applications ranging from flight takeoff, to maintenance tasks, to emergency management (Mosier, Palmer, & Degani, 1992; Roth, Bennett, & Woods, 1987).

This chapter summarizes a recent naturalistic decision making study that examined the impact of highly prescriptive procedures on the decision-making process (Roth, Mumaw, & Lewis, 1994). The particular domain of application was nuclear power plant operations. When an emergency arises in a power plant, operators are required to follow paper procedures step by step. The procedures provide detailed guidance on what plant parameters to check, how to interpret the symptoms observed, and what control actions to take. Given that operators utilize such highly prescriptive procedures, a question arises regarding the nature and extent of cognitive activity required of operators to adequately handle emergencies. One view is that all that is needed of operators is that they understand and follow the individual procedure steps. An alternative view is that higher-level cognitive activities, such as situation awareness, response planning, and mental models of system function and structure, continue to be important for successful operator performance. These two alternative views have very different implications for the kinds of training, procedures, displays, and decision aids that need to be provided.

In this chapter, we attempt to show through illustrative examples taken from Roth et al. (1994) that situation awareness and response planning continue to be important elements of naturalistic decision making even in cases where detailed procedures are provided. We provide evidence that operators actively construct a mental representation of plant state and use this mental representation to identify malfunctions, anticipate future problems, evaluate the appropriateness of procedure steps given the situation, and redirect the procedure path when judged necessary.

STUDY OF OPERATOR DECISION MAKING

An empirical study was conducted examining nuclear power plant operator performance in two cognitively complex simulated emergencies that included aspects that were not fully addressed by the procedures. Data on operator performance were collected using training simulators at two plant sites. Up to 11 crews from each plant participated in the simulated emergencies for a total of 38 cases analyzed. Crew performance was videotaped and transcripts of crew dialogue were produced. The transcripts documented observations made, hypotheses considered, situation assessments made, and actions taken. Particular attention was paid to documenting instances where operators summarized their situation assessment; reflected on the goals or objectives of the procedures; or engaged in extraprocedural activity. Detailed descriptions of the data collection and analysis method can be found in Roth et al. (1994).

Analyses revealed two strands of parallel activity. One strand was governed by the procedures. Crews went through the procedures to insure that all major safety functions were maintained, and that actions required to maintain safety functions were performed. At the same time there was a second strand of activity that was self-initiated and cognitively based. Operators performed additional monitoring, situation assessment, and response-plan monitoring activities. Operators acknowledged alarms and assessed their import, assessed and monitored situation state, and monitored whether the strategies embodied in the procedures were appropriate to the situation.

The analysis identified a number of situations where crews exhibited evidence of situation awareness and response planning that enabled them to handle aspects of the situation that were not fully covered by the procedures. Following, I provide illustrative examples of the types of cognitive activity that were observed. A fuller description of the results can be found in Roth et al. (1994).

Situation Awareness

One aspect of naturalistic decision making that has been found in a number of domains is that individuals actively construct a mental representation of the situation. This has been variously described as "story building" where a person attempts to synthesize observations into a causal explanation (Klein & Crandall, 1995), development of a "mental model" of the situation (Lipshitz & Ben Shaul, chapter 28, this volume), situation assessment (Orasanu, Dismukes, & Fischer, 1993), and situation awareness (Endsley, 1995).

We found evidence that operators actively constructed a mental representation of the factors known or hypothesized to be influencing plant state at any given point in time. They used this mental representation to explain observations, identify and pursue unexpected findings, and anticipate future problems. I draw on examples from one of the events we ran in the study, a simulated interfacing system loss of coolant accident (ISLOCA), to demonstrate this point.

The ISLOCA involved a leak from the reactor coolant system (RCS), which is inside containment, into the Residual Heat Removal (RHR) System, which is outside of containment. One of the factors that made this event complex is that many of the early symptoms (e.g., radiation symptoms inside containment) strongly suggested a leak inside of containment. Further, because of the particular dynamics of the scenario, the emergency procedures directed the operators to a procedure intended to deal with a leak inside containment. Although a procedure for dealing with the ISLOCA case existed, it could not be reached given the procedure transition flow. As a result, ability to detect and respond to the leak into the RHR depended on the success of the operators' situation assessment.

Eventually all nine crews that participated in this scenario identified the leak into the RHR, and attempted to take action to terminate the leak. The cues that enabled the crews to recognize that the event was not a simple leak inside containment, and the activities they engaged in to identify and terminate the leak into the RHR, provided evidence of active construction of a mental representation of plant state on the part of the operators.

Five of the nine crews recognized early that it was not a straightforward leak inside containment based on *a failure to observe expected symptoms*. Given the rate of decrease in the pressurizer, the operators formed expectations with regard to containment symptoms that should appear if it were a simple leak inside containment. When containment pressure did not increase as quickly as they expected, they began to suspect a leak outside containment.

This example illustrates several points. First, operators *actively generate situation assessments* that include possible explanations for the set of symptoms they observe. In this case they began by postulating a leak inside containment. Second, operators *form expectations with regard to additional symptoms they should observe if their explanation is correct*. In this case they expected to see a rapid increase in pressure in containment. Third, *when their expectations are violated, they search for additional (or alternative) factors (influences) to explain the behavior observed*. In this case they began suspecting some kind of leak outside containment. This led them to actively search for a leak pathway.

Operators also used their knowledge of the physical interconnections of plant systems, and the size and temporal characteristics of plant parameter behavior, to *generate and narrow potential explanations of plant symptoms*. For example, in the ISLOCA case, three of the nine crews identified the leak into the RHR by applying their knowledge of the physical interconnections among systems. The leak from the RCS caused an increase in pressure in the RHR. There is a pressure relief valve in the RHR system that relieves into a pressurizer relief tank (PRT) inside containment. Level and pressure in the PRT built up, and it eventually ruptured. When operators noted the PRT symptoms, they considered all the plant components that relieved into the PRT, they narrowed it down to the RHR relief valve, because this was the only one that could relieve sufficient quantities of water to account for the rapid increase in level and subsequent rupture of the PRT.

Operators also showed evidence of *active search for a coherent explanation of multiple diverse symptoms*. In the ISLOCA case, the buildup of pressure in the RHR eventually led to a leak in the heat exchanger between the RHR and the Component Cooling Water (CCW) system. This produced high coolant level and radiation symptoms in the CCW. Although the CCW symptoms were consistent with the hypothesis of a direct leak from the reactor coolant system to the CCW system, that explanation could not account for the symptoms in the RHR, nor for the symptoms in the PRT. Search for coherent explanation that minimized the number of separate faults that need to be postulated led the crews to narrow in on the possibility of a break in the RHR to CCW heat exchanger. Thus the crews were able to come up with a coherent causal sequence that simultaneously explained the symptoms in the RHR (i.e., a leak from the reactor coolant system to the RHR), the PRT (i.e., the RHR relieves into the PRT), and the CCW (i.e., the high pressure buildup in the RHR led to a break in the CCW Heat Exchanger).

There was also evidence that situation assessment is a dynamic activity that involves *keeping track of multiple influences on plant state and their expected impact*. In several instances operators had to determine whether observed plant behavior was the result of known influences on the plant, such as manual and automatic actions and known malfunctions, or was unexpected and signaled an unidentified plant malfunction. This depended on an assessment of the size and direction of plant behavior that would be expected, given known plant influences. An example is a case where we introduced a leak in a pressurizer relief valve while operators were performing a maneuver that was expected to produce disturbances in pressurizer level and pressure. Identification of the leaking relief valve depended on recognizing that the magnitude of observed changes in pressurizer behavior could not be entirely explained as side effects of the ongoing maneuver.

Formulating/Monitoring Response Plans

The emergency procedures dictate what actions operators should take during emergencies. In the simulated emergencies, operators consistently followed the emergency procedures in handling the events. Nevertheless, the operators also monitored plant state and assessed the appropriateness of the procedures they were following at that point in time, given their understanding of plant state. Operators were observed to discuss among themselves the appropriateness of particular procedure steps, as well as whether the procedure path they were following would lead them to the actions needed to achieve safety goals. We found examples in our data of cases where monitoring the response plan imbedded in the procedures enabled crews to:

- Catch and recover from errors—both operator errors and errors in the procedures.
- Assess whether the procedure path they were following was correct, or whether they had missed an important transition.
- Fill in gaps and adapt procedures to the situation.
- Deal with unanticipated situations that went beyond the available procedural guidance.

One of the behaviors we observed in the simulated emergency events is that operators occasionally took actions in anticipation of a later procedure step. These actions were based on an understanding of both plant state and the content and sequence of procedures. They allowed operators to move through the procedures in a more timely manner. As an example, in the ISLOCA case, once they suspected an ISLOCA, several of the crews immediately called down to the auxiliary building to search for evidence of a possible leak outside of containment. The procedure included a step to call the auxiliary building, but that step appeared late in the procedure. By anticipating that step, operators were able to initiate the call early so that they would have the results of the auxiliary building search already available at the point where the procedure called for it.

Operators were also observed to monitor whether actions indicated in procedure steps made sense in the context of the particular event, as well as in the context of the overall response strategy embodied in the procedure as a whole. This is an important factor in helping operators detect and recover from errors—both operator errors as well as errors in the procedure. We observed operators detect instances where they had inadvertently skipped critical steps in the procedure. They recognized their error when they started reading a subsequent step and realized that the actions were not appropriate for the situation. We also found cases where operators detected faulty step logic by recognizing that the actions indicated by the procedure step as written were inappropriate.

Operators were also observed to monitor the appropriateness of the procedure path they were following. In our simulated events, we saw repeated cases where operators stopped to discuss as a group whether the procedure path they were following would eventually lead them to take the actions they recognized to be important for safe recovery of the plant. Because the emergency procedures involve a series of transitions from procedure to procedure, it was not always a trivial task to ascertain whether a given procedure would eventually lead, through a series of transitions, to the specific emergency procedure that contained the relevant guidance. The task was made that much more difficult by the fact that due to event dynamics, the symptoms required for a transition step sometimes did not appear until after the transition step was passed, or other symptoms arose first that caused a transition to a different procedure, from which the desired procedure could not be reached.

CONCLUSIONS AND IMPLICATIONS

The study illustrates two main points. First, it argues for the central role that active construction of a mental representation of the situation plays in naturalistic decision making, at least in this domain. In this study we found little evidence of pure recognition-primed decision making of the sort reported in some domains (e.g., Klein & Calderwood, 1991). Operator decision making was based on active construction of a model of the factors known or hypothesized to be influencing plant state at any given point in time. These results are consistent with the elaborated version of the RPD model that incorporates diagnostic and "story building" elements (Klein, chapter 27, this volume).

Second, the study points to the limitations of preplanned procedures. Although preplanned procedures may reduce the need for operators to develop diagnostic and response strategies on their own in real time, they do not entirely eliminate the need for active situation assessment and response plan monitoring. In our scenarios a number of situations arose that were not fully covered by the procedures. In these situations operators exhibited:

- Knowledge-driven monitoring.
- Search for coherent explanations for plant symptoms.
- Formation and use of expectations based on situation assessment.
- Monitoring of procedure effectiveness in achieving safety goals.
- Adaptation of the procedures to the situation.

These results have implications for the development and evaluation of training, and control room aids (e.g., procedures, displays, decision aids).

In our study we found evidence of situations where crews needed to utilize mental models of physical plant systems and to reason qualitatively about expected effects of different factors influencing plant state in order to localize plant faults and identify actions to mitigate them. There is a need to foster and support accurate mental models through training and control room displays and decision aids (cf. Vicente, Christoffersen, & Pereklita, 1995).

Another type of needed knowledge concerns important plant goals and means to achieve them. Our study found evidence that operators needed to reason about plant goals, and evaluate alternative means to achieve them. Operators also needed knowledge of the procedures, which includes not only knowledge of how to follow the individual procedure steps, but also knowledge of assumptions and logic that underlie the procedures. This includes knowledge of the goal prioritization inherent in the procedures, knowledge of the response plans embodied in the procedures and their rationale, and knowledge of the procedure transition network. The results of the study suggest value in explicitly addressing these types of knowledge through training, background documents, and the structure and formatting of procedures.

ACKNOWLEDGMENTS

The study of operator performance in simulated emergencies was sponsored by the Office of Nuclear Regulatory Research, U.S. Nuclear Regulatory Commission. The views expressed in this chapter are the author's and do not necessarily reflect those of the U.S. Nuclear Regulatory Commission.

REFERENCES

Endsley, M. R. (1995). Towards a theory of situation awareness in dynamic systems. *Human Factors, 37*, 32–64.

Klein, G. A., & Calderwood, R. (1991). Decision models: Some lessons from the field. *IEEE Transactions on Systems, Man, and Cybernetics, 21*, 1018–1026.

Klein, G. A., & Crandall, B. W. (1995). The role of mental simulation in naturalistic decision making. In P. Hancock, J. Flach, J. Caird, & K. Vicente (Eds.), *Local applications of the ecology of human–machine systems* (Vol. 2, pp. 324–358). Hillsdale, NJ: Lawrence Erlbaum Associates.

Mosier, K. L., Palmer, E. A., & Degani, A. (1992). Electronic checklists: Implications for decision making. *Proceedings of the Human Factors Society 36th annual meeting* (pp. 7–10). Santa Monica, CA: Human Factors and Ergonomics Society.

Orasanu, J., Dismukes, R. K., & Fischer, U. (1993). Decision errors in the cockpit. *Proceedings of the Human Factors and Ergonomics Society 37th annual meeting* (pp. 363–367). Santa Monica, CA: Human Factors and Ergonomics Society.

Roth, E. M., Bennett, K., & Woods, D. D. (1987). Human interaction with an "intelligent" machine. *International Journal of Man–Machine Studies, 27*, 479–525.

Roth, E. M., Mumaw, R. J., & Lewis, P. M. (1994). *An empirical investigation of operator performance in cognitively demanding simulated emergencies* (NUREG/CR–6208). Washington DC: U.S. Nuclear Regulatory Commission.

Vicente, K. J., Christoffersen, K., & Pereklita, A. (1995). Supporting operator problem solving through ecological interface design. *IEEE Transactions on Systems, Man, and Cybernetics, 25*, 529–545.

Chapter 18

Aeronautical Decision Making, Cue Recognition, and Expertise Under Time Pressure

◆

Alan F. Stokes
Florida Institute of Technology

Kenneth Kemper
United States Air Force

Kirsten Kite
Florida Institute of Technology

Pilot decision-making prowess has often been viewed as a somewhat occult skill, a recondite and elusive element of "professional seasoning" or the "right stuff." Most literature on the subject has tended to derive from one of three sources: from nonscientific anecdotal and advisory accounts (e.g., *Flying* Magazine's book series, "I Learned About Flying From That"); from intuitively based judgment-training programs (e.g., Telfer, 1987); or from analyses of aircraft accident and incident reports (see, for example, Giffen & Rockwell, 1987). Prevalent themes in this third body of literature include the observations that faulty pilot judgment and poor decision making are causal or contributing factors in the majority of "pilot-error" accidents; that most of these accidents involve time pressure or time criticality; and that more research on pilot judgment and decision making is needed.

Despite these conclusions, few researchers have attempted to bring a theory-based experimental (or at least empirical) approach to bear on the issue of describing the cognitive processing activities that take place as decisions are carried out in the cockpit. In the absence of actual knowledge, standard analytical models of decision making have simply been assumed and imposed on observed behavior. For example, Besco et al. (1994) asserted:

> Decision making in the cockpit follows traditional views of decision making . . . in which the decision maker, i.e. Captain, is 1) presented with a situation that requires a decision; 2) the nature of the situation is assessed by the decision maker; who 3) determines the availability of alternative outcomes to respond to the situation, and 4) after evaluating the risk and the benefits of each alternative; 5) selects an alternative in response to the needs of the situation. (p. 43)

The study reported here is one of a series of experiments that used desktop flight simulation and cognitive (psychometric) testing to examine the adequacy of traditional views of decision making within the context of aviation.

BACKGROUND

According to the traditional view, both experienced and inexperienced aviators presumably rely to a significant extent on analytical, utility-based decision strategies. These in turn necessitate a very comprehensive weighing of situational factors and a full consideration of viable alternatives (Janis & Mann, 1977). However, studies have found little relationship between aeronautical decision performance and the information-processing skills thought necessary for this kind of strategy, such as working or short-term memory (STM) and logical reasoning (Barnett, 1989; Wickens, Stokes, Barnett, & Davis, 1987). This has proven to be particularly true of more experienced pilots, which would suggest that exhaustive situation analysis requiring extensive processing in STM may not in fact be a prescription for (or a description of) optimal decision making in the cockpit. These findings are not surprising, insofar as other, nonaviation studies have shown that exhaustively analytical decision strategies are ineffective in situations where time limits constrain the amount of analysis and cognitive processing that can be carried out (Howell, 1984; Rouse, 1978; Zakay & Wooler, 1984). Nor is it always cost effective for the decision maker to invest the resources necessary to identify the very "best" course of action. For example, protocols obtained from fireground commanders have shown that these experienced firefighters do not expend valuable time weighing alternative options, calculating probabilities, and so forth; rather, they report seeking "workable" and "timely" courses of action (Klein, Calderwood, & Clinton-Cirocco, 1986). Similarly, experienced aviators may, instead of trying for an optimal solution, invest the minimum resources consistent with arriving at an adequate (if not necessarily the most elegant) solution to a given problem. This decision strategy is sometimes referred to as *satisficing* (Simon, 1955), and, it has been suggested, is typically accomplished with the use of a nonanalytical pattern-recognition process (Klein, 1989; Klein et al., 1986).

It was hypothesized, therefore, that although experienced pilots are not necessarily characterized by any special cognitive advantages such as superior reasoning ability (Barnett, 1989; Stokes, Belger, & Zhang, 1990), they do differ from inexperienced pilots in terms of access to domain-specific (aviation-relevant) knowledge representations in long-term memory (LTM). Thus, when confronted with situations that could endanger the safety or efficiency of the flight, high-time pilots may more readily recognize cues relevant to the problem and "pattern match" these cues with situational schemata from LTM. Only if unable to match the cues in this way (it is hypothesized) do pilots drop into an alternative strategy using real-time computational and inferential processes heavily dependent on working memory (see Fig. 18.1). Within this model, then, experienced pilots exhibit more efficient decision making because they possess more extensive experiential repertoires and internalized scripts (Stokes, Barnett, & Wickens, 1987). The study reported here examines this hypothesis experimentally.

Another issue addressed in this study concerns early versus late stages of decision making. As noted already, earlier experiments were unable to demonstrate that information-processing skills were important components of decision making, particularly for experienced pilots. However, because these studies were based on a multiple-choice decision format they focused heavily on the option selection stage of the decision process. The present study examines the relationship of information-processing skills and knowledge representation measures to various stages in the decision-making process, including cue recognition and hypothesis generation.

A third issue is that of ecological validity versus experimental control. Specifically, the multiple-choice decision format used in earlier studies may have attenuated variance between experienced and inexperienced pilots by prompting the latter to search for cues that they might otherwise have overlooked. Obviously, however, such prompting does not reflect operational realities. The current study, therefore, included a nonprompted, open-ended decision format designed to accept decision alternatives generated by the pilot (Stokes, 1991). This provided a more ecologically valid context while retaining the controlled environment of an experimental simulation.

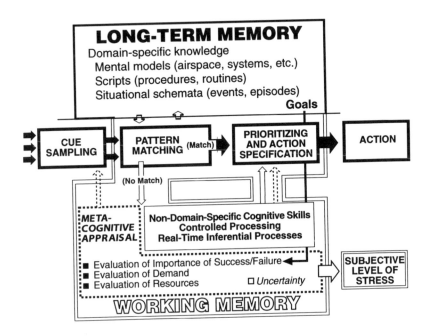

FIG. 18.1. A model of decision making under stress. This figure consists of three major elements. The black-filled outline rectangle at the top represents long-term memory functions. The white-filled outline rectangle at the bottom represents working memory functions. The boldly outlined boxes sequenced left to right represent major stages in the processing of information, including (a) the sampling of (predominantly environmental) cues such as instrument readings or radio messages, and (b) the attempted matching of the pattern of cues with stored information. Only when a pattern match is not obtained does processing "drop down" into a slower, more error-prone working memory intensive mode. In this mode, task-related processing competes for attentional resources with metacognitive processes influencing both task execution and affective state. The dotted-line "balloon" in the diagram may expand to take up more working memory capacity as more attention is diverted to threat. This, in turn, may affect both cue sampling and prioritizing of actions. The diagram is intended to suggest that the pattern-matching mode is the preferred, normal, or default strategy which, when successful, leads most rapidly to action specification. It entails very little of the analytical or deliberative processing of options frequently said to characterize pilot decision making, and it short-circuits mechanisms of stress said to influence such deliberative processes. The pattern-matching strategy places few if any demands upon working memory or basic cognitive processes within it (tests of which nevertheless continue to dominate most fitness-to-fly and aviator selection batteries). There are a number of testable hypotheses implicit in the model, for example, that standard cognitive tests will be poor predictors of aeronautical decision making; that expert pilots will tend to make good decisions without weighing options; and that knowledge (rather than, say, personality variables) will be the prime inoculator against stress-induced performance degradation. To date, the experimental evidence has been consistent with these hypotheses. Adapted from *PC Based Instrument Flight Simulation* (p. 29), by R. Sadlowe (Ed.), 1991, New York: American Society of Mechanical Engineers. Copyright 1991 by American Society of Mechanical Engineers. Reprinted with permission.

METHOD

The general approach was first to evaluate both high- and low-time pilots in terms of STM-based information-processing abilities and LTM-based knowledge representations. The next step involved determining the predictive power of each test on the pilots' performance in a simulated flight in instrument conditions. Subjects consisted of 24 pilots, 12 with considerable flight experience and 12 with relatively little. Specifically, members of the low-time group held a private pilot's license and were either working on their instrument rating or had fewer than 50 hours of total instrument time. The expert pilots had at least 1,500 hours of flight experience as well as instructor, commercial, or higher certificates plus instrument ratings. As a convention (that is, without making any a priori—and inevitably circular—claims about expertise), these groups were labeled instrument-flying "experts" and "novices."

Individual differences in the efficiency of short-term processes in working memory were measured using selected subtasks from the SPARTANS battery (Stokes, Banich, Elledge, & Ke, 1991). This battery assesses elements of cognitive functioning that have been identified as relevant to aviation, including focused and divided attention, memory, psychomotor skills, risk taking, spatial ability, and logical reasoning. All tests were domain-independent and thus did not favor the more experienced group of pilots. (Indeed, previous experiments by Wickens et al. (1987) and Barnett (1989) indicate that few differences can be expected between the two cohorts on these measures.)

The LTM-based knowledge representation tests were domain-specific, and were used to assess individual differences in availability of situational schemata or knowledge representations. First a declarative knowledge measure, the FAA Quiz, was administered to all subjects. The test consisted of 25 standard questions from the FAA Instrument Written Test Manual covering a balanced range of topics (e.g., weather, chart use, systems), but excluding procedural knowledge questions. Procedural knowledge was evaluated using the method developed and used by Barnett (1989) and Stokes et al. (1990), which included two tasks designed to gauge the pilot's ability to construct an accurate mental representation of operations within the airspace. The Situation Recognition Task involved listening to radio exchanges between air traffic control (ATC) and other pilots, building a mental picture of the situation, and selecting from sets of diagrams purporting to represent the scenario. In the ATC Recall Task, subjects listened to radio call sequences and then attempted to reconstruct them from memory. To control for individual differences in working memory capacity, hearing, and the like, half of the radio call trials followed a coherent sequence concerning takeoffs, approaches, or navigation, whereas the other half consisted of arbitrary randomized calls selected from a variety of different flight scenarios; scoring was based not on absolute values but on the difference between the two conditions. This task was designed in direct analogy to the well-known experiments reported by de Groot (1965) and by Chase and Simon (1973), which showed that chess masters were far more capable than novices of reconstructing from memory an arrangement of pieces on a board, but only when the arrangement represented a coherent game sequence. (When pieces were placed randomly, experts were no better than novices at replacing them.) In both domains, presumably, the quality of reconstruction is primarily influenced by the availability of appropriate situational schemata and attendant action scripts.

The simulation, called MIDIS (microcomputer-based decision simulator) was designed from the outset to study inflight decision making rather than stick-and-rudder control. MIDIS included a full, high-fidelity instrument panel with operating attitude, navigational, and engine instruments. The readings on the instrument panel changed throughout the course of the "flight" in synchrony with the prevailing scenario. The object was not to simulate the flight dynamics of an aircraft from control inputs, but rather to impose judgment requirements by presenting a series of time slices or "scenarios" in the course of a coherently unfolding flight. Each situation thus appeared and needed to be addressed in its appropriate chronological and situational context. Scenarios consisted either of a statically configured instrument panel together with a text description of a particular inflight circumstance, or of a dynamic (i.e., moving) instrument panel alone. The latter type made it possible to study decisions involving the detection of changes and the integration of decision cues in real time.

After viewing a scenario, subjects pressed the return key to request the decision options (in multiple-choice scenarios) or input format screens (in open-ended scenarios). Multiple-choice scenarios provided between four and six decision options that had been prerated according to an optimality scale ranging from 1 to 5. The open-ended scenarios, in contrast, required subjects to list the cues identifying the presence and nature of a problem. These took the form of brief descriptors along the lines of "RPM decreased," "preflight weather briefing consistent with icing," or "355 radial unobtainable on either VOR." Subjects were then asked to list all plausible action alternatives (up to 10), and to identify the best of these alternatives, giving a rationale for the choice. These took the form of brief action statements, for example, "Turn on pitot heat and carburetor heat," "Descend to 8000 feet," or "Divert to alternate airport." After indicating their decision, subjects also indicated their level of confidence in that response on a 5-point scale. Evaluations of decision quality were provided by three independent experts, and interrater reliabilities were high (between .82 and .92; p < .001). Although this could justify the collapsing of the different judges' ratings together, they were, in fact, examined separately for each judge to provide a more stringent test of the data as well as a built-in double replication.

The simulated flight was an extended test requiring considerable concentration. Subjects worked alone for about 2 hours in a small, cockpit-like booth with subdued lighting. Subjects were asked to treat the experiment as a real flight and not to leave the booth until after "landing." Altogether, 51 scenarios were presented in the course of the flight. In addition to those static and dynamic scenarios that actually involved a problem, the flight incorporated nonproblem episodes, preserving some of the natural characteristics of a normal flight. Every effort was made to avoid making the flight an implausible assemblage of unlikely catastrophes; rather, the intention was to simulate a realistic (if eventful) experience. Although no artificial time limits or cutoffs were introduced into the pilots' task (see Orasanu & Strauch, 1994), the natural time pressure inherent in flying was enhanced in two ways. First, in addition to the flat fee paid to participants, pilots were made aware of a bonus for efficiency and safety, as indexed by elapsed time and average decision optimality. Second, subjects were made acutely aware of the passage of time (and, therefore, the erosion of the bonus) by the presence on screen of a small stopwatch timer. This automatically appeared as soon as a decision problem was addressed and indicated the number of seconds spent on the problem. The timer both displayed and "beeped" the passing seconds. The timer was reset for each scenario.

RESULTS AND DISCUSSION

On the domain-independent tests of information-processing abilities, initial analyses found, as anticipated, statistically significant differences between the two cohorts on response latencies. After factoring out age, however, the two cohorts failed to differ significantly on any of the information-processing measures—a result consistent with that reported by Stokes et al. (1990). Thus, the evidence favors the view that the two cohorts were essentially equal in basic domain-independent information-processing abilities, and that group differences in criterion performance (that is, the flight simulation) cannot be attributed to age or inherent differences in ability.

As regards the LTM-based knowledge representation measures, consistent with previous studies (Barnett, 1989; Stokes et al., 1990), there were no statistically significant differences between the two groups on declarative knowledge of instrument flight regulations and procedures as measured standard FAA questions. However, high-time pilots scored significantly better than low-time pilots on the two procedural measures: the ATC Situation Recognition Task (7.8 vs. 5.8; $p < .002$) and the ATC Recall Task (see Table 18.1). On the latter test, high-time pilots were able to generate a concept recall score over twice as great as that observed for novices.

TABLE 18.1

ATC Recall Task—Concepts and Words Recalled (Difference) Scores

	Concepts	Words
High-time pilots	1.29	3.25
	[t (24) = 2.89; p < .008]	[t (24) = .27; p < .893]
Low-time pilots	.58	3.01

Interestingly, the word-recall score in this test was similar for both groups. However, the novices recalled a higher proportion of prepositions, conjunctions, and the like, whereas the high-time pilots tended to leave out the less significant "filler" words and remember only content-loaded key phrases. These results lend support to the notion that high-time pilots are better able to make practical use of situational schemata to impose form on sensory data in real time.

On the flight simulation, high-time pilots significantly outperformed the low-time pilots both on multiple-choice and open-ended decision problems. To understand where the difference lay, mean optimalities were calculated and t-tests were performed for each of the three scenario content-analysis categories: memory, declarative knowledge, and spatial problems. Performance by both groups of pilots was best on problems rated as highly dependent on declarative knowledge, second best on scenarios imposing high working memory load, and worst on spatially demanding scenarios. Nonetheless, the between-group differences in each category are dramatic, with the high-time group scoring approximately 15%–16% better (see Table 18.2).

The difference in performance on scenarios rated high in demand for declarative knowledge is somewhat surprising, given the results of the declarative knowledge pretest. It would appear that the application of "ground school" knowledge to a real-time situation is more difficult for the low-time pilot. Considered in conjunction with the other results of this study, this effect suggests that textbook learning may be poorly integrated into the mental models accessed by student pilots as they fly, remaining as "compartmentalized" facts to be invoked by exams rather than events.

The difference in scores on problems rated high in memory demand is

TABLE 18.2

Open-Ended Response Optimalities by Scenario Category

	High-Time Pilots	Low-Time Pilots	Significance
Declarative knowledge	.70	.57	t (24) = 6.25; p < .001
Memory	.67	.51	t (24) = 7.01; p < .001
Spatial awareness	.61	.45	t (24) = 4.96; p < .001

consistent with the hypothesis that low-time pilots are forced to integrate cues in STM, whereas the high-time pilots are able to match the cues rapidly and directly with a familiar situation in LTM. According to Stokes's (1991) model, both groups of pilots attempt to find a pattern match in LTM—the low-time pilots simply fail more often than the high-time pilots. Under this premise, low-time pilots perform less optimally because they are forced to integrate cues and generate an original solution using processes in working memory susceptible to time pressure and perceived stress. Those problem scenarios rated as being highly demanding in spatial skills appear to be the most challenging, especially for the low-time pilots. Although high-time pilots were no different from low-time pilots in non-domain-specific tests of spatial ability, they were clearly better on domain-specific spatial tasks. The results of the ATC situational awareness tests reported earlier are consistent with this finding. Presumably, spatial templates can be recalled from LTM for familiar situational schemata, and high-time pilots have these spatial schemata more readily available.

It was also found, in analyzing the superior performance of high-time pilots, that members of this group identified significantly more cues that were actually relevant to the situation (see Table 18.3). Apart from cohort membership, the best predictor for detection of relevant cues was the total amount of time used to make decisions throughout the flight. Novices, in contrast, reported not only fewer relevant cues but more irrelevant ones, both in absolute terms and relative to the overall number of cues reported. However, the difference was not statistically significant. These results indicate that detection of relevant cues is best accomplished by highly certificated pilots who invest time in reaching their decisions, and that such pilots are also more efficient than novices at confining their attention to relevant considerations.

In addition to identifying more relevant cues, high-time pilots also generated 30% more action alternatives than their less experienced colleagues. Nevertheless, consistent with the satisficing strategy posited for this group, high-time pilots were also far more likely to select and carry out the first alternative that they had considered. Indeed, for those trials where multiple options were considered, high-time pilots chose their first response 71% of the time, whereas low-time pilots selected their first response only

TABLE 18.3

MIDIS Performance—Relevant and Irrelevant Cues

	Relevant Cues	Irrelevant Cues
High-time pilots	15.4	9.9
	[t (24) = 4.20; p < .001]	[t (24) = 1.21; p < 237]
Low-time pilots	8.9	12.8

53% of the time—a very large difference. Although this is itself impressive, we noted that it actually underestimates the extent to which high-time pilots tended toward their first response relative to the low-time group. As we have seen, high-time pilots were much more likely to cite multiple alternatives than the low-time group, and therefore had a greater chance of picking a solution other than the first one listed. Conversely, low-time pilots were sometimes unable to generate more than a single course of action, such that this, their first and only option, was necessarily the one that was selected. A weighting scheme developed to correct for this artifact yielded mean scores for the high-time pilots that were nearly twice that of the low-time group. However, even this method probably underestimated the strength of the effect: The high-time pilots generated 30% more total alternatives than their less experienced colleagues, and it is clearly much more significant if individuals tend to select the first of 10 alternatives than the first of only 2.

In addition to the between-group analyses, a stepwise multiple regression analysis was conducted to determine which variables, including the various psychometric measures, were most predictive of performance on the MIDIS simulated instrument flight. The best single predictor of pilots' decision performance proved to be the level of pilot certificate and ratings held (Private License, Instrument Rating, Commercial License, etc.); this measure alone was so strong that it was able to account for well over half the variance in decision optimality. Within the necessarily limited confines of the experiment, neither age nor hours of experience were found to predict the quality of flight decision making. Thus, this study provided no evidence that better decision making comes automatically with advancing years or the raw accumulation of flight hours.

If knowledge of the pilots' certification level was excluded from the analysis, the best predictor of decision-making performance was the number of relevant cues detected in problem scenarios during the flight. This variable itself was able to account for about half of the total variance in decision-making optimality scores (depending on the judge). However, if only psychological variables were considered, spatial memory and LTM-based knowledge representation measures were the most predictive.

The two procedural knowledge measures—the ATC Recognition Task and the ATC Recall task—correlated both with one another and with overall decision optimality. Performance on the ATC Recognition Task also correlated significantly with the number of relevant cues detected (as noted, the second-best predictor overall), although the ATC Recall Task did not. Finally, the high-and low-time groups were analyzed separately, and it was found that domain-independent spatial memory measures were the best predictors of decision-making proficiency among high-time pilots. Among low-time pilots there was less consistency from judge to judge in the variables that predicted decision-making skill.

CONCLUDING COMMENTS

The following observation rather neatly summarizes the claims embodied in the diagrammatic model of decision making presented in Figure 18.1: "The third revolution [in decision research], which is still in progress, comes from recognition that choices occur relatively rarely; that past experience usually provides ways (policies, habits) of dealing with problems. Decisions are required primarily when these solutions fail" (Beach & Potter, 1992, p. 115).

The outcomes of the experiment are consistent with these hypotheses. They also support the notion that choices may be less ubiquitous than is sometimes assumed. Expert pilots often assert that "there was only one choice," and, of course, there was a clear tendency among expert pilots in this study to persist with the first action alternative identified. There is, however, the potential for conceptual or at least definitional confusion stemming from this work, and from formulations of the problem such as that of Beach and Potter (note especially the last line). Are we to confine "real" decision making only to the selection of choices (via slow inferential processes in working memory), excluding problem solutions more automatically arrived at via LTM-based processes? Although there is no shortage of research to be done on choice, a considerable literature on option selection does exist. By comparison, relatively little is known about how and when experience-based "solutions" succeed or fail, or about what (among apparently equally experienced pilots) determines individual differences in the rapidity and effectiveness with which experiences are applied to novel situations. There are a number of complex possibilities here (the use of analogies, for example) that stand in need of experimental research. However, the experiment reported here casts some light on one simple hypothesis: that proficient "users of experience" abstract more information from their environment and are better able to pattern match the current situation with an appropriate experience.

The idea that experts access more environmental information than novices and that their expertise lies precisely in this quantitative difference is sometimes known as the Information–Use hypothesis. Recently, however, an analysis of five studies (in domains as distinct as livestock judging and auditing) has cast doubt on this hypothesis (Shanteau, 1992). These studies suggest that the most important factor in expert judgment may not be the amount of information that the expert uses but rather the type of information. That is, the important dimension of difference is qualitative and is concerned with the relevance and diagnostic value of the information used. Even though the flight simulation study reported here does not permit the Information–Use hypothesis to be discarded, our results also point to the importance of the qualitative distinction in the type of information utilized by experts and novices.

It has been observed that expert judgments are often characterized by the use of relatively few significant cues (Shanteau, 1992). For example, pathologists have been found to use up to four cues, medical radiologists between two and six, and stockbrokers six or seven. The present flight simulation study indicates that expert pilots, even under time pressure, can and do report rather more cues than this—a mean of over 15 relevant cues per decision problem being recorded. In contrast, novice instrument pilots were only able to cite about nine.

Part of expertise may lie in knowing which data are irrelevant—that is, what to ignore. It has been suggested that judgments may be "watered down" by individuals giving too much attention to extraneous or nondiagnostic data. Certainly, irrelevant data can provide a rich source of information to "feed" biases or inappropriate heuristics (an obvious example being the history of heads and tails in a coin-tossing game). Our results show that expert pilots reported fewer irrelevant cues than novices but not significantly so. Nevertheless, the proportion of total cues identified which were irrelevant was far larger for novices (59%) than for experts.

On the basis of these results we cannot exclude the possibility that expert pilots' superior decision making may be influenced by both quantitative and qualitative advantages in cue recognition. The precise mechanism for this superiority remains unclear, although the predictive power of the domain-independent spatial tests may be an intriguing clue. Certainly, the ability to discriminate relevant from irrelevant cues has been the subject of some speculation—see for example Neale and Northcraft (1989) on "strategic conceptualizations" in rational decisions. We, of course, posit a general mechanism based more on fairly automatic "recognition-primed" LTM processes rather than on precepts of rationality, optimality, or utility. Nevertheless, we anticipate that detailed answers to this question may be highly domain-specific, and suspect that any general conclusions must await a considerable expansion of the research database across domains.

A separate issue concerns the implications of this research for personnel selection and training. The evidence appears to be that expertise (at least in flight-management decision making), which has traditionally been measured in total hours (that is, by quantity of flight time), is better measured in terms of the quality of flight time. Pilot selection decisions should, perhaps, give greater weight to the level of certification or to other qualitative evaluations, and less to total flight hours, as the most important indicator of expertise.

The data from this analysis also suggest that pilot training could be improved by explicitly building the experiential repertoires of students. Ground school training classes could therefore be modified to include or emphasize event-based learning, in conjunction with traditional fact- and rule-based training. In expanding a student's repertoire of situational schemata through vicarious experience, relatively low-cost low-fidelity desktop "event-based simulators" (EBS) could be used. These would not necessarily look like the nonintelligent PC-based flight simulator programs that are now widely available and that concentrate on stick-and-rudder skills and instrument procedures. EBS systems may or (like the MIDIS system used in this series of experiments) may not have conventional controls; in either case they would have to expose students to events or circumstances, and structured classes of events and solutions, which might otherwise take the student years to experience and assimilate.

REFERENCES

Barnett, B. (1989). *Modeling information processing components and structural knowledge representations in pilot judgment.* Unpublished doctoral dissertation, University of Illinois, Urbana–Champaign.

Beach, L. R., & Potter, R. E. (1992). The pre-choice screening of options. *Acta Psychologica, 81,* 115–126.

Besco, R. O., Maurino, D., Potter, M. H., Strauch, B., Stone, R. B., & Wiener, E. (1994). Unrecognized training needs for airline pilots. In *Proceedings of the Human Factors and Ergonomics Society 38th annual meeting* (pp. 41–45). Santa Monica, CA: Human Factors and Ergonomics Society.

Chase, W., & Simon, H. (1973). Perception in chess. *Cognitive Psychology, 4,* 55–81.

de Groot, A. (1965). *Thought and choice in chess.* The Hague: Mouton.

Giffen, W., & Rockwell, T. H. (1987). A methodology for research on VFR flight into IMC. In R. S. Jensen (Ed.), *Proceedings of the Fourth International Symposium on Aviation Psychology.* Columbus: The Ohio State University.

Howell, G. E. (1984). *Task influence in the analytic intuitive approach to decision making: Final report* (Office of Naval Research contract N00014–82 C–001 Work Unit; NR197–074). Houston, TX: Rice University.

Janis, I. L., & Mann, L. (1977). *Decision making: A psychological analysis of conflict, choice, and commitment.* New York: The Free Press.

Klein, G. A. (1989). Recognition-primed decisions. In W. Rouse (Ed.), *Advances in man–machine systems research* (Vol. 5, pp. 47–92). Greenwich, CT: JAI.

Klein, G. A., Calderwood, R., & Clinton-Cirocco, A. (1986). Rapid decision making on the fire ground. In *Proceedings of the Human Factors Society 30th annual meeting* (pp. 576–580). Santa Monica, CA: Human Factors Society.

Neale, M. A., & Northcraft, G. B. (1989). Experience, expertise and decision bias in negotiation: The role of strategic conceptualization. In B. Shepperd, M. Bazerman, & R. Lewicki (Eds.), *Research on negotiations and organizations, Vol. 2.* Greenwich, CT: JAI.

Orasanu, J., & Strauch, B. (1994). Temporal factors in aviation decision making. *Proceedings of the Human Factors and Ergonomics Society 38th annual meeting* (pp. 935–939). Santa Monica, CA: Human Factors and Ergonomics Society.

Rouse, W. B. (1978). Human problem solving performance in a fault diagnosis task. *IEEE Transactions on Systems, Man and Cybernetics, 4* (SMC–8), 258–271.

Shanteau, J. (1992). How much information does an expert use? Is it relevant? *Acta Psychologica, 81,* 75–86.

Simon, H. A. (1955). A behavioral model of rational choice. *Quarterly Journal of Economics, 69,* 99–118.

Stokes, A. F. (1991). Flight management training and research using a micro-computer flight decision simulator. In R. Sadlowe (Ed.), *PC based instrument flight simulation—A first collection of papers* (pp. 47–52). New York: American Society of Mechanical Engineers.

Stokes, A. F., Banich, M. T., Elledge, V., & Ke, Y. (1991). Testing the tests: An empirical evaluation of screening tests for the detection of cognitive impairment in aviators. *Aviation, space and environmental medicine, 62*(7), 783–788.

Stokes, A. F., Barnett, B., & Wickens, C. D. (1987). Modeling stress and bias in pilot decision-making. In P. Rothwell (Ed.), *Proceedings of the Human Factors Association of Canada, annual conference* (pp. 45–48).

Stokes, A. F., Belger, A., & Zhang, K. (1990). *Investigation of factors comprising a model of pilot decision making: Part II. Anxiety and cognitive strategies in expert and novice aviators* (Tech. Rep. No. ARL–90–8/SCEEE–90–2). Savoy: University of Illinois Aviation Research Laboratory.

Stokes, A. F., & Kite, K. (1994). *Flight stress: Stress, fatigue and performance in aviation.* Aldershot, UK: Ashgate.

Telfer, R. (1987). Pilot judgment training: The Australian study. In R. S. Jensen (Ed.), *Proceedings of the Fourth International Symposium on Aviation Psychology* (pp. 265–293). Columbus: The Ohio State University.

Wickens, C. D., Stokes, A., Barnett, B., & Davis, T., Jr. (1987). *A componential analysis of pilot decision making* (Tech. Rep. No. ARL–87–4/SCEEE–87–1). Savoy: University of Illinois, Aviation Research Laboratory.

Zakay, D., & Wooler, S. (1984). Time pressure, training and decision effectiveness. *Ergonomics, 27,* 273–284.

Chapter 19

Capturing and Modeling Planning Expertise in Anesthesiology: Results of a Field Study

◆

Yan Xiao
University of Toronto
and
University of Maryland School of Medicine

Paul Milgram
University of Toronto

D. John Doyle
The Toronto Hospital

Naturalistic studies of problem-solving behavior have several specific missions in comparison with their counterparts carried out in simulated environments such as laboratories. As an end, they are to provide knowledge to designers on what strategies are likely to be adopted in the real world, what problems actually face practitioners, and what training and cognitive supports would be desirable (Bainbridge, 1983; Klein, 1989; Reason 1987). As a means, they are to provide laboratory investigators needed guidance to formulate experimental designs that cover the range of problems facing practitioners (Sheridan & Hennessy, 1984).

Although a list of comparisons can be easily made between a typical laboratory experiment on human problem-solving and what people typically do in a given real-world situation (e.g., Meacham & Emont, 1989), inevitably one would find that two of the critical elements characterizing naturalistic behavior are the *integration* of problem-solving with other activities, and the *continuity* of activity over time: (a) problem-solving activities rarely occur on their own right, but are embedded in some larger context, and are in integration with other activities, most notably physical and perceptual contact with the work environment; and (b) problem-solving activities occur continually in time, with cognitive efforts at any time related to the anticipation of future events and activities.

This paper describes a field study in which special attention was paid to these two attributes of natural problem-solving behavior. We investigated how practitioners planned, in terms of how they prepared for future events and tasks, and how such preparations influenced subsequent activities. Our emphasis on behavior over a continuum of time was not only the result of seeking ecological validity[1], but also the result of the exploratory phase of the field study.

The field study took the domain of anesthesiology as a "laboratory" for studying how proficient practitioners solve problems in real-life work settings. In the following, we focus on the findings of the field study and the question of how to capture and model the expertise involved in preparatory activities. For treatments of other issues related to the field study, refer to Xiao (1994).

METHOD

Domain description

Anesthesia is a procedure to suppress patients' abilities to feel, to move, and/or to remember. It is used in a wide range of medical procedures, mostly to provide suitable patient conditions for surgery. In the majority of surgical cases, anesthesiologists induce patients into an unconscious state prior to the beginning of the surgery, maintain that state during the course of the surgery, and then terminate the process at the end. During anesthesia, it is primarily the anesthesiologist's responsibility to ensure the patient's well-being and to compensate for the effects of surgery and of the anesthesia itself, be they intentional or accidental.

---

[1] As Neisser argued, temporal continuity is one of the aspects essential for ecological validity.

Data collection and analysis

The field study was conducted in a teaching hospital and spanned a period of four years. A staff anesthesiologist acted as the main informant. During the study about 40 cases were observed, 8 of which were audiotaped. A total of 30 hours of audio-recordings were transcribed, and over 800 episodes were extracted and analyzed. For those audiotaped cases, the anesthesiologists were requested to give a report before the case, to "think aloud" and answer probing questions during the case, and to comment on the transcription made from the on-line recordings afterwards, usually within a week. Each episode was analyzed in terms of what was the situation, what was done, and what happened. Then the physical and inferred mental activities were classified, and further inferences about the associations among events, mental states, and activities were drawn (Xiao, 1994).

Discussions of 10 previous cases during four case rounds were audiotaped and analyzed. These case rounds were part of the resident training program and were attended by staff anesthesiologists and residents.

RESULTS

The initial phase of the field study brought to light an important class of problem solving activities, which occurred when anesthesiologists actively anticipated problems and took measures either to prevent troublesome situations from arising, or to prepare for them. In the following, an example is used to illustrate some of the preparatory activities observed in the field study.

In one case, the patient was found to have a history of uncontrolled hypertension prior to the surgery, and the anesthesiologist had anticipated large swings of blood pressure during and after the induction (the process of putting the patient into an unconscious state). She planned to use lidocaine to block the patient's response to the induction, and to use nitroglycerin if the blood pressure was still too high after using lidocaine. She proceeded to prepare the nitroglycerin infusion bag and set the bag up so that, if she needed to use the drug, all she had to do was to plug it into an intravenous port. Note that all these activities occurred *before* the surgery had started. During the course of induction, the patient's blood pressure was high despite a larger than usual dose of anesthetic. The patient was treated with lidocaine and responded well to the treatment. The nitroglycerin infusion bag was never used.

In this example, the prior preparation had provided the anesthesiologist with guiding rules and the needed materials. Even though there were many other ways (e.g., using vasodialating drugs) in which she could have treated the same problem, in the end the task to respond to the anticipated problem of hypertension turned out to be a matter of "match-and-trigger." The cues she looked for were the easily available blood pressure readings, and the needed materials and access were all ready. The scale of the decision problem was considerably smaller than it would have been if the anesthesiologist had to evaluate factors such as drug complications, patient allergies, the medication history, cost and availability of drugs, and so forth, all under time pressure. She would also have had to evaluate readings in addition to the blood pressure monitor to make a situational assessment. Thus the seemingly simple, reactive response was in fact the result of preparation. Further, the response to the problem of high blood pressure observed was in fact the result of preparation, and the response had started before the patient was even brought into the operating room, long before the "problem" actually occurred.

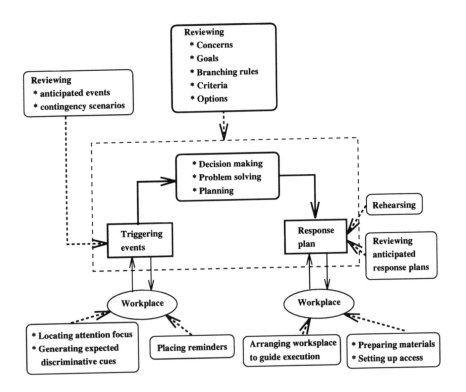

FIG. 19.1. A framework for preparatory activities. Labels outside the dashed rectangle in the center indicate eight categories of preparatory activities summarized from the behavioral patterns observed in the field study. Previous research has paid attention mostly to how people *react* to events after problematic situations arise, and to the cognitive activities inside the dashed rectangle.

During the course of attempting to understand the data collected from the field study, a framework (see Fig. 19.1) was devised for capturing various strategies associated with preparatory activities. Note that some of these preparatory activities are physical, while others are mental. The framework encompasses the major findings and highlights preparatory activities, along with their impact on the practitioner's performance.

In the following, the major findings of the field study are summarized in terms of two conceptual workspaces: mental and physical, which the practitioners can tag and re-organize.

Preparing the Mental Workspace

The activities comprising preparation of the mental workspace were not directly observable. However, the results of such activities can be viewed as a mental state of readiness for responding. One type of preparatory activity was associated with identifying a list of *points for consideration* (PFCs). This type of activity was most obvious during the pre-case interviews, at which time the anesthesiologists reported major potential obstacles to the success of a case. (Anesthesiologists term such PFCs *anesthetic considerations*) Concerns include special conditions in the patient's physiology, expected or potential non-routine events during the surgery, expected or likely troublesome scenarios, obstacles to routine procedures, and, perhaps most importantly, errors that one is prone to commit. The concern list functions as a set of "warning flags" that guide the anesthesiologist's attention in a dynamic, multi-task environment and for achieving multiple and sometimes conflicting goals. The preparation of the concern-list often led to the formulation of contingency response plans for those troublesome situations which were either anticipated or likely to occur. It could also lead to decisions about the general approach to a case, about situations to be avoided, and about solutions which should not be applied or were not implementable.

During the surgical operation, in comparison, the scope of anticipation and preparation was on a short-term basis, involving the monitoring and controlling of the patient's status, and the execution of planned actions. One type of preparatory activity was associated with the formulation and use of local rules, which were ad hoc rules that could be used to control and monitor the patient's status. For example, in one case of a bone marrow harvesting procedure, the anethesiologist used blood pressure readings to guide fluid infusion, although more invasive monitoring techniques (e.g., central venous pressure monitoring) are usually used for this task.

Preparing the Physical Workspace

In many situations, the contingency responses require not only mental readiness, but also readiness in terms of materials, because a good response plan may not be realizable, simply because of a lack of materials. Some materials (e.g., intravenous drugs that require dilution) may be available but require complicated and time consuming setup procedures before use.

At the beginning of each case, the anesthesiologist configures the workplace by assembling and arranging materials needed (e.g., syringes, endotracheal tube, laryngoscope, etc.) for busy periods and potential emergencies. Anesthesiologists were also found to place "triggering cues" for planned actions into their workplaces, such as placing unusual but easily forgotten drugs in a prominent place on the drug cart as reminders.

Properties of the Preparatory Activities

Some general properties of anesthesiologists' preparatory activities were identified in the field study. Three prominent ones are summarized in the following.

First, evidence from the pre-operative interviews and the case round discussions suggested that only a small number (often three or four) of aspects of a case were examined and prepared for. The mental preparation seemed to be *fragmentary*, not systematic, and dealt with only a small fraction of all possible aspects of a case. In addition, the decisions made at the time of preparation were frequently non-specific (e.g., using less anesthetic without deciding on the exact dosage).

Second, not only did the preparation seem fragmentary, but it also appeared to be concerned more with the *identification* of problems than with the *solution* of problems. In case round discussions, for example, it was often true that proper identification of concerns alone (e.g., difficult intravenous access) was sufficient, without actually offering any solutions. How these types of mental preparation affect behavior in the operating room thus became an interesting question for the field study. One of the subjects offered the following comments during a post-case interview: "If you know the patient has aortic stenosis, if you observed hypotension, you'll give epinephrine faster. You'll be very concerned and intervene early. On the other hand, I might have a plan to prevent in the case to develop hypotension." According to this subject, preparation primed him for quick responses and might lead him to prevent anticipated troublesome situations from occurring.

Third, the case rounds (attended by both staff and resident anesthesiologists) showed that anesthesiologists could quickly identify potential obstacles and prescribe precautions for executing selected action plans. This finding suggested that, in examining an action plan, experienced practitioners could identify likely trouble spots in the planned solutions effectively, and probably without detailed mental simulations.

Findings represented at different levels of abstraction

The major findings of the field study can be represented at three different levels of abstraction (see Table 19.1; Hollnagel, Pedersen, & Rasmussen, 1981). At the bottom level, the findings are represented by domain strategies (or "performance fragments," as defined by Hollnagel et al., p. 10), which can be detected directly in the data collected from the field. However, these findings are difficult to generalize to domains outside anesthesiology, and they do not reveal the underlying cognitive activities.

After removing the domain context and adding cognitive descriptions, the domain strategies are represented by specific strategies. These strategies are no longer context specific, although they can be "illuminated" by examples in empirical data. Findings represented at this evel have direct implications in terms of design and training.

At the top level, specific strategies can be synthesized into a single generic strategy. It represents the essence of the preparatory activities observed in the field study, and reveals a fundamental characteristic of the interaction between a proficient worker and a complex, dynamic task environment. Identification of such a generic strategy can also direct the search for other types of specific strategies.

MODELING PREPARATORY ACTIVITIES

As noted earlier, the mental preparation observed in the field study seemed fragmentary and occurred in various forms, ranging from general decisions, concerns, and overall action plans, to specific procedures and actions. How

TABLE 19.1

Findings Represented at Different Levels of Abstraction

Levels of Abstraction	Findings
Generic strategy	• Reduce response complexity through anticipating future situations, mental preparation, and reorganizing the physical workspace.
Specific strategies	• Schedule tasks (off-load). • Build local models and rules. • Be preventive: Think of probable side effects, pitfalls, and predictors of these side effects. • Prepare necessary materials and access. • Rehearse impending procedures (mental simulation)
Domain strategies	• Prepare induction and emergency syringes. • Pay more attention to muscle relaxation. • If blood pressure fluctuates too wildly, start nitroglycerin infusion. • Prepare nitroglycerin whenever there is a chance. • Tape vaporizers to prevent the use of vaporizers (in a total intravenous anesthetic). • Use only short-acting drugs if surgery duration is uncertain.

then does mental preparation influence subsequent interactions with the environment?

To explain the role of preparatory activities, we outline a conceptual model based on Rasmussen's (1976) action control framework, the decision ladder. The decision ladder framework captures a central characteristic of human information processing: an experienced practitioner can utilize experience to "short-circuit" information processing paths. High-level goals, for example, are not always involved in making decisions. This flexibility in the process of decision making implies that, whenever a solution is not readily available, high-level goals have to be consulted and options have to be evaluated in some way. All of the components in the decision ladder (e.g., short-cuts, options, goals, branching rules, procedures, etc.) are thus important or potentially important, in making decisions and controlling actions, and they can be viewed as *action resources* for the practitioner to use in the interaction with the environment. The decision ladder can therefore be utilized as a way to represent action resources. Preparatory activities, or planning, can thus be viewed as the process of preparing action resources (as opposed to that of generating sequences of actions), and the role of planning is to reduce the response complexity (as opposed to determine action sequences) in the subsequent interactions with the task environment.

DISCUSSION

In the management of complex systems, unexpected crises that require on-site "inspiration" do occur. However, "preparation" precedes most crisis situations. How well one anticipates and prepares for problematic situations has significant impact not only on the frequency and the nature of crisis situations, but also on ways of handling them. The field study reported here showed that, when behaving in a naturalistic environment and when the consequences of performance failure are costly, practitioners *can* and *do* anticipate future situations and prepare for them. These findings were also supported by previous studies in complex work domains (e.g., Amalberti & Deblon, 1992).

To date, most training research has been concerned primarily with procedures and diagnosis (see the review by Cannon-Bowers, Tannenbaum, Salas, and Converse, 1991). Similarly, research on the design of decision aids also reveals a lack of attention to preparatory activities (e.g., Wiener & Curry, 1980). The current field study indicates that timely identification of concerns is often more valuable to practitioners than solutions, and that practitioners need more support in identifying potential problems than in solving them. Further studies should therefore be conducted to investigate what types of critical skills and knowledge are involved in anticipating and preparing.

In summary, the field study highlighted the role of anticipation. It showed that key components in expertise are anticipation and preparation. It is expected that the principal findings of this field study, as listed in Table 19.1, will provide preliminary guidelines, both general and specific, on how to support and train practitioners.

ACKNOWLEDGMENTS

The study was supported by a doctoral fellowship from IBM Canada awarded to the first author and research grant #OGP0004101 from NSERC awarded to the second author. The revision of this paper was partially supported by a grant from the Office of Naval Research, #N00014-91-J-1540.

REFERENCES

Amalberti, R., and Deblon, F. (1992). Cognitive modeling of fighter aircraft process control: A step towards an intelligent onboard assistance system. *International Journal of Man-Machine Studies, 36,* 637–671.

Bainbridge, L. (1983). Ironies of automation. *Automatica, 19,* 775–779.

Cannon-Bowers, J. A., Tannenbaum, S. I., Salas, E., & Converse, S. A. (1991). Toward an integration of training theory and technique. *Human Factors, 33,* 281–292.

Hollnagel, E., Pedersen, O. M., & Rasmussen, J. (1981). *Notes on Human Performance Analysis.* Tech. Rep. No. Riso–M–2285. Riso National Laboratory, Roskilde, Denmark.

Klein, G. A. (1989). Recognition-primed decisions. In W. B. Rouse (Ed.), *Advances in Man-Machine Systems Research* (Vol. 5.; pp. 47–92). Greenwich, CT: JAI.

Meacham, J. A., & Emont, N. C. (1989). The interpersonal basis of everyday problem solving. In J. D. Sinnott (Ed.), *Everyday problem solving.* New York: Praeger.

Neisser, U. (1976). *Cognition and reality—Principles and implications of cognitive psychology.* New York: W. H. Freeman.

Rasmussen, J. (1976). Outlines of a hybrid model of the process plant operator. In T. B. Sheridan & G. Johannsen (Eds.), *Monitoring behavior and supervisory control.* New York: Plenum.

Reason, J. (1987). Cognitive aids in process environments: prostheses or tools. *International Journal of Man-Machine Studies, 27,* 463–470.

Sheridan, T. B., & Hennessy, R. T. (Eds.). (1984). *Research and modeling of supervisory control behavior: Report of a workshop.* Washington, DC: National Academy Press.

Wiener, E. L., & Curry, R. E. (1980). Flight deck automation: Promises and problems. *Ergonomics, 23,* 995–1012.

Xiao, Y. (1994). *Interacting with complex work environment: A field study and a planning model.* Unpublished doctoral thesis, University of Toronto, Toronto, Ontario, Canada.

Chapter 20

Decision-Making Errors Demonstrated by Experienced Naval Officers in a Littoral Environment

◆

Susan G. Hutchins
Naval Command, Control and Ocean Surveillance Center
and
Naval Postgraduate School

Despite the fact that the primary design criteria for U.S. warships are based on supporting a major conflict on the high seas, littoral (i.e., near land) scenarios are expected to represent a major portion of future Naval operations. Such situations are characterized by unique problems and system stresses. Such operations must be conducted in spite of: uncertain motivations of various forces in an area; complex rules of engagement; and the presence of friendly, neutral, and hostile shipping and aircraft. Moreover, the risk of damage to friendly and neutral ships and aircraft must be held very low, despite possible ongoing hostile activities. The confined and congested water and air space occupied by friends, adversaries, and neutrals can make identification profoundly difficult. Often, the decision maker has to make a high-risk decision based on ambiguous information in a volatile, uncertain environment under tight time constraints.

Decision-making errors in stressful, complex environments are often due as much to ill-designed human–system interfaces as to human error. This has been well documented in other fields by both the popular and technical literature (Buck, 1989; Casey, 1993; Norman, 1988; Perrow, 1984; Wilson & Zarakas, 1978). "Human errors account for a growing share of death and misfortune in our skies, waterways, workplaces, and hospitals" (Casey, 1993, p. 9). Numerous examples have been documented of instances where "the very system designed to increase the safety, reliability, or effectiveness, (e.g., a flight protection system for an aircraft, or oversensitive alarms) inspire overconfidence or complacence and contribute to the probable cause of accidents" (p. 104). New technologies will succeed or fail based on our ability to minimize these incompatibilities between the characteristics of the things we create and the way we use them. There are many well-documented instances of critical systems or parameter changes going unnoticed or unheeded because the operating procedures, or the human machine interface, provided no historical trace. For example, an unnoticed increase in an aircraft's altitude contributed to the shoot down of an Iranian airbus—when the team mistakenly believed the aircraft to be descending—because there was no historical trace to make the aircraft's actual *increasing* altitude apparent (Roberts & Dotterway, 1995).

The Tactical Decision Making Under Stress (TADMUS) program is being conducted to apply recent developments in decision theory and human–computer interaction technology to the design of a decision support system (DSS) for enhancing tactical decision making under highly complex conditions. Data was collected in the Decision-Making Evaluation Facility for Tactical Teams Laboratory to establish a baseline on command-level decision-making performance. During the baseline phase of testing, a detailed understanding was developed of the cognitive processes underlying the various tasks involved in tactical decision making—and where the bottlenecks occur. This understanding was then used to design the way the information is presented to the user in order to facilitate performance of the required tasks. Our goal was to design a DSS based on (a) an understanding of the cognitive strategies people use when dealing with the types of decisions required in tactical decisionmaking and (b) incorporating human–computer interface design principles that are expected to help compensate for human cognitive processing limitations.

A case has been made that previous generations of decision support systems, which focus primarily on solution optimization and base decision support on normative models of human decision making, are less applicable than a decision support system that parallels the cognitive strategies used by domain experts in situations characterized by time-constrained situations with uncertain and ambiguous data (Smith & Grossman, 1993). Additionally, traditional decision support systems were primarily used for option generation and evaluation, rather than in situation assessment. The TADMUS program has adopted the position that decision aiding systems should assist in the decision-making *process*, and focus on aiding the situation assessment portion of the decision-making task.

METHOD

Subjects

This study focused on the command-level decision makers of an antiair warfare team—the commanding officer and the tactical action officer. Subjects in the study were 12 active duty Naval personnel; some were from training commands; whereas others were from operational commands aboard ship or assigned to group staffs. Rank ranged from LT ($n = 3$), LCDR ($n = 3$), CDR ($n = 2$), through CAPT ($n = 4$).

Materials

Two weeks prior to the experimental session, a preexperimental package was mailed to the subjects to allow review of the following scenario materials: (a) overview of the experiment; (b) political/military background for the scenarios; (c) intelligence summary; and (d) rules of engagement. A quantified set of specific rules of engagement was developed for use by teams engaging in TADMUS scenarios. These rules of engagement allow us to meet potentially contradictory goals of achieving experimental control while maintaining operational realism. The rules of engagement provide time and accuracy measures against which observed performance can be compared with and without the decision support system tools in place. Some subset of the rules of engagement apply to each of the contacts of interest in all scenarios.

A situation awareness data collection form was administered once during each scenario at a predetermined, unannounced time. Data was collected at a point in the scenario when subjects had received most of the information they needed to form their situation awareness. The commanding officer and tactical action officer were asked to list all contacts of interest and to categorize them in terms of identification (i.e., enemy, assumed enemy, unknown, neutral, or friend) and intent (i.e., attack, harass, reconnaissance, no factor, or undetermined), and to list the factors that were most valuable in forming their assessment of each track. Subjects were given 5 minutes to complete the form and were asked to complete the forms independent of one another.

Data collected with this measurement instrument represents a "snapshot" of the command-level decision-maker's assessment of the tactical situation at a critical point in each scenario. For analysis, responses of the commanding officer and tactical action officer were compared with each other within the scenarios and these data were compared with "ground truth" (i.e., the known identification and intent categories for each contact). Types of questions which can be answered with these data include: Do the commanding officer and tactical action officer share the same assessment of what is occurring in the scenario at a given time? How close is this snapshot of the command-level decision-makers' assessment to ground truth? What factors do they use to form their assessments?

Design and Procedure

Data was collected in the Decision-Making Evaluation Facility for Tactical Teams Laboratory, a six-station test-bed environment that simulates console positions in a Navy Aegis cruiser combat information center. Four stations were filled by confederates (active duty Navy personnel) who play antiair warfare support-team member roles. These roles include the antiair warfare coordinator, identification supervisor, tactical information coordinator, and electronic warfare supervisor. (For a detailed description of the Decision-Making Evaluation Facility for Tactical Teams Laboratory, see Hutchins, 1996.) After approximately 1½ hours of orientation to the laboratory, and training in the use of the computer consoles, the subjects engaged in four test scenarios. The scenarios each lasted about 25 minutes and contained between 11 and 18 contacts of interest per scenario, in addition to numerous background contacts.

Team communications were recorded on a multichannel audio recorder; these included all intrateam exchanges, as well as all communications with simulated offship personnel. Audio tapes were used to produce verbatim transcripts of all team communications. A modified version of the TapRoot® Incident Investigation System (Paradies, 1991; Paradies & Unger, 1991) was then applied to identify errors. The TapRoot® System is a systematic technique for identifying anomalies in information flow and it provides an approach for analyzing a team's cognitive processing and decision making under the conditions posed in the TADMUS experiments. The resultant list of types and frequency of different error categories were then used to provide guidance for the development of the decision support system. For a more detailed presentation of the method see Hutchins and Kowalski (1993) and Hutchins and Westra (1995).

The objective was to identify tactically significant errors committed during the scenario. The following criteria were used for counting an error as tactically significant: (a) loss of tactical awareness, (b) failure to take definitive action when within the weapon's range of an approaching contact, or (c) a violation of rules of engagement. Video recordings were made of the commanding officer and tactical action officer computer screens. Detailed analyses of all audio and video recordings were conducted using the TapRoot® System.

RESULTS

These complex, time-constrained, decision-making situations, where many contacts need to be classified and processed under conditions of high time pressure and uncertainty, resulted in a large number of decision errors. The mean number of tactically significant errors documented across six teams and the four scenarios was 16 errors per scenario; the number of errors per scenario ranged from 9 to 22. The standard deviation was 3.7. The ordinal agreement between three raters (navy subject matter experts) on error counts from TapRoot® analysis was computed. Results showed a high degree of agreement with the Kendall's W of .93 indicating that 93% of the possible rank variance is accounted for.

Categorization of Errors

The antiair warfare tactical decision-making process has been conceptualized as involving the five decisions listed in Table 20.1 (Kaempf, Wolf, Thordsen, & Klein, 1992). These decisions are not always performed in the order listed; rather, they are usually iterative. Table 20.1 depicts the percentage of different types of errors, which were documented by using the TapRoot® analysis process, categorized by the type of decision involved. An example of an error categorized under the category of recognizing a problem involved a failure to take appropriate actions. Two contacts were within the 30-nautical-mile rules of engagement limit, yet no action had been taken. An example of an error categorized under the category of determining intent involved a situation where the tactical action officer lost the tactical picture. He failed to stop the tactical information coordinator from initiating tornado procedures (an unnecessary and potentially provocative act) because he was unable to recall prior information related to this contact (e.g., it had been evaluated as a probable commercial aircraft earlier). This task requires memory of past, as well as current events, to maintain situation awareness.

TABLE 20.1

Percentage of Tactically Significant Errors
Categorized by Type of Difficult Decision

Decision Category	Percent of Total Errors
Recognizing a problem	15
Situation assessment	
Track identification	26
Determining intent	20
Course of action selection	
Avoiding escalation	24
Engaging a track	16
Totals	100

Situation Awareness Data

Table 20.2 presents the mean percentage match between the commanding officer and tactical action officer, and the mean percentage match with "ground truth" for the commanding officer and tactical action officer, for contacts regarding identification and intent. The situation awareness form was open-ended; that is, the subjects listed contact numbers and the associated information for contacts which they considered to be "contacts of interest." This resulted in situations where subjects sometimes listed different contacts on their individual forms. This explains cases where the commanding officer and tactical action officer have a fairly high match with "ground truth" yet do not have a high percentage match with each other. The most critical result to note is the low percentage of matches between the commanding officer and tactical action officer for their situation awareness. The implication is that when these two command-level decision makers do not share the same view of the tactical situation, it could result in a delay in taking action or, if the person with the less accurate view makes the final decision regarding course of action, and it turns out to be the wrong decision, the effects could be grievous. Due to the ambiguous nature of the contacts, it was decided to use a "reasonable decision maker" approach in scoring; in many cases two or three identification and intent categories were accepted as "correct." When a more stringent approach was used, scores were extremely low.

DISCUSSION

In these scenarios a large number of contacts are monitored for changes in any of several key parameters. When the decision maker is faced with several concurrent contacts of interest, all of which have numerous associated data

TABLE 20.2

Situation Awareness Scores: Means and Standard Deviations for Percentage Match for Track Identification (ID) and Determination of Intent

| | Match Between CO/TAO | | Match with Ground Truth | | | |
| | | | CO | | TAO | |
Scenario	ID	Intent	ID	Intent	ID	Intent
B	53 (27)	47 (18)	94 (7)	88 (16)	83 (11)	96 (8)
C	45 (30)	40 (25)	70 (17)	96 (8)	85 (13)	91 (11)
D	82 (11)	58 (23)	74 (22)	80 (25)	100 (0)	96 (6)
E	79 (11)	71 (17)	78 (16)	87 (19)	84 (13)	80 (19)
Overall mean	65 (20)	54 (21)	79 (16)	88 (9)	88 (9)	91 (11)

items (i.e., as many as a dozen), some or all of which may change over the course of the scenario (e.g., intelligence, active radar emitters, various kinematic parameters, etc.) memory load easily exceeds human capacity. Moreover, changing parameters will impart different interpretations to what is occurring. The cause of many of the tactically significant errors is attributed to the extremely high task demands levied on the decision maker by the scenario and the human decision maker's limited attentional resources. Many cases are also attributed to working memory limitations. Maintaining an awareness of the status of each contact and the status of many actions to be taken by the antiair warfare team—which actions have been taken and what the contact's response to the action was—severely taxes the decision maker's working memory (Hutchins, in press). The high workload entailed in the scenarios produces a high time-compressed decision-making situation.

These time-compressed decision-making situations—where attentional resources and working memory capacity are limited—do not allow the decision maker to maintain an accurate situation awareness for all tracks at any given time. The low-percentage match scores for situation awareness may be explained by the large degree of ambiguity inherent to these scenarios (albeit a very realistic level) and the limited time available to process the contacts. In the antiair warfare problem, threat assessment is particularly difficult because the available information is often incomplete or ambiguous. If the decision maker had access to all data about a contact approximately twelve variables would be used to determine identity and to infer intent. Two or three of these items, alone, do not provide definitive answers because, in many cases, these parameter values do not fall within clear-cut ranges for a particular assessment category (i.e., threat, non-threat). Thus, a single time slice of information provides an incomplete picture of the situation. In the dynamic, ambiguous conditions characteristic of littoral operations, the rate and direction of change (data history) can help one better assess the threat and predict the future state of the situation. When the incoming information changes over time, the integration of information as it changes can help the user extract the message (Kirshenbaum, 1992). The DSS was designed to do precisely this: to facilitate the integration process and present a synthesized picture of the situation to the user in a format that can be quickly assimilated (Hutchins, Kelly, & Morrison, 1996).

CONCLUSIONS

I conclude by briefly discussing remedies for the errors documented in this study. They take the form of DSS modules that are being developed under the TADMUS program to aid the tactical decision-making process. In recognition of the limited cognitive resources of the decision maker, several modules were designed to reduce working memory requirements. The track history module is one example. This module was designed to support the recognition-primed decision model of decision making. For tasks involving rapid decision making (e.g., several seconds to one or two minutes) a recognitional strategy appears to be highly efficient (Klein, 1993). The objective for this module is to facilitate the track identification process by providing information that is integrated in a way that supports a recognitional decision strategy. A large amount of parametric data is synthesized and a coherent picture is portrayed graphically for rapid assimilation by the user. This module includes a geometric representation of both the contact's worst case weapon release envelope and own-ship's weapons coverage. Changes in the contact's speed, course, altitude or range are immediately apparent with this graphical depiction of the track history. This graphical representation was hypothesized to be particularly useful as in previous systems the user had to remember the previous parameter values (e.g., altitude) and compare them with the current values for those parameters. However as short-term memory degrades under stress, the user may not be able to accurately perform this function. Empirical evaluation of the DSS indicates that the TADMUS system can support users by enabling them to access critical data easily, to visualize the integral relations among data, and to manage complex response actions (Kelly, Hutchins, & Morrison, 1996; Kelly, Morrison, & Hutchins, 1996).

Features offered by the DSS to address errors attributed to limited attentional resources include focusing attention on (1) high priority contacts (i.e., track priority list and alerts), as well as on (2) missing data (e.g., basis for assessment), and (3) enabling the decision maker to use more data than is typically used in current systems (e.g., track history, comparison to normal values). Current systems require the user to retain previous contact data in memory to compare with current values for critical parameters. Current systems also require the user to rely on recall of vast amounts of information from training and experience. Presenting all known data on a contact in a synthesized way should reduce working-memory requirements and facilitate recognition. Presentation of trend and history data, as well as the basis for assessment and comparison to normal values modules, should also mitigate cognitive biases, such as "tunnel vision," where the decision maker attends to a smaller number of cues when under stress.

In summary, the decision support modules included in the TADMUS system (a) use graphics to support intuitive processes and reduce cognitive processing requirements, (b) present data organized in some form—such as in the form of the track history module to support a recognitional decision strategy, (c) construct a set of explanations for all the available evidence, (d) prompt the user regarding rules of engagement and required responses, (e) provide the ability to compare multiple hypotheses, and (f) structure and present information in a format that parallels the decision maker's cognitive strategies.

ACKNOWLEDGMENTS

Funding for the research cited in this chapter was received from the Cognitive and Neural Science and Technology Division of the Office of Naval Research. I gratefully acknowledge the many contributions of Steve Francis, Brent Hardy, Pat Kelly, C. C. Johnson, Ron Moore, Connie O'Leary, Mike Quinn, and Will Rogers in data collection, interpretation, DSS development, and technical expertise.

REFERENCES

Buck, L. (1989). Human error at sea. *Human Factors Bulletin, 32*(9), 12.

Casey, S. (1993). *Set phasers on stun and other true tales of design, technology, and human error.* Santa Barbara, CA: Aegean.

Hutchins, S. G. (in press). *Decision Making Evaluation Facility for Tactical Teams* (Tech. Rep. 1). San Diego, CA: Naval Command, Control, and Ocean Surveillance Center, RDT&C Division.

Hutchins, S. G., Kelly, R. T., & Morrison, J. G. (in press). *Decision support for tactical decision making under stress.* Proceedings of the Second International Symposium on Command and Control Research and Technology.

Hutchins, S. G., & Kowalski, J. T. (1993). Tactical decision making under stress: Preliminary results and lessons learned. In *Proceedings of the Symposium on Command and Control Research* (pp. 85–96). McLean, VA: SAIC.

Hutchins, S. G. & Westra, D. P. (1995). Patterns of Errors Shown by Experienced Navy Combat Information Center Teams. In *Proceedings of the 39th annual meeting of the Human Factors and Ergonomics Society,* (pp. 454–458).

Kaempf, G. L., Wolf, S., Thordsen, M. L., & Klein, G. A. (1992). *Decision making in the Aegis Combat Information Center* (Tech. Rep. No. 1). Fairborn, OH: Klein Associates.

Kelly, R. T., Morrison, J. G., & Hutchins, S. G. (1996). Impact of naturalistic decision support on tactical situation awareness. *Proceedings of the 40th annual meeting of the Human Factors and Ergonomics Society* (pp. 199–203).

Kirshenbaum, S. S. (1992). Influence of experience on information-gathering strategies. *Journal of Applied Psychology, 77,* 343–352.

Klein, G. A. (1993). A recognition-primed decision (RPD) model of rapid decision making. In G. A. Klein, J. Orasanu, R. Calderwood, & C. E. Zsambok (Eds.), *Decision making in action: Models and methods* (pp. 138–147). Norwood, NJ: Ablex.

Norman, D. A. (1988). *The psychology of everyday things.* New York: Basic Books.

Paradies, M. (1991). Root cause analysis and human factors. *Human Factors Society Bulletin, 34*(8), 1–4.

Paradies, M., & Unger, L. (1991). *TapRoot Incident Investigation System Manual* (Vols. 1–7). Knoxville, TN: System Improvements, Inc.

Perrow, C. (1984). *Normal accidents: Living with high risk technologies.* New York: Basic Books.

Roberts, N. C., & Dotterway, K. A. (1995). The Vincennes Incident: Another Player on the Stage? *Defense Analysis* Vol. 11, No. 1, pp. 31–45.

Smith, D. E., & Grossman, J. D. (1993). *Understanding and aiding decision making in time-constrained and ambiguous situations.* Manuscript submitted for publication.

Wilson, G. L., & Zarakas, P. (1978, February). Anatomy of a blackout. *IEEE Spectrum,* 339–346.

Chapter 21

Evidence of Naturalistic Decision Making in Military Command and Control

◆

Raphael Pascual
Simon Henderson
UK Defence Research Agency

For the past 4 years, The UK Defence Research Agency (DRA)–Fort Halstead has been conducting research into various aspects of Command and Control (C^2) decision making, examining how decision strategies are implemented by C^2 staff in a number of controlled simulated environments. One major study (Henderson, Pascual, & Fernall, 1993) involved a comprehensive survey of existing models of human decision making and potential impacting variables, to identify requirements for future research relating to C^2 decision making and assessment. That work, together with observations from previous exercises, provided the foundation for the empirical study reported here.

The program sought to demonstrate how previously identified critical factors (both human characteristics and external variables) modify process utilization within different decision-making strategies. In addition, the work sought to assess the validity and relevance of existing models of human decision making. The identified models reflected a broad range of theoretical viewpoints, including both naturalistic and classical analytical paradigms.

THE EXPERIMENTAL PROGRAM

Two experimental sessions, Exercises Shere Khan I and II, were run using commissioned and noncommissioned officers from the British Army. Fifteen subjects, who had varying degrees of experience in C^2 Operations Headquarters tasks, took part in the study. Individual decision makers were assessed in a controlled, yet militarily realistic, simulated environment. This was reflected in the provision of all the usual HQ facilities, maps, reference material, communications networks, surrounding support personnel, and in the quality and quantity of military information supplied to the subjects.

All the subjects completed three scenarios, undertaking a variety of typical problem-solving and decision tasks drawn from the C^2 Operations domain. The first scenario was a training session, to enable subjects to become accustomed to the HQ cell configuration and task regime. This was followed by one planning and one dynamic scenario. In dynamic scenarios, friendly forces are under continual enemy harassment and there exists a constantly changing battlefield situation which must be responded to rapidly. Planning scenarios involve less interaction and are characterized by a plethora of resource planning and utilization tasks, requiring significant spatial and temporal coordination under time-pressured circumstances.

In one example of a dynamic scenario, a subject was placed in command of a reserve demolition in a screen force battle. Typical decision tasks included deciding when to blow enemy bridges, considering possible routes for withdrawal, and responding to predicted enemy helicopter intentions. The decision themes usually evolved over an elapsed period of time and reflected multiple aspects of the decision process—including situation assessment, option evaluation, course of action selection, risk analysis, and so forth. These identified critical decision themes subsequently provided the focus for exercise monitoring and debriefing activities.

DATA COLLECTION TECHNIQUES

All experimental sessions were video and audio recorded, including all communications traffic between the cell under test and the control role players. Written outputs were collected, including all tables, schemata, and plans produced. This provided valuable evidence of the physical manifestations of working practices and problem-solving activities. Individual's subjective perceptions of workload were captured online, using Instantaneous Self-Assessment (ISA) workload recording devices.

During a session, subjects were requested to verbalize out loud what actions they were taking as the scenario unfolded. For example, subjects highlighted when they were sending or receiving information, planning, setting goals, or processing information. Subjects were not asked to explain the reasoning behind any of their actions or judgements. All recordings of exercise sessions were subsequently transcribed for further analysis.

The exercise sessions were monitored by teams of scientific and military staff, who filled in video event diaries, logging military events and decision-making activity of interest. These diaries were subsequently utilized in conjunction with the video recordings of exercises in postexercise debrief sessions. The aim of the debrief sessions was to understand the subjects' rationale underlying decision making and problem solving. Thus, detailed information was obtained on decision strategies used, errors made, and on the contextual factors that influenced their method of working. A modified version of the Critical Decision Method (Klein, Calderwood, & MacGregor, 1989) provided the foundation for the knowledge elicitation process. All debrief sessions were recorded, transcribed, and coded, providing a rich source of information not only on C^2 decision-making behavior, but also capturing opinions about the nature of possible computer support tools, training, and adherence to Standard Operating Procedures (SOPs).

Finally, data were also collected on individual differences, using a range of validated psychometric tests, together with a specifically designed military experience proforma. These captured information on personality, problem-solving and learning styles, visiospatial skills and individual experience levels, among many other dimensions.

FINDINGS

Coding

One of the clearest findings that emerged from the process of coding the transcripts was the difficulty in rigorously attributing a particular model to the processes described. This problem resulted from uncertainty in what to look for in subjects' responses to probes, and was compounded by the fact that many models in fact describe the same processes using alternative concept descriptors. A solution to this problem was found in the development of a Concept Translation Meta-model (Pascual, Henderson, & Fernall, 1994).

Strategy Attribution

A primary finding from the analysis of the coded responses across all scenarios was the dominance of naturalistic strategy utilization (87%) over classical (2%), hybrid (3%), and other (8%) strategies. There are a number of issues that must be taken into account when considering this finding. The scenarios were all familiar to the subjects (i.e., terrain, force structures, incursions from Eastern Europe, etc). It is believed that a more novel scenario (e.g. Bosnia, or the Gulf) would result in much greater utilization of classical or hybrid strategies, due to the lack of experience with the situation (i.e., little or no prototypicality from which to draw). Second, the subjects were largely quite experienced, and were readily able to draw on analogues, past working practices and problem-solving techniques, and so forth—the cornerstones of naturalistic decision making. Finally, it may be easier to detect naturalistic strategy utilization than classical strategies. For example, people verbalize serially, and thus cannot easily describe utilizing concurrent evaluation processes.

Model Attribution

Figure 21.1 shows the proportion of model attribution across all subjects in Shere Khan II. Recognition-Primed Decision making (RPD; Klein, 1989) clearly dominated, followed by attributions of Search for Dominance Structure (SDS; Montgomery, 1989). Although RPD-like decision-making behavior may indeed be easier to identify, code, or both, than other models, it is felt that the RPD model provides the most appropriate, accurate, and utilitarian concepts for describing a broad range of C^2 decision-making behavior, particularly for those subjects with considerable military experience.

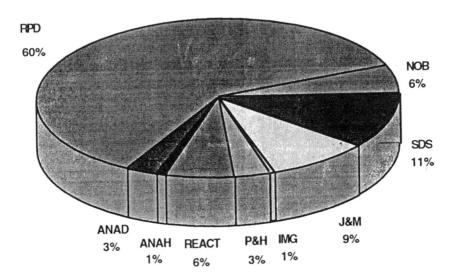

Key to model attribution

RPD—Recognition Primed Decision Making (Klein, 1987)
NOB—Schmata Model (Noble, 1986)
SDS—Search for Dominance Structure (Montgomery, 1989)
J&M—Coping Model (Janis and Mann, 1978)
IMG—Image Theory (Beach and Mitchell, 1987)
P&H—Story Model (Pennington and Hastie, 1986)
REACT—Risk Assessment Model (MacCrimmon and Wehrung, 1986)
ANAH—Analytical Holistic (e.g., Rapoport and Wallsten, 1972)

FIG. 21.1.   Model attribution.

External Environmental Factors

One measure used to examine the impact of external environmental variables was the ratio of incoming to outgoing calls. Results showed that the lowest ratios were obtained for dynamic scenarios and the highest for planning scenarios, with the training scenarios (involving both planning and dynamic elements) in between. It is hypothesized that these results are attributable to the planning scenario necessitating the proactive coordination and allocation of resources by the commander; the dynamic scenarios result in the commander being more reactive to the unfolding enemy situation. Thus, information gathering, monitoring, expectancy checking, and the application of a course of action increased in planning environments. Dynamic environments reduced the *opportunity* for proactivity in the conduct of such processes.

Workload

In studying workload, a key interest was whether particular decision problems gave rise to high or low workload, and also to examine the impact of individual differences. This was done by combining all the workload responses across several subjects for each scenario, and then examining (through video analysis) the context under which responses varied greatly from the average workload profile. Figure 21.2 shows an example of this analysis. It was discovered that the clusters of high or low responses often referred to particular individuals who, for example, did not have adequate experience of conducting a particular task component, or were perhaps more susceptible to workload because of risk aversion in certain decision situations, such as when deciding whether or not to blow a key demolition. On other occasions, common high workload points were common to many of the subjects—such as for the task of determining enemy objectives in

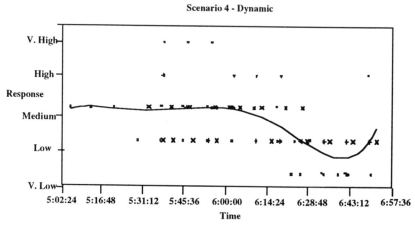

FIG. 21.2. Scenario average workload profile.

complex, time-pressured situations. This finding appeared to remain extant, largely irrespective of an individual's level of experience.

Proportional Model Attribution Per Subject

Figure 21.3 illustrates the individual differences found in model utilization on Exercise Shere Khan II. Although naturalistic theories predominated, there were clear individual differences with Officer F, for example, having a high attribution of Janis and Mann's (1977) hypervigilance in decision making observed.

Individual Differences

Thirty-three individual difference dimensions were identified and formed the basis for detailed personality profiles that were developed for each subject. Key differences were highlighted between operations and intelli-

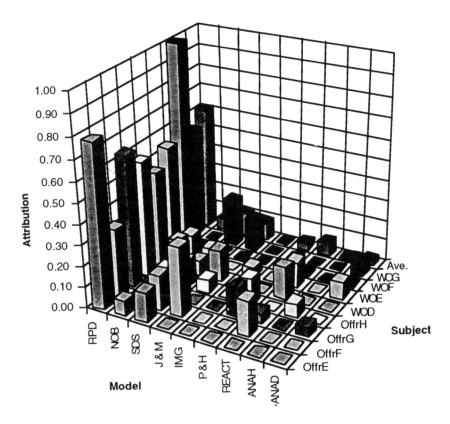

FIG. 21.3. Proportional model attribution per subject.

gence C^2 staff, and also between commissioned and noncommissioned officers. Four primary individual differences have been found to impact on strategy usage: Experience (related to experiential learning of task-specific and task-analogous methods, training, and "Life Experience"); Pragmatism (related to Learning Style); Activism (related to Learning Style); and Independence (as measured in the California Psychological Inventory).

Standard Operating Procedures (SOPs)

Throughout all scenarios, subjects rarely considered the Tactical Aide Memoire or the Staff Officers' Handbook as useful aids. In summary, inexperienced subjects did not know how to use them or what might be found in them, whereas experienced subjects made much more use of personalized aide memoirs, containing information on critical task components, potential "short-cuts" and "tricks of the trade." One training implication is that those who could benefit most from using the wealth of information in the army handbooks do not know that it is there or how to find it. Although the SOPs enable the principles underlying decision making to be proceduralized, they are perceived as too rigid and time-consuming to be practically applied in time-pressured, high-workload operating environments. It is under these conditions that decision makers need to be trained to craft workable and effective solutions, often those previously learned and applied by experienced colleagues.

Working Practices

One of the most significant (and unanticipated) factors that emerged in the study of C^2 decision making was the clear difference in working methods adopted by different decision makers. Use of these practices was found to be strongly related to experience. In all, 22 practices were identified, as summarized in Table 21.1. These practices, or behaviors, are significant because they affect directly the particular strategies adopted by a decision maker. Decision making is difficult to study without the context in which it takes place, and thus for decision making to be properly understood, the broader context in which problem solving occurs must also be taken into account. The working practices in Table 21.1 have subsequently been related to four different levels of expertise, based on a detailed analysis of the empirical data gathered, providing a foundation for the characterization and understanding of decision-making behavior in C^2 environments (Pascual et al., 1994).

TABLE 21.1

List of Identified C^2 Working Practices Modified by Expertise

No.	C^2 Working Practices Affected by Level of Expertise
1	Initial organization of the information environment
2	Initial definition of objectives, information requirements, and priorities
3	Initial assessment of the situation—decision-making requirements
4	Priority / order of DM task / problem solution
5	Use of mental imagery and simulation
6	Use of analogues / "mlitary principles"
7	Use of standard operating procedures and tactical aid memoires
8	Development of a structured checklist (outside of SOPs)
9	Exploitation of interpersonal relationships
10	Delegation of tasks
11	Development and refinement of task plan
12	Development of "satisfactory" versus "optimal" decision solutions
13	Response to interrupts
14	Monitoring of activities set in motion
15	Anticipation of problems / changes in the information environment
16	Provision for errors and approximations in decisions / actions taken
17	Implementation of "short-cuts"
18	Identification of anomalies / gaps in the information environment
19	Perception and control of concurrent activity
20	Selection and evaluation of a course of action
21	Identification and exploitation of new opportunities
22	Predictability of decision making

WHAT IS BEING DONE WITH THESE FINDINGS?

These findings are having a significant impact on a number of research programs at Fort Halstead. One project, Command Decision Making Aids, is developing computer support tools to assist commanders in their decision making. These tools are taking into account the emerging principles from Naturalistic Decision Making (for example, the tools concentrate on assisting the development and refinement of a single, workable Course of Action, rather than forcing the commander to generate and choose between three, as is the norm in military doctrine). In addition, the Working Practices identified in Table 21.1 are being exploited within tool development. Thus the tools assist with identification of objectives and priorities by organizing the information environment, developing tailored checklists, and so forth.

The tool supports not only the commander's decision making, but also his or her understanding of the context in which it takes place.

Another research program is developing evaluation methods for assessing the extent to which proposed military information systems meet the decision-making requirements of its users, under a range of operational conditions. A key input into this work is the understanding obtained on the sources of decision-making errors. This is providing valuable information for developing criteria to assess whether proposed information systems provide support to areas where decision making is weak (e.g., handling temporal and spatial relationships).

CONCLUSIONS

This chapter has conveyed the primary findings from a major experimental program to investigate decision making in command and control. The results that have emerged to date show clearly the utility of naturalistic decision making in command and control. These findings are being rigorously exploited in a raft of application areas, including the development of computer support tools; information system evaluation; training, recruitment, and selection; and computer simulation of decision making. It is also imperative that further empirical research be conducted to define those C^2 operating conditions when alternative approaches to problem solution, such as the use of hybrid decision-making strategies, are likely to be most effective. These findings must also be subsequently reflected in any future approaches developed to support C^2 decision making. Finally, the authors believe that although pure military decision-making research must advance further, it must do so in a manner which produces useful, exploitable products as early as possible if military customers are to maintain confidence in the field.

REFERENCES

Beach, L. R., & Mitchell, T. R. (1987). Image theory: Principles, goals and plans. *Acta Psychologica, 66,* 201–220.

Henderson, S. M., Pascual, R. G., & Fernall, R. P. (1993). *Decision making and human computer interaction models—Final report.* Kent, UK: DRA Fort Halstead.

Janis, I. L., & Mann, L. (1977). *Decision making: A psychological analysis of conflict choice and commitment.* New York: The Free Press.

Klein, G. A. (1989). Recognition-primed decisions. In W.B. Rouse (Ed.), *Advances in man-machine system research* (Vol. 5, pp. 47–92). Greenwich, CT: JAI Press.

Klein, G. A., Calderwood, R., & MacGregor, D. (1989). Critical decision method for eliciting knowledge. *IEEE Transactions, Systems, Man & Cybernetics, 19*(3), 462–472.

MacCrimmon, K. R., & Wehrung, D. A. (1986). *Taking risks: The management of uncertainty.* New York: The Free Press.

Montgomery, H. (1989). The search for a dominance structure: Simplification and elaboration in decision making. In D. Vickers & P. L. Smith (Eds.), *Human information processing: measures, mechanisms and models* (pp. 471–483). Amsterdam: North-Holland.

Pascual, R. G., Henderson, S. M., & Fernall, R. P. (1994). *Strategic research package* (SRP) ASO1BWO5—Final report. Kent, UK: DRA Fort Halstead.

Pennington, N., & Hastie, R. (1986). Evidence evaluation in complex decision making. *Journal of Personality and Social Psychology, 51*, 242–258.

Rapoport, A., & Wallsten, T. S. (1972). Individual decision behaviour. *Annual Review of Psychology, 23*, 131–176.

Tversky, A., & Kahneman, D. (1972). Availability: A heuristic for judging frequency and probability. *Cognitive Psychology, 4*, 207–232.

Chapter 22

Discovering Requirements for a Naval Damage Control Decision Support System

◆

Jan Maarten Schraagen
TNO Human Factors Research Institute

Damage control on board naval frigates is a complex and time-critical task, and the Royal Netherlands Navy has ordered an investigation into the possibilities of decision support systems for supporting Damage Control (DC) officers in carrying it out. The first stage in this investigation was analysis of the functional requirements. Obtaining requirements for decision support systems involves cognitive task analysis, because the tasks to be supported draw on the task performers' knowledge instead of their physical dexterity. This chapter describes the results of the cognitive task analysis carried out in the DC domain. First, a brief description of the DC task itself is presented.

THE DC TASK

Basically, the DC task is a Command and Control task, consisting of a cycle of processes: (a) gather information, (b) assess situation, (c) plan and select actions, and (d) execute and monitor actions.

For instance, in case of a fire, the DC officer *gathers information* on where the fire is located, the time at which it was located, what initial actions have already been taken, and the extent of the fire.

227

The DC officer next *assesses the situation* with regard to the ship's mission, as determined by the commander-in-chief. For instance, if the ship's mission is to fight and particular weapon systems are crucial for fulfilling this mission, these weapon systems should be safeguarded at all costs. If a fire constitutes a threat to these weapon systems because its location is close to that of the weapon systems, then a high priority should be given to fighting this fire.

Next, countermeasures are *planned* in order to fight the fire. The means with which the fire is attacked are determined, as are the direction from which the fire is to be fought and who will command the operation. When a final plan has been made, the DC officer gives orders to *execute the plan* and *monitors plan execution* in order to observe whether the fire is contained.

PERFORMING THE DC TASK

The conditions under which the DC officer has to make decisions are clearly suboptimal: Information presented to the DC officer is often incomplete, uncertain, or just wrong and information can be presented after considerable delay. This is because information is gathered in extreme circumstances (e.g., rooms filled with smoke) and passes along several other persons before it reaches the DC officer. Also, the DC officer has to deal with a number of highly complex interacting systems, such as the electrical, propulsion, ventilation, firemain, and dewatering systems.

The question arises: How is it possible for DC officers to perform their task? The way DC officers are trained is basically by teaching them standard procedures to be carried out when certain well-defined calamities occur, such as the crash of a helicopter. Also, on-board exercises are carried out regularly, but these tend to be quite routine. In real-life situations, such as in wartime, calamities can be worse than expected. The most common way of dealing with nonroutine situations is by recognizing these situations as instances of general prototypes. Indeed, this is the way Klein, Calderwood, and Clinton-Cirocco (1986) assumed their Fire Ground Commanders to operate. However, an important difference between Fire Ground Commanders and DC officers is their experience with actual firefighting events. The Fire Ground Commanders studied by Klein et al. had an average of 23.2 years of experience. As is common with military officers, DC officers do not actively practice their job for more than 3 years, and even in this 3-year period they do not encounter the same number of firefighting events that civil Fire Ground Commanders would encounter. Therefore, we may not expect the DC officer to rely as much as the Fire Ground Commander on an extensive set of firefighting events stored as prototypes in memory.

However, the DC officer has the advantage of fighting fires in a known location, namely one particular ship, whereas the Fire Ground Commanders have to fight fires in novel locations every time. Therefore, we may assume

DC officers to make use of this ship knowledge when fighting fires, for instance, by being able to anticipate a fire spreading to a dangerous location such as an ammunition storage compartment, by choosing the most efficient routes for fire and repair parties, or by intelligently reconfiguring the system when parts of it are out of order. The importance of knowing the layout of the installation and processes was also demonstrated in a series of interviews with Offshore Installation Managers who had experienced a variety of incidents (Slaven & Flin, 1994).

The knowledge actually used by DC officers was investigated in a study comparing more and less experienced DC officers (Schraagen, 1989). The experienced DC officers had carried out the DC task for at least 1 year, whereas the less experienced DC officers knew what the task entailed because they had recently completed a Damage Control course, but had not actively practiced it. Two subjects in each group participated. The basic philosophy behind the study was to use accepted laboratory techniques often used in expert–novice studies, such as card sorting, thinking aloud while generating actions in a scenario, and free recall of the scenarios. In the sorting task, the subjects had to sort cards with information about reports of calamities (location, nature of calamity, way of reporting) printed on them. Reports on which the subjects would undertake the same action had to be placed in the same pile. Identical scenarios were presented to the DC officers and objective ways of scoring their responses could be employed (for more details, see Schraagen, 1989).

RESULTS

The thinking aloud results showed that, when confronted with a particular scenario, experienced officers generated more actions, made more predictions (consequences for nearby compartments), and omitted fewer correct actions than the less experienced officers. The predictions made by experienced officers could largely be attributed to their specific knowledge of potentially dangerous compartments near the calamity. Thus, if there was a fire in place A, the experienced officers could imagine what consequences this fire would have for place B (an ammunition depot, for instance). The less experienced officers mentioned fewer of those consequences, not because they did not know where A and B were (in case they did not know, the location was pointed out to them on a map), but presumably because they did not regard A and B as belonging together. In other words, they had not drawn the connection: "If there is a fire in A, then be aware of its spreading to B, C, ... " Only the experienced officers mentally simulated the fire spreading to other potentially dangerous locations. This was apparent from statements in the protocols, such as:

There is a missile impact in the officers' mess. OK, let me think, this is close to the NATO Sea Sparrows ammunition storage depot. There will certainly be a fire in the officers' mess, so this fire may spread to the ammunition storage depot, or at least, because of the heat, it may cause the ammunition to explode. So there is a high priority to contain this fire.

On the other hand, one less experienced officer said:

A missile impact in the officers' mess? Then there will be a small fire because there is a frying pan in the officers' mess that may fall and catch fire.

No further consequences were elaborated by this officer.

The card-sorting results confirmed the importance of ship knowledge for the experienced DC officers. The experienced DC officers sorted more according to location, whereas the less experienced DC officers sorted more according to calamity. These results further confirm that ship knowledge is more important in this domain than experience with many types of calamities.

The recall protocols were analyzed by measuring pauses, with longer pauses indicating retrieval of different cognitive units (cf. Reitman, 1976). The results showed that the experienced DC officers recalled information faster and better than the beginners. The experienced officers had stored the information from the scenarios in causal form; that is, the initiating calamity was recalled first, then the secondary, ensuing calamities were recalled, then the final consequences, and then the actions taken. The causal schema was duplicated for each part of the ship: Pauses between causes and consequences in one part of the ship were significantly shorter than pauses between switching from the back to the front of the ship. The less experienced DC officers did not recall the information in causal, but rather in temporal form: They tried to recall what was the first alarm presented to them, what was the second, and so forth.

One implication of these results is that a decision support system should show the DC officer important compartments that may be affected by a calamity near those compartments. This may be particularly important for young DC officers who may lack detailed ship knowledge, but is also important for experienced DC officers who may not have proceduralized this knowledge to such an extent that it could be applied under all circumstances, including high-time-pressure situations.

A second implication, derived from the results of the recall protocols, is that a decision support system should present information on calamities and actions taken in a format compatible with the DC officer's mental model. This means that the front and the back of the ship should be represented as distinct units and that information should be presented in the format: primary calamity, secondary calamity, final consequences, actionstaken.

THE DECISION SUPPORT SYSTEM

Both the time pressure and the number of interacting systems involved make the damage-control task very exacting. For instance, in case of damage to the firemain system, the system needs to be reconfigured. This involves knowledge of what valves to close in case of, for example, a leaky duct. This often is a time-consuming task in which the DC officer has to search through manuals and graphs of the firemain system, and make correct deductions on the consequences of closing particular valves on the rest of the system.

By modeling the various subsystems and the knowledge on how to manage these subsystems, the decision support system developed in this project could carry out the necessary calculations, thereby alleviating the DC officer's task substantially. Furthermore, the system alerts the DC officer to potentially dangerous compartments near the calamity. In order to accomplish these tasks, the system has at its disposal knowledge about the structure of the vessel and its subsystems, knowledge about the actual state of the vessel, knowledge about the desired state of the vessel, knowledge about the working of subsystems, knowledge about procedures to be followed, and knowledge based on experience of DC officers.

The interface of the decision support system has been given much attention. Color graphic displays were used in order to present images of underlying technical ship systems. These images consist of two-dimensional representations of pipe networks containing all relevant information required for monitoring and control of these systems. In addition, a three-dimensional image of the ship with overlays of technical systems is used to convey spatial information. The graphic displays aid the DC officer in visualizing the impact of various calamities on the ship's systems and compartments. Online manipulations can be made to simulate the consequences of decisions (a realization of the mental simulation strategy used by experienced DC officers). This is realized by reasoning on the basis of structural and functional information about the system components. Further details on the system can be found in Perre, Rutten, and Mols (1990) and Van Leeuwen (1993).

CONCLUSIONS

Use of standard knowledge elicitation techniques is an important component of cognitive task analysis. The present study has shown the feasibility of employing these techniques for discovering requirements for a naval damage control decision support system. Generally, it is clear that if we can identify knowledge that is not readily available when it should be, either decision support or training may be recommended. In the damage control case, knowledge of potentially dangerous compartments near the calamity and knowledge of what valves to close in what situation were identified as candidates for implementation in a decision support system, and these types of knowledge have been implemented in the final system.

The exact timing and potential overlap of subtasks was not systematically studied in the task-analysis phase. However, various evaluations with prototypes of the system have indicated that the time spent in searching for the correct valves to close can be drastically reduced by using a decision support system instead of searching through manuals and graphs by hand.

This leaves the DC officer with more time for planning actions.

REFERENCES

Klein, G. A., Calderwood, R., & Clinton-Cirocco, A. (1986). Rapid decision making on the fire ground. In *Proceedings of the Human Factors Society 30th annual meeting* (pp. 576–580). Santa Monica, CA: The Human Factors Society.

Perre, M., Rutten, J. C. R., & Mols, D. L. (1990). DAMOCLES: An expert system for damage control management aboard standard frigates. In *Proceedings of the Ninth Ship Control Systems Symposium* (Vol. 2, pp. 101–109). Washington, DC: Naval Sea Systems Command, Department of the Navy.

Reitman, J. S. (1976). Skilled perception in Go: Deducing memory structures from inter-response times. *Cognitive Psychology, 8,* 336–356.

Schraagen, J. M. C. (1989). Requirements for a damage control decision support system: Implications from expert–novice differences. In *Proceedings of the Second European Meeting on Cognitive Science Approaches to Process Control* (pp. 315–322). Ispra, Italy: Commission of the European Communities Joint Research Centre.

Slaven, G., & Flin, R. (1994). Learning from offshore emergencies. *Disaster Management, 6,* 19–22.

Van Leeuwen, E. W. A. (1993). ANDES: A real-time damage control expert system. In *Proceedings of the Tenth Ship Control Systems Symposium* (Vol. 2, pp. 17–28). Ottawa: Canadian Department of National Defense.

Chapter 23

The Decision-Making Expertise of Battle Commanders

◆

Daniel Serfaty
Jean MacMillan
Elliot E. Entin
Eileen B. Entin
ALPHATECH, Inc.

In the fields of decision making, planning, and problem solving, *expertise* has been one of the most difficult concepts to understand, capture, and quantify. The challenge is perhaps even greater in complex domains such as military command and control, where decision tasks reflect the high levels of complexity, uncertainty, and high tempo inherent in tactical and operational environments, and where there is often no single correct answer. An essential component of expertise in military command and control is the ability to make and implement decisions in a timely, efficient, and effective manner, most often with very limited information, in an increasingly fluid and multidimensional battlespace. We call this ability *battle command decision-making expertise.*

A significant challenge for research on battle command decision-making expertise is the difficulty associated with assessing the degree of expertise in an individual. The domain of the military commander is highly uncertain, multidimensional, affected by extraneous, uncontrollable factors (e.g., enemy, weather, and terrain), and dangerous. This is in contrast to many domains for which significant research in expertise has been conducted, such as chess. In chess the board ("ground truth") is always visible, the pieces have fixed capabilities and limited moves, the goal is unambiguous, and games can be played inexpensively and, if desired, repeated from any point. Whereas in chess a system of national and international rankings of expertise is available, no similar approach for rating battle command decision-making expertise is practical. Easily available metrics such as military rank and years of service have not proven to be reliable predictors of military command performance. On the other hand, the evidence for expertise associated with victory in an actual combat situation is debatable because of the many possible extraneous factors involved. Thus, a method for reliably assessing the battle command decision-making expertise of an individual is a necessary starting point for meaningful research in this area.

A second major challenge in the study of battle command decision-making expertise is how to design and conduct rigorous experiments that allow for systematic testing of theoretical hypotheses about the nature of expertise without sacrificing the complex, dynamic, and uncertain setting in which true expertise emerges. There is a large body of empirical research in behavioral decision theory indicating that experts do not always perform impressively (Arkes & Hammond, 1986; Hammond, 1987; Shanteau, 1987). Research seems to show that expert judgment is seldom better than that of novices, and that it is often improved on when replaced with simple mathematical models relying on outcome predictions in various domains (Sage, 1981). Although there are major domain differences, a simple explanation for this disturbing conclusion might be the fact that in several of the studies cited, the tasks given to the experts and the questions asked were, to a large degree, simplistic and artificial. The task was often stripped of its natural context; thus the problems faced by the experts did not emerge naturally and were not presented in their natural language (Zakay & Wooler, 1984).

The importance of studying experts in their working environments has lead to the systematic study of naturalistic decision making. Such studies have lead to significant progress in modeling expert decision processes such as the Recognition-Primed Decision (RPD) model (Klein, 1988; Klein, Calderwood, & MacGregor, 1989). The naturalistic study of experts insures a valid problem domain and a valid context, but it does not necessarily provide a setting that is conducive to the *systematic empirical investigation of hypotheses*. In our research into the nature of battle command decision-making expertise, we have attempted to bridge the gap between naturalistic observation and systematic laboratory research. Based on an extensive literature review, interviews with commanders, and field observations of military exercises, we have developed a theoretical framework and hypotheses on expertise. We have tested our theory in a rigorous, yet realistic, experiment. Our research design combines the unique insights of military experts with the systematic control inherent in scientific experimentation.

THEORETICAL FRAMEWORK

Mental Models and Expertise

What is the precise nature of battle command decision-making expertise? The underlying premise of the theoretical framework is the cognitive-science concept of *mental models*. Mental models are our internal representation of the external world. The literature on mental models suggests that such models can provide a mechanism for representing the expert's understanding of the situation. The expert's memory consists of an array of "patterns," with information items grouped and indexed by their relevance for problem solving in the domain of expertise.

We suggest that an expert commander has a mental model of the tactical situation that differs in measurable ways from that of a novice. The expert's pattern-indexed memory supports the construction of a better initial mental model of the situation. The expert can retrieve a problem representation structure from memory that is similar to the problem at hand in a way that facilitates a solution. In addition to allowing the expert to organize the available information in a meaningful way, the model allows the expert to detect what essential information is missing.

The available literature does not tell us what, exactly, is contained in the battle command decision-making expert's mental model of the situation that makes it superior. The literature also does not provide insight into how military commanders use their mental models to deal with the uncertainty of tactical situations. To gain insight into these issues, we conducted a series of preliminary interviews (Serfaty, MacMillan, & Deckert, 1991; Serfaty & Michel, 1990) in which military commanders were probed with concrete instances of tactical scenarios (vignettes). These "critical incident" interviews provided us with more specific material on the nature of decision-making expertise in a *battle command context*, allowing us to identify its similarities to and differences from expertise in other fields. They suggested a new framework for the study of expert behavior.

Overview of the Theoretical Framework

The framework for studying battle command decision-making expertise is based on an *integration* of the theories of expertise and the observations made in the preliminary interviews. The theory is derived from several models of cognition (such as Duncker's theory described by Newell, 1985) and is based on a three-stage *hourglass* model (recognition–exploration–matching) of the process by which mental models are developed and used by the expert (see Fig. 23.1).

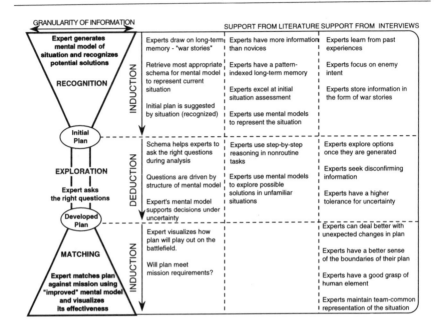

FIG. 23.1. "Hourglass" framework for battle command decision-making process.

Recognition. We suggest that expert tacticians organize their knowledge base in order to store a large amount of information in their domain, and they are able to "chunk" information by grouping details together into patterns. They store and retrieve information about their domain in a different way than novices, and are able to very quickly bring to mind relevant experience and information. Cognitive science research indicates that human beings encode a variety of information about an experience, and can access their memory of it along many different dimensions. Johnson-Laird (1983) argued that parallel processing occurs in memory retrieval, precluding the possibility of conscious awareness. Thus expert commanders may not be aware of how they access their collection of war stories, just that certain circumstances bring certain previous experiences to mind.

How does the expert's retrieval of information from memory support the construction and use of mental models? We suggest that the expert commander builds a mental model of the current situation and the appropriate plan for dealing with it. The first step is to retrieve relevant experiences, which are very specific, from memory. None of them will match the current situation exactly. Taken together, however, they give the decision maker a "schema" or skeleton that indicates which aspects of the current situation are most important. The decision maker can then proceed to fill in the empty "slots" of the schema through information gathering, analysis, progressive deepening, and so forth.

Exploration. After expert commanders orient themselves through retrieving relevant instances from their previous experiences, they need to explore the details of the current situation and the plan being considered. This is a conscious analytical process, and may be supported by computers and other planning tools. What is the role of the mental model at this stage of the process? We believe that the decision maker's model, based on experience, suggests what kinds of important information are missing and must be acquired. The model keeps the decision maker from being overwhelmed by details and "lost in the trees." Information gathering and an analytic, deductive effort fill in the *critical* empty slots in the initial model.

How is battle command decision-making expertise manifested at this stage? Experts may have a better set of analysis techniques at their disposal. Also, a better model for the situation and the important inconsistencies and missing links may lead to better questions, that is, better use of the available information-gathering resources.

Matching. Once a decision maker has developed a mental model of the situation and filled in its gaps through analysis and information gathering (to the extent possible), the plan of action can be played out with checks for feasibility and mission effectiveness. How is battle command decision-making expertise manifested at this stage? We hypothesize that the expert has a "better" model of the tactical situation and of the plan and therefore that the expert can do a better job of visualizing outcomes, problems, etc. The expert's model may be better along a number of different dimensions.

A set of major components of battle command decision-making expertise emerge from the proposed theoretical framework (see Fig. 23.2). Based on the theory, detailed hypotheses were formulated in Serfaty, et al. (1991). In summary, we believe that experts keep an extensive store of specific experiences in memory, supported by high-level principles, that allow them to develop very quickly a rough, still incomplete mental model for a new situation. These high-level principles (Serfaty & Michel, 1990) are usually doctrine-driven and based on offensive and defensive tactical theory. Their purpose is to "constrain" the space of significant alternatives. This initial representation is associated with possible courses of action (COA) and, as such, helps the expert to focus immediately on the most critical aspects of the situation, to ask the right questions, and to gather the most relevant information. This information is then used to build a richer mental model that captures the situational dynamics in both space and time, and enables the expert to visualize the outcomes of possible courses of action.

The use of this richer mental model allows the expert commander to develop a course of action that is both flexible and robust. The COA's flexibility is the result of contingency planning and a careful exploration of the options and the associated "branches." A richer mental model allows

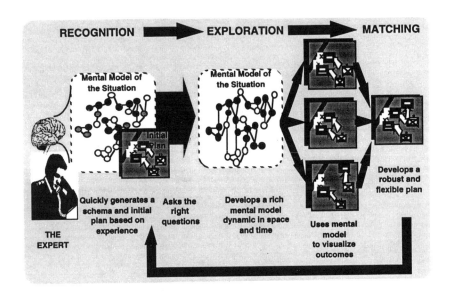

FIG. 23.2. Components of battle command decision-making expertise.

experts to simulate mentally the different ways their decisions may affect the situation ("what if" questions) and to prepare for deviations from the main COA. The COA's robustness reflects its resilience against uncertainty. The combined qualities of flexibility and robustness result in the expert's taking effective action in the situation.

EXPERIMENTAL DESIGN

To meet the challenges of eliciting and assessing expertise in a laboratory setting, and to test the concepts proposed in the theoretical framework, we designed an experiment that posed realistic, nontrivial problems, simulated the procedure and materials used in real-world tactical situations, and involved a significant number of military officers.

Methodology

Forty-six military officers served as subjects in the experiment. Based on their rank, which ranged from Captain to General, all of the officers were expected to have some level of expertise that would vary as a function of ability and experience. Subjects were presented with written materials and maps that described a battlefield scenario set in the Persian Gulf. The materials included a statement from the commanding general describing the overall mission and descriptions of four tactical situations based on the general scenario. Acting as division or brigade commanders, subjects were asked to recommend appropriate action in each of the tactical situations. After providing an initial verbal response to the tactical situation, subjects were permitted to ask questions about the situation. These questions were answered by a military-knowledgeable experimenter who role-played the subordinate and lateral commanders to whom the questions were directed. After obtaining responses to their questions, the subjects completed their written commander's intent and orders/messages. The subjects then explained the rationale for the chosen course of action, responded to questions posed by the experimenter, and filled out a written questionnaire. The experiment sessions were videotaped in order to capture the subjects' verbal responses.

The Assessment of Expertise

Expertise Assessment by Judges. The battle command decision-making expertise shown by each subject in each situation was judged independently by two or three masters, or "superexperts," in the domain. These judges were retired three- and four-star generals with extensive tactical experience, who are widely recognized by the military community as experts in their domain. By design, the judges were not supplied with a definition of expertise. For each expertise rating, they provided an expertise score which could range from 1 (*novice*) to 7 (*expert*). In addition, each judge provided—in writing—supporting evidence for his decision. The judges made three sets of expertise ratings. First, they rated the written materials produced by the subjects (commander's intent statements and orders/messages). Then, using the videotapes, they rated the degree of tactical expertise displayed in the subjects' decision-making process. The written and process ratings were done separately for each tactical situation. Finally, considering both the written and verbal responses in all of the situations, the judges rated each subject's overall level of expertise.

Assessment of Behavioral Components of Expertise. Using a set of behavioral measures (Deckert, Entin, Entin, MacMillan, & Serfaty, 1994) derived from our theoretical hypotheses, two observers who were not experts in tactical decision making also rated the videotaped responses. The theory predicts that domain experts would manifest certain types of behaviors as they reacted to the situation and developed their plans (e.g., develop initial plan with built-in contingencies, ask critical questions to fill in information gaps, consider show stoppers while developing a COA, etc.) These behaviors, described in detail in Deckert et al. were captured in a set of measures that could be observed and coded by military nonexperts. It is important to note that the ratings made by the observers did not require them to judge the "quality" of the subjects' responses and plans, and that the observers had no access to the judges' ratings.

RESULTS AND DISCUSSIONS

Did We Measure Expertise?

A key premise of this research is that battle command decision-making expertise can be identified and reliably assessed, even though there may be no universal definition of this expertise. A second premise was that expertise could be elicited in a laboratory situation. If these premises are correct, then the judges should be able to assess levels of expertise, using their own definitions and standards of expertise, and their overall expertise assessments should show high rates of agreement. If there is a recognizable and measurable skill called battle command decision-making expertise, it should be assessed similarly by multiple judges; that is, the judges' overall expertise assessments should be highly correlated. The coefficient alpha, a measure of internal consistency and an important form of reliability (Cronbach, 1970), was found to be 0.81 (out of a max. of 1.0) for the overall expertise rating, implying high reliability or consistency among the judges. All other detailed components of expertise—both output and process—were found to achieve similar levels of consistency.

The judges provided three ratings of expertise for each individual: an evaluation of the expertise skills expressed in the individual's written output; an evaluation of the expertise process expressed in the individual's verbal output; and an overall expertise rating that expresses the judge's overall assessment of the individual's expertise level. If expertise is a coherent, measurable skill, judges should be able to recognize and assess it in terms of both product and process components. Figure 23.3 shows the relationship

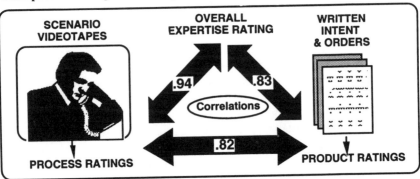

FIG. 23.3. Relationship among product, process, and overall ratings of expertise.

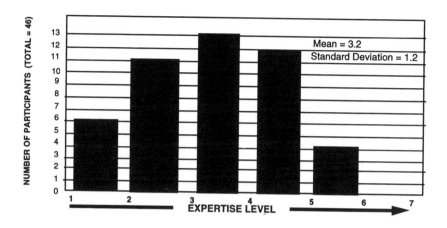

FIG. 23.4. Distribution of expertise for experiment subjects.

among these three indices of expertise. The high correlation between the process and product ratings suggests the existence of a measurable underlying quality.

Are there varying degrees of expertise that can be reliably assessed? Given the range of backgrounds of the 46 subjects in our sample, we expected that there would be a range in the subjects' command decision-making expertise (Fig. 23.4). The average expertise scores assigned by the judges are almost symmetrically distributed around the mean. The range and shape of the score distribution suggest that the method discriminated among subjects with varying levels of expertise.

Finally, we asked whether battle command decision-making expertise is directly related to rank or years of service, or whether there is a component of expertise that is independent of experience. The correlations between rank and expertise and between time in service and expertise (.20 and .25, respectively) were not statistically significant, indicating no strong relationship between years of military service and battle command decision-making expertise.

Hypothesis Testing

The relationship among the major components of battle command decision-making expertise suggested by the theoretical framework and the behavioral measures used in the experiment is illustrated in Fig. 23.5. It also shows which of the measures is significantly correlated with expertise level as assessed by the judges, as well as the direction (positive or negative) of the correlation.

According to the theory, the expert is able to rapidly retrieve from memory an initial schema and a possible corresponding plan of action. This

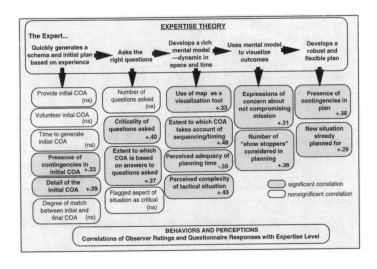

FIG. 23.5. Relationships between theoretical framework and behavioral measures.

hypothesis is based on other research, specifically Klein's observation of decision making in naturalistic settings, and Shank's theory of case-based reasoning (Riesbeck & Schank, 1989). We found no evidence that more expert subjects were more likely than less expert subjects to provide an initial COA. On the other hand, we did find evidence that when subjects provided an initial COA, subjects with higher expertise provided a more detailed COA with more contingencies.

The finding that the high-expertise subjects' initial COAs contained more detail is consistent with the idea that experts are able to draw on previous experience to generate a more complete schema for the tactical situation and therefore can produce a more detailed plan for action early in the planning process. The hypothesis that the expert is able to focus immediately on critical unknowns and to ask the right questions was supported by a significant correlation between the criticality of questions asked and the use of the answers to questions in developing the COA. (The criticality of questions was based on ratings of a list of possible questions by superexpert judges.)

The hypothesis that the expert builds and uses a richer (more complex) mental model was supported by all five behavioral measures derived from that hypothesis. The more expert subjects were more likely to develop COAs that took account of the sequencing and timing of events, indicating that they had a better mental representation of these complex relationships than less expert subjects. They were also more likely to use the map as a visualization tool, indicating that they were considering the effects of terrain and distance on possible movements, another element of complexity. Based on questionnaire responses from the subjects, the more expert subjects perceived the initial situation as more complex than did the less expert subjects, and felt that the time allocated for planning was less adequate. We conclude that, based on the same information, the expert "sees" more complexity in the situation than the novice.

The theory suggests that the expert uses the mental model to visualize outcomes, and that the mental model allows the expert to act under uncertainty. Two measures support this prediction of the theory. More expert subjects were more likely to mention their concern about possible compromises to the overall mission outcome, and when asked, volunteered more "show stoppers" that they had considered when formulating their plans. Experts seem better able to "see" what could go wrong with their plans, which is also consistent with the idea of the creation and use of a richer mental model to visualize outcomes. Experts are able to produce a more robust and flexible plan of action because they have already visualized what could go wrong, and have planned for it. This behavior was supported in the experiment by two measures: the presence of contingencies in the plans developed by the more expert subjects and the extent to which changes in the tactical situation had already been planned for in the expert's COA.

CONCLUSIONS

Researchers in naturalistic decision making are often concerned with defining and evaluating expertise in complex domains where there are no operationalized expertise levels, no single correct answers to problems, and where the observation and measurement of real-world expert performance is difficult. The research effort described in this chapter builds on a mental model theory of expert decision making to design an experiment in which expertise is assessed in an extremely complex and demanding domain: battle command decision making in tactical warfare. The experiment demonstrates that command decision-making expertise can be recognized by domain experts, differences in the command decision-making expertise of individuals can be identified even under conditions that do not fully replicate the real world, and observers who are not domain experts can recognize the expert behaviors predicted by a mental-model theory about the nature of expertise. We proposed an experimental methodology that could simultaneously satisfy two apparently contradictory requirements: (a) decision-making expertise must emerge in a manner natural to the battle commanders, and (b) a rigorous experimental design must allow for measurement of key behaviors and the testing of hypotheses suggested by the theory.

Bridging the Laboratory–Field Gap: Experimental Innovations

There were three major innovative aspects of this study's design. The first one is the use of quasi-realistic scenarios to elicit the subjects' battle command decision-making expertise. The use of tactical situations based on actual battlefield scenarios allowed us to capture in a laboratory setting some of the uncertainty and complexity inherent in combat situations. The use of large wall maps like those actually used in combat situations enhanced the fidelity of the situation. The presence of a role player who responded to questions posed by the subjects elicited the aspect of verbal interchange between commander and supporting staff that occurs in a command headquarters.

The second major innovation was the use of masters, or "superexperts" who, without using an explicit definition of expertise, provided quantitative evaluations of the degree of battle command decision-making expertise exhibited by the subjects and qualitative evidence supporting the quantitative evaluations. By using multiple experts making multiple evaluations of expertise in multiple situations, we were able to assess both the reliability and validity of our approach.

The third major innovation was the use of nonexpert observers to rate behaviors suggested by the theory as observable components of expertise. Observers who are not domain experts cannot be charged with evaluating the quality of expertise exhibited by experiment subjects, but they are able to look for the presence/absence of observable behaviors or to rate the extent to which particular behaviors were observed.

Testing the Theory of Expertise

Although it is always a challenge to prove that a theory is correct, there is much empirical evidence to support the proposed theory of expertise. We found substantial evidence in the data to support our mental-model theory of expertise, both in the judges' comments and, quantitatively, in the relationship between the behavioral measures and expertise. The theory was able to suggest behavioral measures of expertise that are not dependent on the observer's domain knowledge.

Several theoretical issues remain unanswered and should provide an exciting challenge for subsequent research in this domain. Although the evidence obtained in this research partially supports the mental-model construct of expertise, other systematic methods are needed to access the expert's mental model, especially if it is to be a useful abstraction that can be used in other domains of expertise. We also did not find direct support in the data for the aspect of our theory, based on the RPD model, suggesting that an expert should be more likely to produce an initial plan of action in response to a problem posed in a tactical situation, and that the expert should do this in a more spontaneous and rapid manner than the novice. As we have already suggested, the lack of support may be a function of our methodology, or it could suggest a flaw in our theoretical framework.

Another question not answered by our study is how the knowledge structure of experts is organized and how experts retrieve knowledge. Is the information stored on a case-by-case basis, and, if so, how are these cases organized for retrieval? The war stories told by the military experts that we interviewed and the recognition-primed decision behaviors observed in naturalistic settings suggest that the commander's experiences are compiled into a set of rules. How do experiences get translated into rules?

Implications of the Research

Command decision making on the battlefield is a critical skill for military officers, but it is a skill that is difficult to define, assess, and train. Opportunities for real-life experience are rare. Large-scale realistic exercises have been developed in part to address the commanders' need to acquire battle-command experience and to hone their decision-making skills in a realistic setting. These programs are expensive to conduct and time-consuming for participants, and therefore must use tools to assess their training value to the commander's decision-making skills. Relevant indicators of command decision-making expertise such as the ones developed in this research effort must guide this assessment and target the training of specific battle-command skills. The results of this experiment should provide encouragement for researchers and practitioners working to understand, measure, and enhance expert performance in complex domains.

ACKNOWLEDGMENTS

This research effort was supported by the U.S. Army Research Institute, Fort Leavenworth Field Unit, KS. The authors would like to thank Mr. Rex Michel (ARI) for his energetic direction and hands-on participation and Dr. Stanley Halpin for his continuous guidance. We also acknowledge the mentorship of the retired Army General officers who helped us design the tactical scenarios and who served as "superexperts" in the project. Finally, we sincerely thank the Army officers who volunteered to participate in this study and share their battle-command expertise with the research team.

REFERENCES

Arkes, H. R., & Hammond, K. R. (1986). *Judgment and decision making: An interdisciplinary reader.* Cambridge, England: Cambridge University Press.

Cronbach, L. J. (1970). *Essentials of psychological testing* (3rd ed.). New York: Harper & Row.

Deckert, J. C., Entin, E. B., Entin E. E., MacMillan J., & Serfaty, D. (1994). *Military command decision-making expertise* (Final rep. TR–631. Burlington, MA: ALPHATECH, Inc.)

Hammond, R. (1987). Toward a unified approach to the study of expert judgment. In J. L. Mumpower (Ed.), *Expert judgment and expert systems* (pp. 1–17). New York: Springler-Verlag.

Johnson-Laird, P. N. (1983). *Mental models.* Cambridge, MA: Harvard University Press.

Kahan, J. P., Worley, D. P., & Stasz, C. (1989). *Understanding commander's information needs* (TR–3761–A), Santa Monica, CA: RAND Corporation.

Klein, G. (1988). Naturalistic models of C^3 decision making. In S. Johnson & A. Levis (Eds.), *Science of command and control: coping with uncertainty* (pp. 86–92). Fairfax, VA: AFCEA International Press.

Klein, G., Calderwood, R., & MacGregor, D. (1989). Critical decision method for eliciting knowledge. *IEEE Transactions on Systems, Man, and Cybernetics, 19*(3), 462–472.

Newell, A. (1985). Duncker on thinking: An inquiry into progress in cognition, in S. Koch & D. Leary (Eds.), *A century of psychology as science.* New York: McGraw-Hill.

Riesbeck, C. K., & Schank, R. C. (1989). *Inside case-based reasoning.* Hillsdale, NJ: Lawrence Erlbaum Associates.

Sage, A. (1981). Behavioral and organizational considerations in the design of information systems and processes for planning and decision support. *IEEE Transactions on Systems, Man, and Cybernetics, 11*(9), 640–678.

Serfaty, D., MacMillan, J., & Deckert, J. C. (1991). *Toward a theory of tactical decision-making expertise* (TR–496–1, Burlington, MA: ALPHATECH, Inc.

Serfaty, D., & Michel, R. R. (1990). Toward a theory of tactical decision making expertise In I. Gravitis (Eds.), *Proceedings of the 1990 Symposium on Command and Control Research* (pp. 257–269).

Shanteau, J. (1987). Psychological Characteristics of Expert Decision Makers. In J. L. Mumpower (Ed.), *Expert judgment and expert systems* (pp. 289–304). Springer-Verlag.

Zakay, D., & Wooler, S. (1984). Time pressure, training, and decision effectiveness. *Ergonomics, 27,* 273–284.

Chapter 24

Situation Assessment and Decision Making in Skilled Fighter Pilots

◆

Wayne L. Waag
Herbert H. Bell
United States Air Force Armstrong Laboratory

This chapter presents preliminary findings from an attempt to identify situation assessment and decision making processes that account for differences in observed mission performance among skilled fighter pilots. The impetus for the study came directly from the U.S. Air Force Chief of Staff. In 1991, he posed a number of questions concerning situation awareness (SA) for the F-15 fighter world. First, what is SA? Can it be objectively measured? Is SA learned, or does it represent a basic ability? In response to the question, "What is it?," a working group at the Air Staff produced the following operator's definition of SA: "a pilot's continuous perception of self and aircraft in relation to the dynamic environment of flight, threats, and mission, and the ability to forecast, then execute tasks based on that perception" (Carroll, 1992). Although definitions of SA in the research literature focus primarily on processes underlying the assessment and resulting knowledge of the situation (Endsley, 1988), our working definition also included forecasting, decision making, and task execution. From a naturalistic, as well as an operational Air Force perspective, SA is more than simply knowledge and understanding of the environment.

The Armstrong Laboratory subsequently initiated a research investigation that had three goals: first, to develop and validate tools for reliably measuring SA (Waag & Houck, 1994); second, to identify basic cognitive and psychomotor abilities that are associated with pilots judged to have good

SA (Carretta, Perry & Ree, 1996); and third, to determine whether SA can be learned, and if so, to identify areas where cost-effective training tools might be developed and employed (Waag, 1994).

To develop measurement tools, it was first necessary to identify and describe critical behavioral indicators of the fighter pilot's ability to maintain good SA and successfully complete his mission. Earlier, Houck, Whittaker, and Kendall (1993) conducted a cognitive task analysis of a typical F-15 air combat mission. The resulting analysis identified the significant types of decisions required of the flight members, the information required for making these decisions, and the observable activities the flight members performed to acquire this information.

Using the taxonomy, a number of SA rating scales were developed to measure SA in operational units. Data were collected on 238 mission-ready F-15 pilots. From these data, a composite measure of SA was derived that was found to be highly related to previous flight experience (Waag & Houck, 1994). These measures also served as a means of selecting a sample of pilots to participate in a simulation phase of the effort, in which performance was observed under realistic combat conditions. During this phase, simulated air combat mission scenarios were developed for assessing SA, and a variety of performance measures were gathered in an attempt to determine those underlying situation assessment and decision making processes that distinguish pilots with good SA.

METHOD

Subjects

Forty F-15 pilots, who were flight lead qualified, served as subjects. An additional 23 F-15 pilots served as wingmen.

Equipment

The Armstrong Laboratory multiship simulation facility was used for data collection. This facility permits pilots to fly realistic air combat missions in a multibogey, high threat environment. The facility consists of a number of independent simulations operating as part of a secure distributed simulation network. For this study, the facility consisted of two high fidelity F-15C simulators, two F-16 simulators used as adversary stations, an automated threat engagement system, and an exercise replay system. This local simulation network located in Arizona was connected to an air weapons controller simulator at Brooks Air Force Base, TX by a dedicated telephone line. Additional details concerning the basic simulation architecture and components are available in Platt and Crane (1993).

Scenario Design

The approach taken toward the measurement of SA was through scenario manipulation and observation of subsequent performance as recommended by Tenney, Adams, Pew, Huggins, and Rogers (1992). Other approaches such as the use of explicit probes (Endsley, 1988) were considered and rejected due to their lack of face validity for the study participants. Because we were using mission-ready F-15 crews, it seemed essential that we provided a simulation experience as realistic as possible. A week-long SA "evaluation" exercise was constructed that consisted of 9 sorties, with 4 engagements per sortie. Over the week, engagements increased in complexity in terms of numbers of adversaries, enemy tactics, lethality of ground threats, type of controller support, etc.

A typical engagement scenario is presented in Fig. 24.1. This depicts an air combat mission in which the objective of the two F-15s is to defend the home airfield. In this case, the attackers consist of two bombers accompanied by two fighters. The engagement begins at 80 nautical miles (nm) separation in which the fighters are flying at 20,000 ft and the bombers at 10,000 ft. At 35 nm, the fighters begin a corkscrew type of maneuver in which they rapidly descend to 3,500 ft. At this time, they drop off of the F-15's radar screen. Upon completion of the maneuver, the fighters will trail the bombers and will be at a much lower altitude. Although F-15s can easily continue tracking the bombers, it requires the crew to "predict" the actions of the fighters so that they may be quickly reacquired on radar. At 15 nm, the bombers do a hard right turn and descend to 2500 ft. At this time, the bombers momentarily drop off the radar screen. Because the range is very close (10–12 nm), this situation requires the crew to accurately "predict" the actions of the bombers and correctly use their radar so that they may be quickly reacquired. The problem is further complicated in that the bombers and fighters now "merge" in roughly the same airspace. If the fighters are ignored, then they can launch their weapons against the F-15s. If the F-15s "lock" their radar on the fighters, which is usually the case at this point, then the bombers can continue toward the airfield "untargeted." Once the fighters are engaged, it is very difficult to reacquire the bombers since they are low and will be flying away from the F-15s. If the F-15s fail to kill the fighters, the problem will only be compounded.

This example not only shows the approach taken toward the design of the mission scenarios, but also serves to illustrate our contention that SA is more than knowledge of the current situation. In naturalistic environments, situation assessment and decision making are viewed as tightly coupled and are often difficult to separate. For fighter pilots to be successful, they must not only be able to "build the big picture," but they must also translate that assessment into a decision to employ weapons. Often, the inability to make these critical employment decisions may lead to mission failure, despite a correct assessment of the situation. In the sample scenario, the key to success

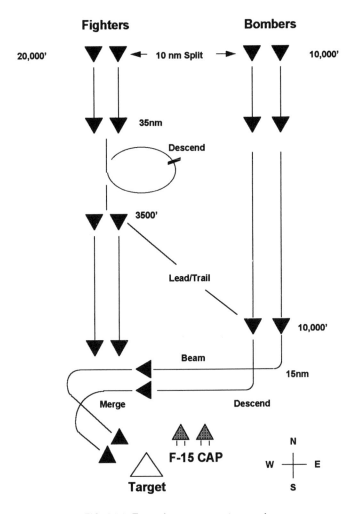

FIG. 24.1. Example engagement scenario.

is to target and destroy the bombers prior to 15 nm and then target the fighters. If the ranges become so close that all four threats must be dealt with simultaneously, then the mission is likely to fail. It is through the careful design of such mission scenarios that the failure to correctly assess the situation or to make correct employment decisions can be successfully inferred by observing pilot performance in the mission scenario.

Data Sources

In his discussion of process-tracing methodologies, Woods (1993) described the use of behavioral protocols. Essentially, behavioral protocols require that one gather as much data as possible so the investigator can piece together the data in order to infer what underlying processes most likely occurred. Although a variety of objective data was gathered (Waag, 1994), the most important data sources, in our view, were the judgments and observations of two retired fighter pilots, who possessed an in-depth understanding of the air combat domain. The same two subject matter experts (SMEs) were used throughout the data collection effort. For each mission, one of the SMEs first attended the mission briefing session conducted by the crew. Both SMEs then observed each engagement and independently completed a checksheet to record pertinent events, notes, and outcomes. On completion of each mission, one of the SMEs accompanied the crew and observed the debriefing. The two SMEs then discussed each engagement and completed a consensus performance rating scale consisting of 24 behavioral indicators of SA related to F-15 mission performance. The SMEs also produced a written critical events analysis for each mission, which attempted to identify those events that, in their opinion, affected the outcome of the mission and were indicative of the crew's SA.

RESULTS

Given the voluminous amount of data gathered (1440 engagements), a major concern was how to select a subset of engagements that could be evaluated "in-depth" in an attempt to infer the processes that likely occurred. First, a composite score was computed for each pilot based upon the average performance rating assigned by the SMEs across all engagements. Based on this score, participants were rank ordered. The top 4 performing and bottom 4 performing pilots were then selected for exploratory analysis to identify whether or not there were consistent differences in certain classes of behaviors. The average number of flight hours in the F-15 for the top 4 was 1030 compared with the 719 for the bottom 4. All missions flown by each pilot were included.

The written critical event summaries were reviewed for the purpose of identifying comments indicative of both poor and good performance in hopes that they might differentiate the top performing and bottom performing pilots. Initially, five categories, loosely based on the taxonomy and the categories from the SA ratings scales, were defined—mission planning and tactics, system operation, communication, situation assessment, and weapons employment decisions. A sixth category, flight leadership, which represents the ability of the flight lead to effectively manage available resources, that is, the wingman and air weapons controller, was added due to the frequency of written comments by the observers.

Comments indicative of good performance tended to be fairly evaluative and nonspecific. The major discriminator between the top and bottom performing pilots was simply the frequency of positive comments (70 vs. 8). Frequency of comments falling in each category were then tallied. It is of interest to note that the most frequently cited categories for the top performing pilots were flight leadership (23) and mission planning and tactics (15). Conversely, the least frequently cited categories were systems operations (0) and weapons employment decisions (1).

Comments indicative of poor performance were also placed in the six categories and frequencies were computed. These results are presented in Table 24.1. Several things should be noted. First, even the top group of pilots in the sample performed poorly at times. This is not surprising given the fact that the scenarios were designed to increase in complexity to a level with which the pilots were unfamiliar. Toward the end of the week, most of the engagements required the F-15 pilots to successfully defeat 6 enemy fighters plus 4 bombers. During operational training using the aircraft, such scenarios are extremely rare due to resource constraints. Second, and as expected, the bottom performing group produced about twice as many written comments indicative of poor performance.

Also of interest was the fact that neither of the groups produced any written comments by observers regarding system operations. In other words, at this level of competency, procedural operation of the onboard systems is mastered and not considered a problem. Of the six categories, weapons employment decisions clearly produced the highest number of "poor performance" comments, in fact, about twice as many compared with situation assessment. Usually, such comments focused on the failure to employ weapons, that is, take a shot, during the premerge phase of the engagement. Thus it would appear that weapons employment decisions may be more problematic than "building the big picture." It is also of interest to note that poor flight leadership comments were reported only for the bottom group. For the top rated performers, there were few comments regarding mission planning and tactics and none for flight leadership. For this group, poor weapons employment decision accounted for over half of the comments.

TABLE 24.1

Frequency of Poor Performance Comments Across Categories

Category	Top Rated Performers		Bottom Rated Performers	
	N	%	N	%
Planning and Tactics	8	22.8	17	27.4
Systems Operations	0	0	0	0
Communication	2	5.7	4	6.4
Situation Assessment	7	20	11	17.7
Employment Decisions	18	51.4	21	33.9
Flight Leadership	0	0	9	14.5
Totals	35	100	62	100

DISCUSSION

Although labeled differently, most cognitive models of information processing refer to at least three closely coupled and continuous components: perception and its meaning (situation assessment), decisions, and actions. For naturalistic settings, only actions are directly observable, the situational assessment and decision components must be inferred. Additionally, behavior in naturalistic environments is purposeful, and thus characterized by a structured goal hierarchy that gives direction to the behavior.

In the fighter environment, high level goals such as *destroy the enemy* and *avoid being destroyed* are fairly obvious. However, the mission plan, with tactics to be employed is less apparent and must also be inferred. It is often found that the mission plan is simply not working and that a change is necessary. The ability to successfully make such changes is a characteristic of the skilled fighter pilot. Again, it is the task of the observer to infer that such changes have occurred. And finally, the fighter pilot works in a team environment. The team in this case is very structured with flight lead in command of the other members including his wingman and air weapons controller. The ability to effectively lead must also be inferred by the observer, although in this instance the task is greatly simplified in the sense that verbal communication is the primary means whereby he exercises his leadership role.

The data gathered in the current study yielded some working hypotheses about the salience of each of these components, and how they change with the development of skill and proficiency. First, consider the basic assessment–decision–action cycle. The results suggest that task execution in terms of the actions involved in flying the aircraft and operating its systems is not a problem at the levels of competence in our sample. By the time the pilot is flight lead qualified, action has been mastered. Instead, most problems are manifested in the assessment and decision portion of the cycle. The results further suggest that problems involving employment decisions are twice as likely to be observed as problems involving situation assessment.

The results also point to the significance of decisions involving mission planning and tactics. Both groups produced comments indicative of tactical decisions that resulted in poor performance mission performance. These data, combined with data involving assessment and employment decisions, suggest that cognitive errors rather than procedural errors are the most likely cause of performance problems among competent fighter pilots.

The data also indicate that highly skilled performers are more likely to exercise effective flight leadership. As the flight leads' skill improves, they are better able to plan and adjust their tactics according to the demands of the scenario. They also improve in their ability to effectively manage all of the resources they have available. Thus, the most frequently cited comments regarding the top rated performers concerned their excellent flight leadership.

Taken as a whole, the results suggest that flight leadership, tactical mission planning, situation assessment, and decision making are critical for developing competent fighter pilots. The quality of such skills becomes manifested in the actions that lead to mission success. As Welford (1968) pointed out, these mental skills are critical components of skilled performance. From an operational perspective, the question becomes how to effectively and efficiently develop these skills. One approach is through multiship combat simulation such as those used in this study. In fact, the opinions of study participants were quite positive toward the use of simulation, with many citing its potential value for learning tactics and flight leadership skills. The value of such a medium lies in the fact that it provides real-world uncertainty which in turn forces the planning, assessment, decision, and action cycle within a naturalistic, team environment. Although mastery and practice of the individual component processes may provide some training value, it is our contention that such benefits will be limited. The development of competency in the fighter pilot community can be maximized to the extent that the training environment presents a realistic depiction of the richness and diversity of the naturalistic environment.

REFERENCES

Carretta, T. R., Perry, D. C., & Ree, M. J. (1996). Prediction of situational awareness in F-15 pilots. *International Journal of Aviation Psychology, 6,* 21–41.

Carroll, L. A. (1992). Desperately seeking SA. *TAC Attack* (TAC SP 127-1) 32: 5–6.

Endsley, M. R. (1988). Design and evaluation for situation awareness enhancement. In *Proceedings of the Human Factors Society 32nd annual meeting* (pp. 97–101). Santa Monica, CA: Human Factors Society.

Houck, M. R., Whittaker, L. A., & Kendall, R. R. (1993). An information processing classification of beyond-visual-range intercepts (Rep. AL/HR-TR-1993-0061). Brooks Air Force Base, TX: Armstrong Laboratory.

Platt, P., & Crane, P. (1993). Development, test, and evaluation of a multiship simulation system for air combat training. In *Proceedings of 15th Interservice/Industry Training Systems and Education Conference* (pp. 629–639). Orlando, FL: National Security Industrial Association.

Tenney, Y. J., Adams, J. J., Pew, R. W., Huggins, A. W. F., & Rogers, W. H. (1992). A principled approach to the measurement of situation awareness in commercial aviation (NASA Contractor Rep. 4451). Langley, VA: National Aeronautics and Space Administration.

Waag, W. L. (1994). Multiship simulation as a tool for measuring and training situation awareness. In *Proceedings of 16th Interservice/Industry Training Systems and Education Conference* (No. 5-3). Orlando, FL: National Security Industrial Association.

Waag, W. L., & Houck, M. R. (1994). Tools for assessing situational awareness in an operational fighter environment. *Aviation Space and Environmental Medicine. 65* (5, Suppl.) A13–19.

Welford, A. T. (1968). *Fundamentals of Skill.* London: Methuen & Co., Ltd.

Woods, D. D. (1993). Process-tracing methods for the study of cognition outside of the experimental psychology laboratory. In G. A. Klein, J. Orasanu, R. Calderwood, & C. E. Zsambok (Eds.), *Decision making in action: Models and methods* (pp. 228–251). Norwood, NJ: Ablex.

Part IV

Methodological
and Theoretical
Considerations

Chapter 25

Training the Naturalistic Decision Maker

◆

Marvin S. Cohen
Jared T. Freeman
Bryan B. Thompson
Cognitive Technologies, Inc.

The naturalistic approach to training decision making takes as its starting point the way real decision makers make decisions. Unlike normative approaches that are based on decision theory or other formal models, naturalistic training is in no rush to replace ordinary decision processes with methods that may appear to be more theoretically justified, but which are qualitatively different from the methods that proficient decision makers use. If a house is basically sound, it can be better to renovate than to tear down and rebuild from scratch.

This chapter addresses a specific set of skills that are utilized by experienced decision makers in novel situations. Naturalistic models of decision making have tended to emphasize recognitional processes; however, recognition is inadequate when no familiar pattern fits the current situation. Although recognition is at the heart of proficient decision making, other processes may often be crucial for success. For example, Klein (1993) discussed mental simulation of a recognized option. We argue that many of these processes, which verify the results of recognition and improve situation understanding in novel situations, are *metarecognitional* in function.

Training metacognitive skills may be applicable across a variety of domains as part of a naturalistic approach to improved decision making. Here we briefly discuss a framework for naturalistic decision making, called the

Recognition/Metacognition (R/M) model (Cohen, Adelman, Tolcott, Bresnick, & Marvin, 1993; Cohen, Freeman, & Wolf, 1994), that highlights
metarecognitional skills. We then turn to our research on training metarecognitional skills in Army battlefield situation assessment.

METARECOGNITION IN DECISION MAKING

Metarecognition is a cluster of skills that support and go beyond the
recognitional processes in situation assessment. They are analogous in many
ways to the *metacomprehension* skills that proficient readers use when they
construct a mental model based on the information in a text. For example,
according to Baker (1985), skilled readers test and evaluate the current state
of their ongoing comprehension, and they adopt a variety of strategies for
correcting problems that are found, such as inconsistency or gaps in their
understanding. More formally, according to Nelson and Narens (1994),
metacognition involves splitting cognitive processes into at least two levels,
whose relationship is characterized by the direction of information flow and
control. The metalevel monitors the object level, maintains a model or
description of the object level, and modifies object-level activity, but not
vice versa.

In the R/M model, the object level consists of recognitional processes that
activate schemas in response to internal and external cues. It also includes
a situation model that integrates these recognitional schemas under the
influence of metalevel control. The metalevel includes (a) processes of
critiquing that identify problems with the recognitional schemas and the
evolving situation model and (b) processes of *correcting* that may instigate
additional observation, additional information retrieval, reinterpretation of
cues in order to produce a more satisfactory situation model and plan, or
any combination thereof. These two processes are controlled in turn by (c)
a higher level process called the *quick test*, which considers the available
time, costs of an error, and degree of uncertainty or novelty.

The following example is based on think-aloud problem-solving sessions
with active duty Army officers who were presented with a battlefield
scenario. A division plans officer is trying to predict the location of an enemy
attack. The enemy has had the greatest success in the south, which the
enemy is likely to want to exploit; its most likely goal, city Y, is in the south;
it has the best supplies in the south; and the best roads are in the south. The
planner concludes that the attack will be in the south.

The normal, recognitional meaning of each cue (prior success, a lucrative
goal, supplies, and roads) is to expect attack in the sector associated with
the cue. If time is limited or the consequences of being wrong about the
location of attack are not great, the planner will not consider the issue
further. However, when the stakes are high, time is available, and the

situation is not completely routine, he may not be content with this initial recognitional response; he may critique it.

Critiquing can result in the discovery of three kinds of problems with an assessment: *incompleteness*, *unreliability*, or *conflict*. An assessment is incomplete if key elements of a situation model or plan based on the assessment are missing. In order to identify incompleteness, the recognitional meanings of the cues must be embedded within a story structure. Story structures depict causal and intentional relations among events and have characteristic sets of components (Pennington & Hastie, 1993). In particular, the main components of stories concerned with assessments of enemy intent are *goals*, *capabilities*, and *opportunities* (that elicit) the *intent* to attack at a particular place and time (that leads to) *actions* (that result in) *consequences*. For example, an officer might conclude that the enemy's intent to attack in the south was adopted because of higher level goals such as capturing city Y and exploiting prior success in the south, superior capabilities in the south by virtue of better supplies, and superior opportunity via better roads. Future actions that would be expected include removing obstacles in the relevant sector, massing artillery, and moving up troops.

In our example, the officer looks for an argument supporting the conclusion that the enemy will attack in the south based on *each component* of the story structure. He finds the story to be incomplete because none of the enemy actions expected to occur prior to an attack have yet been observed. More subtly, the story may also be incomplete because the officer has not fully considered the factor of capability. What about the relative strength of artillery, armor, and leadership in the north versus the south? Moreover, he has not fully considered the factor of accessibility. What about mountain or river crossings required in a southern versus a northern attack? Correcting steps may generate the information required to complete this story by directing the retrieval of prior knowledge, the collection of new observations or analyses, or the revision of assumptions.

Another function of critiquing is to find conflict—new arguments whose conclusions contradict the conclusions of existing arguments. In our example, the officer's further consideration of enemy capabilities produced an assessment that both troop strength and leadership were superior in the north. The normal, recognitional meanings of these assessments are that the enemy intends to attack in the north. Moreover, fleshing out the accessibility component of the story produced another conflicting argument: The northern forces had superior river-crossing skills, making the northern route easier on the whole.

Critiquing can also expose unreliability in a situation model or plan. Understanding and planning is unreliable if the argument from evidence to conclusion, or from goals to action, is conditioned on doubtful assumptions. For example, taken by itself, troop movement toward the south is an unreliable indicator of attack in the south because there may be even more troops moving north, or the enemy may intend to move the observed troops north at the last minute. Unreliability is different from conflict, because here critiquing neutralizes the argument for attack in the south based on troop movements but does not provide an argument *against* attack in the south.

Critiquing and correcting for one problem may lead to the creation and detection of other problems. In this example, efforts to create a complete story led to discovery of the conflict between better capabilities and accessibility in the north versus more plausible goals in the south. The officer resolved this conflict by rejecting the normal, recognitional meaning of the evidence favoring attack in the north. He generated an alternative interpretation of these same data, that the *main* attack will be in the south but that a *diversionary* attack is planned for the north. This resolution of the conflict, however, opened the door to a new problem: unreliability of the assumption about a diversionary attack in the north.

Figure 25.1 summarizes how steps of critiquing and correcting can be linked in the R/M framework. The three types of problems explored by critiquing are shown as three points on a triangle, representing model incompleteness, unreliable assumptions in arguments for the key assessment (e.g., intent to attack in the south) or in rebuttals of arguments against the

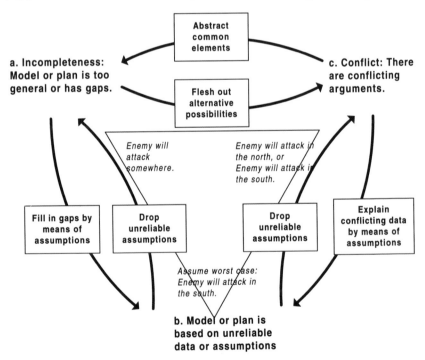

FIG. 25.1. Ways in which correcting steps can lead to new problems.

key assessment, and the existence of conflicting arguments that contradict the key assessment. The arrows showing transitions from one corner of the triangle to another represent correcting steps. It is these correcting steps that may sometimes, but not always, produce new problems. For example, correcting incompleteness in the situation model by retrieving or collecting data or by making assumptions can lead either to unreliable arguments or to conflict with other arguments. Resolving conflict by critiquing a conflicting argument can lead to unreliable assumptions in rebuttals. Dropping or replacing unreliable assumptions can restore the original problems of incompleteness or conflict. These new problems may then be detected and addressed in a subsequent iteration of critiquing.

Our analysis of 34 critical incident interviews with Army command staff suggests an important feature of naturalistic decision making related to Fig. 25.1. Proficient decision makers first try to fill gaps and explain conflict, and only then assess the reliability of assumptions. Thus they tend to advance from the upper right and left corners of the triangle down to the bottom, converting problems of incompleteness and conflict into problems of unreliability. In short, they try to construct *complete and coherent situation models*. They do this if possible by means of newly collected or retrieved information, but if necessary by adopting assumptions. Success in filling gaps and resolving conflict does not mean that decision makers accept the resulting situation model. However, it does tell them what they must believe *if* they were to accept it. This process facilitates evaluation of a model by reducing all considerations to a single common currency: the reliability of its assumptions. If unreliability is too great, a new cycle of critiquing will hopefully expose it and trigger efforts to construct a new story.

The officer in our example continued to monitor the situation, but new observations conflicted with his hypothesis of a main attack in the south. Each time, however, the officer generated a modification of the story that accommodated the new evidence. These modifications involved assumptions that neutralized the impact of the conflicting argument. For example, the enemy's deep interdiction efforts destroyed southern bridges, making it difficult for the enemy to advance along the expected attack route. The officer rejected the normal, recognitional interpretation of this observation, which supports attack in the north. He explained the observation by supposing that the enemy might have more bridging equipment than previously known, or that the enemy was more concerned about reinforcements of our forces in the south than rapidity of advance. Another piece of potentially conflicting evidence involved movement of enemy artillery to the north. The officer generated possible explanations of this as well: The artillery might be moved at the last minute to the south; it might have a longer range than supposed; it might have moved north by mistake.

In creating a coherent and complete story, the officer appears to be guilty of the *confirmation bias*, reinterpreting conflicting data to make it consistent with a favored hypothesis. However, we are not convinced that this "bias" is always an error. As discussed in Cohen (1993), it can also be an error to abandon a hypothesis in the face of one apparently conflicting observation. It is possible that the normal meaning of the conflicting observation is not correct. It can be perfectly reasonable for decision makers to use conflict with a larger or more trusted set of arguments as a prompt to look for alternative explanations of conflicting cues.

While metarecognitional processes help trigger the search for alternative interpretations of cues, they also guard against excesses. They monitor for unreliable assumptions required by explaining away too much. For example, in order to hold onto the view that the main attack will be in the south, the officer must now accept at least some of the assumptions mentioned earlier, or other equivalent ones: for example, that a diversionary attack is planned for the north, that the enemy has better bridging equipment than expected, and that the artillery will be moved prior to the attack. He might be expected to become increasingly uncomfortable as the number of conflicting observations grows, along with the unsubstantiated assumptions required to explain them. Each new explanation is like stretching a spring. At some point, the extended spring snaps back. At this point he drops the assumptions, and decides to focus seriously on the alternative possibility—a main attack in the north. He then explores a new situation model.

TRAINING METARECOGNITION

The R/M model describes a set of skills that supplement pattern recognition in novel situations. These skills include identifying key assessments and the recognitional support for them; checking stories and plans based on those assessments for completeness; noticing conflicts among the recognitional meanings of cues; elaborating stories to explain a conflicting cue rather than simply disregarding it; sensitivity to problems of unreliability in explaining away too much conflicting data; attempting to generate alternative coherent stories to account for data; and a sensitivity to available time, stakes, and novelty that regulates the use of these techniques.

These skills are neither as domain-specific as simple pattern recognition, nor as general-purpose as analytical methods. Like analytical tools, metarecognitional skills may be applicable with minor adaptations across a wide range of domains. Unlike analytical skills, however, their use requires a relatively strong base of familiarity in a domain. They build on the knowledge embedded in recognitional skills, but do not by any means replace it.

We have prepared and tested a training program on metarecognition skills in two domains: Army command staff battlefield situation assessment and Navy ship-based antiair warfare. In both cases, the existing training method consists of several segments, each of which contains (a) reading material,

(b) some brief explanations and examples by the instructor, and (c) classroom discussion and interactive exercises. Examples for exercises and discussion are drawn from trainees' own experiences, from the pretest scenario that is run prior to the training, and from modified case studies in the training materials.

The Army training has two major segments. One focuses on situations that the decision maker feels relatively certain of his or her conclusions, and focuses on critiquing and correcting for unreliability. The other focuses on critiquing and correcting for conflicting observations. (The Navy training also addresses critiquing and correcting for incompleteness and the quick test.) We very briefly convey some of the flavor of each of the two Army segments.

Handling "Certainty"

We begin the discussion by asking officers for a personal experience that they felt completely certain of some assessment. We then show how that certainty could be questioned. This method forces the officers to generate an alternative story that covers all the evidence. In doing so, it exposes assumptions underlying the story that they currently accept, and helps them evaluate the story for reliability. These assumptions can be evaluated and, if time and stakes warrant, can be checked. Appropriate correcting steps can be taken when weaknesses in the story are found. In the end, even if officers retain the original story, their confidence in it will have been earned.

The following example of certainty was volunteered in one class. A battalion officer, facing an enemy across the river, predicted that they would cross the river at point X. Point X was relatively close to the enemy's present position, the river at point X was relatively shallow, and a combination of vegetation and terrain there would provide concealment. He recommended concentration of friendly forces in the vicinity of point X.

The method for finding hidden assumptions consists of four steps:

1. Select a critical assessment, no matter how confident you are that it is true (e.g., that the enemy will cross the river at point X).
2. Imagine that a perfect intelligence source, such as a crystal ball, tells you that this assessment is wrong.
3. Explain how this assessment could be wrong.
4. The crystal ball now tells you that your explanation is wrong and sends you back to step 3.

After each new exception was mentioned, the crystal ball told the trainee, "No, that's not the reason why the assessment is wrong. Come up with another explanation." In this particular case, the crystal ball elicited a number of ways this "certain" assessment might fail: (a) The enemy might anticipate that our force will be at point X and decide not to cross there. (b) The enemy might detect the movement of our force to point X and decide not to cross there. (c) There are good crossing sites that we missed. (d) The enemy does not know how good a location point X is. (e) The enemy does not have any river-crossing assets. He cannot cross the river at all. (f) The enemy's river-crossing assets are so good that he can cross elsewhere. (g) The enemy has a large enough force that they can accept casualties in crossing elsewhere. (h) The enemy's objectives are not what we thought. He does not need to cross the river. (i) The enemy will use air assault forces to get across the river.

Usually, trainees are surprised at the quantity and the plausibility of the exceptions that the crystal ball elicits from them. They now realize that the original assessment rested on the assumption that none of these exceptions was true. However, the existence of these possible exceptions is not adequate cause to abandon that assessment. The next step is to evaluate the exceptions. Each one should be considered, at least briefly. The class is asked how they would handle each one. Some possible exceptions may be implausible—for example, that the enemy can afford large casualties. Some can be tested by data collection or by requesting additional intelligence—for example, that the enemy has superior river-crossing assets. Other exceptions may motivate a change in plans to make them less likely. For example, to avoid anticipation or detection of our forces at point X, we might position our forces elsewhere, then move to point X later. Other exceptions may cause adjustments in planning to handle them in case they turn out true. For example, we might place reserves on paths behind the river in case we missed some sites or the enemy missed point X. Exceptions may also cause the adoption of a contingency plan. For example, if the enemy's objective turns out to be on the other side of the river, we might prepare to cross the river ourselves. Finally, some exceptions might have to be accepted as known risks—for example, if the enemy uses air assault.

Handling Conflicting Data

In the second unit of training, we ask officers to describe personal experiences that they were surprised; for example, the enemy attacked in an unexpected sector. We then ask if any cues or indicators had been observed that, in hindsight at least, could have served as a warning. Typically, such cues are clearly remembered, but were disregarded at the time.

A common response to observations that conflict with a previous conclusion is to disregard or discount them. Another response to conflict, which may be equally bad, is to lose confidence and immediately abandon the original assessment. An unexpected event means that situation understanding is imperfect, but the fault may not lie in the original assessment. It may lie in an incorrect interpretation of the new event. In situations where no patterns fits all the data, the correct assessment *must* involve some "explaining away" of conflicting data.

In this training segment, officers learn to monitor or critique for conflicting evidence and learn how to handle conflicting observations when they occur. When conflict is detected, they begin by modifying the current story to explain the surprising events in terms of their original assessment. They then evaluate the reliability of the resulting story. If the explanations prove to be implausible, officers alter the assessment itself and create a new story. The procedure consists of these steps:

1. Notice unexpected events.
2. Explain an unexpected event in terms of your current assessment. If there have been previous unexpected events, try to find the simplest reliable explanation covering all of them.
3. Evaluate the reliability of your account of all the unexpected events.
4. If the explanation is not reliable, change your assessment and return to step 1.

The crystal ball technique can be useful in generating explanations of conflicting data. Now, however, the crystal ball tells you that the original assessment is *correct*, despite the conflicting observation, and asks you to explain how this could be. The crystal ball rejects each explanation the trainees generate and asks them to find another. For example, the destruction of southern bridges might be consistent with a main attack in the south, if: (a) The bridge was destroyed to prevent our troops from being reinforced; (b) the enemy has better bridging equipment than we thought; (c) destruction of the bridge was a mistake; (d) it was part of a deception; or (e) the bridge was destroyed by our own troops rather than by the enemy. The other surprising event, observation of artillery moving north, can be explained similarly by elaborating the story based on the original assessment.

A list of this kind may never be exhaustive. Nevertheless, it provides an understanding of the *kinds* of ways that the current assessment could still be true despite a conflicting observation. To hold onto the assessment, it is not necessary to know which, if any, of these explanations is the case. However, some such story elaboration must be true if the original assessment is to be maintained. Thus, the original assessment is no more reliable than the best of these explanations. These explanations may also point to ways that the assessment can be tested.

The training proceeds to illustrate and discuss the dangers of explaining away too many conflicting cues. We have compared this process of adopting new assumptions to the gradual stretching of a spring until finally it snaps back, the assumptions are dropped, and a new assessment is explored. The spring stretches less if explanations of new conflicting events do not introduce new assumptions. The most reliable overall account must consist of individually plausible explanations, but it should also be simple. It should not require a large number of separate assumptions. For example, an account of events that relies heavily on enemy mistakes to explain all of the unexpected events is maximally simple, but not necessarily reliable. Nevertheless, it may be worthwhile to look for an explanation that accounts for a variety of unexpected observations in the same way, such as a single unreliable source or a single enemy strategic concept.

EXPERIMENTAL TESTS

The training methods have been tested in a pretest/posttest control design with active duty Army officers. We have space here only for the most cursory summary of the results. Participants varied in rank from Captain to Lieutenant Colonel. Twenty-nine received 90 minutes of training while 8 controls were engaged in a discussion of traditional situation assessment techniques. All participants were given a pretest and posttest consisting of a military scenario followed by a set of information updates pertaining to the scenario. Each update consisted of two parts. The first part presented observations concerning enemy activity, terrain, attitudes of the local populace, international diplomatic developments, and other issues and gave an assessment based on those observations. The second part gave a new set of observations, which supported or disconfirmed the original assessment. Participants were asked to evaluate the assessment twice: first, after receiving the original observations, then after receiving the new ones.

Trained participants tended to consider more factors in their evaluations of the assessments than did controls ($F_{1,31} = 3.67$, $p = .065$). More importantly, training increased the value of the factors that officers considered ($F_{3,27} = 3.166$, $p = .04$). Training caused participants to consider fewer factors that were neutral (or irrelevant) with respect to the assessment and to notice more factors that conflicted with the assessment. Moreover, the increase in the number of factors that trained officers considered did not occur at the expense of quality. On the contrary, the issues presented by trained participants were judged in a blind rating by a subject matter expert to be of higher quality on average than those presented by control participants ($F_{1,31} = 3.255$, $p = .081$).

CONCLUSION

Based on critical incident interviews in two domains, the R/M framework appears to capture the way experienced decision makers test and improve the results of recognition. Moreover, it explains how critical thinking can be an effective component of situation assessment without the need for inflexible analytical methods. For example, the R/M model offers a very different approach to situation assessment than Bayesian decision theory. In the latter, every possible implication of every observation must be identified and quantified at the beginning of an analysis. To avoid intractability, the assessment of the impact of a given piece of evidence must be independent of the context of other evidence. This requires a difficult counterfactual mind-set: What would this piece of evidence mean if I did not know about all the other things that have occurred and if I had never thought about them? The product of all this work is a set of probabilities over statements rather than a coherent picture of the situation.

According to the R/M framework, by contrast, proficient experienced decision makers work with evolving situation models or stories. They approach new data in terms of these models, while at the same time looking for gaps and conflicts and being prepared for surprises. When an unexpected or conflicting event occurs, they elaborate the model to take it into account. However, they maintain an awareness of their elaborative efforts and stay alert to the danger of going too far. Alternative stories are investigated when the current story has been stretched too thin (as in the training for handling conflict) or even if a story appears sound, when time is available and the stakes are high (as in the training for handling "certainty").

Based on our very preliminary tests, these metarecognitional skills can be enhanced by training. The result may be more effective situation assessment across a very wide range of decision domains.

ACKNOWLEDGMENTS

This research was funded by Contract No. MDA903–92–C–0053 from the Army Research Institute. We are grateful to Jon Fallesen of the Fort Leavenworth Field Unit for his insightful assistance at all stages of the project. We also thank Martin Tolcott, Leonard Adelman, Terry Bresnick, Freeman Marvin, and Gen. Charles Ottstot (U.S. Army Ret.) for their invaluable help.

REFERENCES

Baker, L. (1985). How do we know when we don't understand? Standards for evaluating text comprehension. In D. L. Forrest-Pressley, G. E. MacKinnon, & T. G. Waller, (Eds.), *Metacognition, cognition, and human performance* (pp. 155–205). New York: Academic Press.

Cohen, M. S. (1993). The naturalistic basis of decision biases. In G. A. Klein, J. Orasanu, R. Calderwood, & C. E. Zsambok (Eds.), *Decision making in action: Models and methods* (pp. 51–99). Norwood, NJ: Ablex.

Cohen, M. S., Freeman, J. T., & Wolf, S. (1994). Meta-cognition in time-stressed decision making: Recognizing, critiquing, and correcting. *Journal of the Human Factors and Ergonomics Society.* (Submitted for publication)

Cohen, M. S., Adelman, L., Tolcott, M. A., Bresnick, T. A., & Marvin, F. F. (1993). *A cognitive framework for battlefield commanders' situation assessment* (Tech. Rep. No. 93–1). Arlington, VA: Cognitive Technologies, Inc.

Klein, G. A. (1993). A recognition-primed decision (RPD) model of rapid decision making. In G. A. Klein, J. Orasanu, R. Calderwood, & C. E. Zsambok (Eds.), *Decision making in action: Models and methods* (pp. 138–147). Norwood, NJ: Ablex.

Nelson, T. O., & Narens, L. (1994). Why investigate metacognition? In J. Metcalfe & A. P. Shimamura (Eds.), *Metacognition: Knowing about knowing.* Cambridge, MA: MIT Press.

Pennington, N., & Hastie, R. (1993). A theory of explanation-based decision making. In G. A. Klein, J. Orasanu, R. Calderwood, & C. E. Zsambok (Eds.), *Decision making in action: Models and methods* (pp. 188–201). Norwood NJ: Ablex.

Chapter 26

The Role of Situation Awareness in Naturalistic Decision Making

◆

Mica R. Endsley
Texas Tech University

Naturalistic decision making (NDM) is emerging as a field of research providing a descriptive view of how people make decisions in actual settings that often feature unstructured problems imbedded within complex and dynamic systems. Decision making in these settings tends to differ significantly from the analytic style inferred from structured laboratory decision tasks that form the basis for traditional decision theory research. A growing body of research indicates that under realistic conditions experts make decisions using a holistic process involving situation recognition and pattern matching to memory structures to make rapid decisions (Dreyfus, 1981; Klein, 1989, 1993; Klein, Calderwood, & Clinton-Cirocco, 1986). Within this framework, a person's situation awareness (SA), an internal conceptualization of the current situation, becomes the driving factor in the decision-making process. For novices as well, who may operate using very different decision strategies, understanding the situation frequently poses the major portion of their task. In most settings effective decision making largely depends on having a good understanding of the situation at hand.

For instance, in a review of National Transportation Safety Board (NTSB) aircraft accident reports across a 4-year period, 88% of the accidents attributed to human error involved SA as a major causal factor (Endsley, 1995a). Other studies have also found SA to be a leading causal factor in aircraft mishaps (Hartel, Smith, & Prince, 1991). Many human errors that are attributed to poor decision making actually involve problems with the SA portion of the decision-making process as opposed to the choice of action portion of the process. Decision makers make the correct decision for their perception of the situation, but that perception is in error. This represents a fundamentally different category of problem than a decision error in which the correct situation is comprehended but a poor decision is made as to the best course of action, and indicates very different types of remediation strategies. In order to understand and positively impact decision making in real-world environments, it is therefore necessary to understand the construct of situation awareness, its role in the decision-making process, and the factors that impact it. To this end, a definition and description of the construct of situation awareness is presented here, a framework model describing the theorized role of underlying processes and mechanisms in achieving SA is summarized, and the relationship between this body of research and naturalistic decision-making research is elaborated.

SITUATION AWARENESS

Concurrent with the growing interest in NDM, situation awareness has developed as a research focus in the past 10 years, largely in the aviation environment, but more recently in many other domains, including the nuclear power industry, automobile driving, air traffic control, medical systems, teleoperations, maintenance, and advanced manufacturing systems. Situation awareness is formally defined as "the perception of the elements in the environment within a volume of time and space, the comprehension of their meaning and the projection of their status in the near future" (Endsley, 1988, p. 97). Situation awareness therefore involves perceiving critical factors in the environment (level 1 SA); understanding what those factors mean, particularly when integrated together in relation to the person's goals (level 2); and at the highest level, an understanding of what will happen with the system in the near future (Level 3). These higher levels of SA allow decision makers to function in a timely and effective manner.

Level 1 SA—Perception of the Elements in the Environment

The first step in achieving SA is to perceive the status, attributes, and dynamics of relevant elements in the environment. A pilot needs to perceive important elements such as other aircraft, mountains, or warning lights along with their relevant characteristics. A flexible manufacturing system operator needs to be aware of the status of various machines, parts, material flows, and backlogs. An automobile driver needs to know where other vehicles and obstacles are, their dynamics, and the status and dynamics of one's own vehicle.

Level 2 SA—Comprehension of the Current Situation

Comprehension of the situation is based on a synthesis of disjointed level 1 elements. Level 2 SA goes beyond simply being aware of the elements which are present, to include an understanding of the significance of those elements in light of one's goals. The decision maker puts together level 1 data to form a holistic picture of the environment, including a comprehension of the significance of objects and events. For example, a military pilot or tactical commander needs to comprehend that the appearance of enemy aircraft arrayed in a certain pattern and in a particular location indicates certain things about their objectives. The operator of a complex power plant needs to put together disparate bits of data on individual system variables to determine how well different system components are functioning, deviations from expected values, and the specific locus of any deviant readings. In these environments, a novice decision maker may be capable of achieving the same level 1 SA as more experienced decision makers, but may fall far short of being able to integrate various data elements along with pertinent goals in order to comprehend the situation as well.

Level 3 SA—Projection of Future Status

It is the ability to project the future actions of the elements in the environment, at least in the very near term, that forms the third and highest level of situation awareness. This is achieved through knowledge of the status and dynamics of the elements and a comprehension of the situation (both level 1 and level 2 SA). For example, knowing that a threat aircraft is currently offensive and is in a certain location allows fighter pilots or military commanders to project that the aircraft is likely to attack in a given manner. This gives them the knowledge (and time) necessary to decide on the most favorable course of action to meet their objectives. Similarly, an air traffic controller needs to put together various traffic patterns to determine which runways will be free and where potential collisions are. A flexible manufacturing system operator needs to predict future bottlenecks and unused machines in order to schedule.

Situation awareness, therefore, involves far more than simply perceiving information in the environment. It includes comprehending the meaning of that information in an integrated form compared to one's goals, and providing projected future states of the environment. These higher levels of SA are particularly critical for effective decision-making in many environments.

Model of Situation Awareness

In order to provide an understanding of the processes and factors that influence the development of SA in complex settings, a theoretical model describing factors underlying situation awareness has been developed (Endsley, 1988, 1990, 1995b). This model brings together a great deal of research on cognition into an organized framework for conceptualizing SA. Even though considerable debate exists within the psychology community on the exact structure and nature of many cognitive mechanisms (an issue that is outside of the scope of the current effort), a useful model showing the role of each mechanism, *as generally understood*, in obtaining SA can do much to guide research in this area. Key features of the model are summarized here and are shown in Fig. 26.1. (The reader is referred to Endsley, 1995b, for a full explanation of the model and supporting research.) Some of the primary features of the model include a consideration of the role of limited attention and working memory, mental models and schema, pattern matching and critical cues, ties between SA and automatic action selection, categorization, data-driven and goal-driven processing, expectations, and dynamic goal selection.

Limited Attention and Working Memory.   In dynamic systems, the development of situation awareness and the decision process are restricted by limited attention and working memory capacity for novices and those in novel situations. Direct attention is needed for perceiving and processing the environment to form SA, for selecting actions and executing responses. In complex and dynamic environments, information overload, task complexity, and multiple tasks can quickly exceed a person's limited attention capacity. Because the supply of attention is limited, more attention to some information may mean a loss of SA on other elements. The resulting lack of SA can result in poor decisions leading to human error. In a review of NTSB aircraft accident reports, poor SA resulting from attention problems in acquiring data accounted for 31% of accidents involving human error (Endsley, 1995a).

Similarly, working memory capacity can act as a limit on SA. In the absence of other mechanisms, most of a person's active processing of information must occur in working memory. New information must be combined with existing knowledge and a composite picture of the situation developed (level 2 SA). Projections of future status (level 3 SA) and subsequent decisions as to appropriate courses of action must occur in working memory as well. For novices, or those dealing with novel situations, working memory may constitute the main bottleneck for situation awareness and can seriously constrain the decision-making process.

Mental Models and Schemata.   In practice, however, experienced decision makers may use long-term memory stores, most likely in the form of

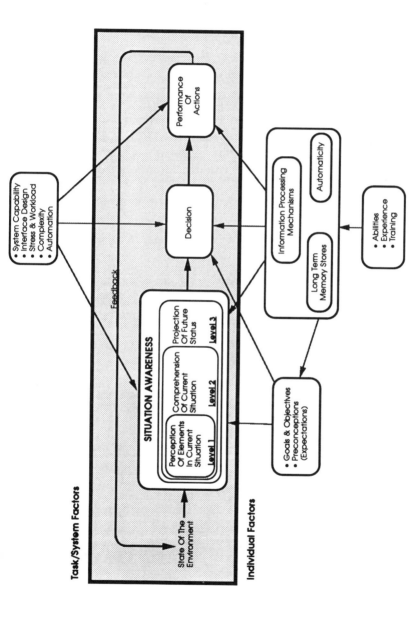

FIG. 26.1. Model of situation awareness in dynamic decision making. Reprinted from *Human Factors*, *37*(1), 1995. Copyright 1995 by the Human Factors and Ergonomics Society. All rights reserved.

schemata and mental models, to circumvent these limits for learned classes of situations and environments, as depicted in Fig. 26.2. These mechanisms provide for the integration and comprehension of information and the projection of future events. They also allow for decision making on the basis of incomplete information and under uncertainty.

Experienced decision makers often have internal representations of the system they are dealing with—a mental model. A well-developed mental model provides: (a) knowledge of the relevant "elements" of the system that can be used in directing attention and classifying information in the perception process, (b) a means of integrating elements to form an understanding of their meaning (level 2 SA), and (c) a mechanism for projecting future states of the system based on its current state and an understanding of its dynamics (level 3 SA). During active decision making, their perceptions of the current state of the system may be matched to related schemata in memory that depict prototypical situations or states of the system model. These prototypical situations provide situation classification and understanding and a projection of what is likely to happen in the future (level 3 SA).

For example, a pilot may perceive several aircraft (considered to be important elements per the mental model) that are recognized as enemy fighter jets (based on critical cues) approaching in a particular spatial arrangement (forming level 1 SA). By pattern matching to prototypes in

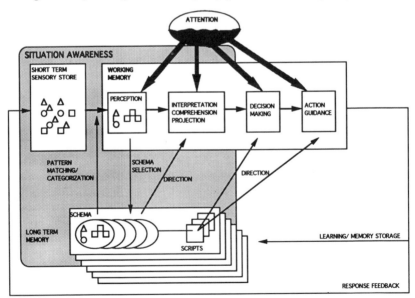

FIG. 26.2. Mechanisms of situation awareness. Reprinted from *Proceedings of the Human Factors Society 32nd Annual Meeting*, 1988. Copyright 1988 by the Human Factors and Ergonomics Society. All rights reserved.

memory associated with the mental model, these separate pieces of information may be classified as a particular recognized aircraft formation (level 2 SA). According to an internally held mental model, the pilot is able to generate likely attack scenarios for this type of formation when in relation to an aircraft with the location and flight vector of his or her own ship (level 3 SA). Based on this high-level SA, the pilot is then able to select prescribed tactics (a script) that dictate exactly what evasive maneuvers should be taken.

A major advantage is that the current situation does not need to be exactly like one encountered before due to the use of categorization mapping (a best fit between the characteristics of the situation and the characteristics of known categories or prototypes). This process can be almost instantaneous due to the superior abilities of human pattern-matching mechanisms. When an individual has a well-developed mental model for the behavior of particular systems or domains, it will provide: (a) for the dynamic direction of attention to critical cues, (b) expectations regarding future states of the environment (including what to expect as well as what not to expect) based on the projection mechanisms of the model, and (c) a direct, single-step link between recognized situation classifications and typical actions.

The use of mental models also provides default information for decision makers. These default values (expected characteristics of elements based on their classification) may be used by individuals to predict system performance under incomplete or uncertain information, unless some specific exception is triggered providing a more refined classification. This allows experts to have access to reasonable default information about system behavior, yielding more effective decisions than novices who will be more hampered by missing data. Default information may furnish an important coping mechanism for experts in forming SA in many challenging domains where information is missing or overload prevents them from acquiring all the information they need.

In addition, people's SA may include information on their degree of uncertainty about their mapping of world information to the internal model and their uncertainty about future projections based on the model. This feature allows people to make decisions effectively, despite numerous uncertainties. Small shifts in these uncertainties, however, can dramatically change resultant conclusions.

Pattern Matching and Critical Cues. According to this model, SA and decision making are dependent on pattern matching between critical cues in the environment and elements in the mental model. The crucial factor in using these models to achieve good SA rests on the ability of the individual to recognize key features in the environment—critical cues—that will map to key features in the model. The model can then provide for much of the higher levels of SA (comprehension and projection) without loading working memory.

Automatic Action Selection.   When scripts have been developed for given prototypical situation conditions, the load on working memory for generating alternative courses of action and selecting between them is even further diminished. Associated goals or scripts can be used to dictate decision-making and action performance. This provides a mechanism for the single-step, "recognition-primed" decision making described by Klein (1989). This process is hypothesized to be a key mechanism allowing people to efficiently process a large amount of environmental information to make rapid and effective decisions in challenging circumstances.

Categorization of Information.   The classification of information into understood representations forms Level 1 SA and provides the basic building blocks for the higher levels of SA. Long-term memory stores play a significant role in classifying perceived information into known categories or mental representations almost immediately in the perception process (Hinsley, Hayes, & Simon, 1977). Categorization is based on integrated information and typically occurs in a deterministic, nearly optimal manner (Ashby & Gott, 1988). The classification of perceived information into components that map to the mental model is very important.

With well-developed memory stores, very fine categorizations may be possible. A highly detailed classification provides the expert with access to detailed knowledge (from long-term memory) based on a small amount of information from the environment. A novice who may not be able to make the same level of classification would have less information from the same data. The cues used to achieve these classifications are very important to SA. With higher levels of expertise, people appear to develop knowledge of critical cues in the environment that allow them to make very fine classifications.

Data-Driven and Goal-Driven Processing.   In a data-driven process, environmental features are processed in parallel through preattentive sensory stores where various signal properties are detected providing cues for further focalized attention. Cue salience has a large impact, therefore, on which portions of the environment are attended to, these elements forming the basis for the first level of SA.

In addition, people can operate in a goal-driven fashion. Situation awareness is impacted by a person's goals and expectations, which influence how attention is directed, how information is perceived, and how it is interpreted. In a top-down decision process, the person's goals and plans direct which aspects of the environment are attended to. That information is then integrated and interpreted in light of these goals to form level 2 SA. Activities are then selected by the decision maker that will bring the perceived environment into line with the person's plans and goals based on that understanding.

On an ongoing basis, trade-offs between top-down and bottom-up processing will occur in dynamic environments. While goal directed processing is occurring, patterns in the environment may be recognized indicating that new plans are necessary to meet active goals or that different goals should be activated. In this way a person's current goals and plans may change to be responsive to events in the environment. Alternating top-down and bottom-up processing allows a person to process information effectively in a dynamic environment.

Expectations. Both working memory and long-term memory have an important role in this process. Information sampling is frequently used to circumvent attention limits, following a pattern dictated by long-term memory concerning relative priorities of information and the frequency with which information changes. Working memory also plays an important role, allowing one to modify attention deployment on the basis of current goals or other information perceived, forming expectations. One's preconceptions or expectations about information can effect the speed and accuracy of the perception of information. Repeated experience in an environment allows people to develop expectations about future events that predisposes them to perceive the information accordingly. They will process information faster if it is in agreement with those expectations and will be more likely to make an error if it is not (Jones, 1977).

Dynamic Goal Selection. Goals and mental models are critical to the formation of SA and are hypothesized to be integrally linked, as shown in Fig. 26.3. A person's current goal(s), selected as the most important among competing goals, will act to direct the selection of a mental model. Plans can then be devised for reaching the goal, using the projection capabilities of the model. A plan will be selected whose projected state best matches the goal state. When scripts are available for executing the selected plan, they will be employed. When scripts are not available, actions will have to be devised to allow for plan completion, using the projection capabilities of the system model.

As an ongoing process, decision makers observe the current state of the environment, their attention directed by the goal activated model and interpreted in light of it. The active model allows for future projections of key environmental elements and expectations concerning future events. When these expectations match with what is observed, all is well. When they do not match because values of some parameter are different, an event

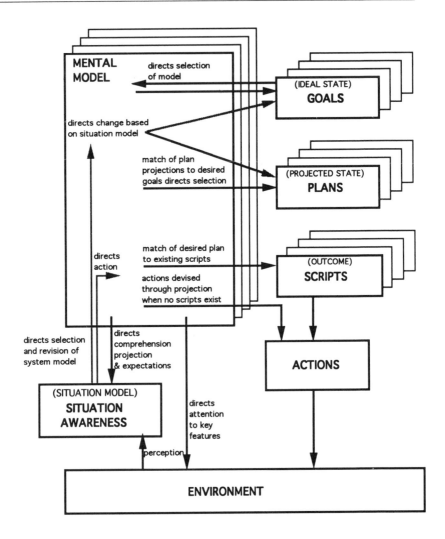

FIG. 26.3. Relationship of goals and mental models to situational awareness. Re-printed from *Human Factors, 37*(1), 1995. Copyright 1995 by the Human Factors and Ergonomics Society. All rights reserved.

occurs that should not, or an event does not occur that should, this signals the individual that something is wrong, and indicates a need for a change in goals or plans due to a shift in situation classes, a revision of the model, or selection of a new model.

This process can also act to change current goal selection by altering the relative importance of goals, as each goal can have antecedent rules governing situation classes in which each needs to be invoked over others. When multiple goals are compatible with each other, several may be active at once. When goals are incompatible, their associated priority level for the identified situation class determines which shall be invoked. Similarly, plans may be altered or new plans selected if the feedback provided indicates that the plan is not achieving results in accordance with its projections, or when new goals require new plans (Beach & Mitchell, 1987). Through learning, these processes can also serve to create better models, allowing for better projections in the future.

Automaticity. SA can also be impacted by automaticity, which may be useful in overcoming attention limits, but may leave the individual susceptible to missing novel stimuli. Developed through experience and a high level of learning, automatic processing tends to be fast, autonomous, effortless, and unavailable to conscious awareness in that it can occur without attention (Logan, 1988). Automatic processing provides (a) good performance with minimal attention allocation, and (b) significant difficulty in accurately reporting on the internal models used for such processing and on the environmental cues that were important. Although automaticity may provide an important mechanism for overcoming processing limitations in achieving SA and making decisions in complex, dynamic environments, it also creates an increased risk of being less responsive to new stimuli as automatic processes operate with limited use of feedback. When using automatic processing, a lower level of SA can result in nontypical situations, decreasing decision timeliness and effectiveness.

IMPLICATIONS FOR NDM

This model of SA provides a description of mechanisms that provide for dynamic goal selection, attention to critical cues, expectancies regarding future states, and ties between SA and typical action, as called for by Klein (1989). More research is needed, however, to fully explore the construct of SA and to develop an understanding of its role in the decision process. This research needs to be fed by an understanding of how people make decisions in realistic settings. Similarly, work on naturalistic decision making needs to take into account developing research on SA in order to fully understand the processes involved in decision making in complex, dynamic environments.

The link between SA and decision making is multidimensional. In addition to forming the basis for decision making as a major input, situation awareness may also impact the types of processes used for decision making. There is considerable evidence that a person's manner of characterizing a situation will determine the decision process used to solve a problem. Through a review of numerous studies, Manktelow and Jones (1987) have shown that the context of a problem largely determines the ability of individuals to adopt an effective problem-solving strategy. Key situational specifics determine the adoption of a mental model that leads to the selection of a problem-solving strategy. In the absence of an appropriate model, people often fail in solving a problem correctly, even when it requires the same logical processes as others they can solve.

In addition, the way a given problem is presented can determine how the problem is solved (Tversky & Kahneman, 1981). Different problem framings can induce different information integration (situation comprehension) and it is a person's situation comprehension that determines the selection of a mental model to use for solving the problem. Thus, it is not only the detailed situational information (level 1 SA), but also the way the pieces are put together (level 2 SA) that directs decision strategy selection.

A major area that needs to be addressed is a determination of when people will use situation-recognition-based decision processes, as described by Dreyfus (1981) and Klein (1989), and when they will use other approaches. Many authors, including Hamm (1988), and Hammond (1986, 1988) have sought to classify decision strategies into a continuum ranging from analytic to intuitive. Hammond (1986), for instance, proposed that task features such as complexity of the task structure, ambiguity, and form of presentation determine whether a person will select an analytic or intuitive decision style for a particular task. Hamm (1988) augmented this work by finding that decision styles can shift between analytic and intuitive many times within a single problem and in recognizing the role of individual differences in decision style selection.

These approaches, however, group all nonanalytic styles into a single category termed *intuitive*. This places decisions that are based on guessing or loose heuristics in the same category with decision strategies that use holistic situation recognition as a basis for decisions, as described here and elsewhere (Klein, 1989). The former indicates a somewhat arbitrary decision style associated with low expertise and effort. The latter draws on a high level of expertise for problem resolution. True, neither style is analytic. To call both intuitive, a term that Hammond (1986) has noted is generally considered inferior, biased, and hazardous, however, does serious disservice to the effectiveness of nonanalytic expert decision making. This style can provide good decisions, very rapidly.

A continuum of decision strategies might more appropriately be represented on a scale ranging from arbitrary to analytic to holistic, as in Fig. 26.4. This classification more accurately reflects the role of expertise in making decisions. Although both arbitrary and holistic styles may be preferred over analytic styles by decision makers in situations involving time pressure, a holistic style can only be used when the decision maker has a sufficiently developed knowledge base. In situations without time pressure, any style could be used. Task factors, such as those noted by Hammond (1986), may

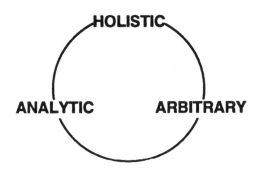

induce a more analytic (as opposed to arbitrary style) for many decisions; however, those with sufficient expertise may still opt for a holistic style when possible due to its efficiency in resource utilization.

A useful point regarding this classification is that it is a continuum. Even within a single problem, a holistic style may only be used for those parts of the problem for which a sufficient knowledge base exists. Other parts of the problem may be solved analytically, if rules and processes are known, or arbitrarily if not deemed important enough to merit the extra effort required by analytic processes. Christensen-Szalanski (1978, 1980) and Shugan (1980) offered an excellent discussion of the effects of time constraints and cost of thinking on decision strategies that is highly applicable.

CONCLUSION

In conclusion, researchers are beginning to explore decision making in realistic settings to find decision processes that are at odds with those predicted by traditional decision theory (Klein, Orasanu, Calderwood, & Zsambok, 1993). In many of these settings, particularly those involving complex and dynamic systems, establishing an ongoing awareness and understanding of important situation components poses the major task of the decision maker. Thus, situation awareness provides the primary input to the decision process and the basis decision strategy selection.

In novel situations, decision makers may be forced to use arbitrary processes of low decision accuracy and reliability, or analytic processes that stress limited internal resources. With increasing experience, they may be able to draw on mental models and schemata of prototypical situations to provide high levels of situation understanding and good decisions without overloading attention and working memory constraints. The key to effective decision making in all of these cases rests in correctly understanding the situation. For those able to operate using a holistic decision style, the ability to correctly classify as situation based on critical cues in the situation is paramount, allowing for the selection of a mental model for problem solving or script for directing the proper response.

Understanding and aiding decision-making in these environments rests on clearly understanding the factors associated with the development of situation awareness. The model outlined here provides an initial attempt at establishing a general theoretical framework for understanding of the processes and factors that impact SA and its role in the dynamic decision-making process.

REFERENCES

Ashby, F. G., & Gott, R. E. (1988). Decision rules in the perception and categorization of multidimensional stimuli. *Journal of Experimental Psychology: Learning, Memory and Cognition, 14*(1), 33–53.

Beach, L. R., & Mitchell, T. R. (1987). Image theory: Principles, goals and plans in decision-making. *Acta Psychological, 66,* 201–219.

Christensen-Szalanski, J. J. J. (1978). Problem solving strategies: A selection mechanism, some implications, and some data. *Organizational Behavior and Human Performance, 22,* 307–323.

Christensen-Szalanski, J. J. J. (1980). A further examination of the selection of problem-solving strategies: The effects of deadlines and analytic aptitudes. *Organizational Behavior and Human Performance, 25,* 107–122.

Dreyfus, S. E. (1981). *Formal models vs. human situational understanding: Inherent limitations on the modeling of business expertise* (ORC 81–3). Berkeley: Operations Research Center, University of California.

Endsley, M. R. (1988). Design and evaluation for situation awareness enhancement. In *Proceedings of the Human Factors Society 32nd Annual Meeting* (pp. 97–101). Santa Monica, CA: Human Factors Society.

Endsley, M. R. (1990). *Situation awareness in dynamic human decision making: Theory and measurement.* Unpublished doctoral dissertation, University of Southern California, Los Angeles, CA.

Endsley, M. R. (1995a). A taxonomy of situation awareness errors. R. Fuller, N. Johnson, N. McDonald (Eds.), In *Human Factors in Aviation Operations: Proceedings of the Western European Association of Aviation Psychology 21st Conference.* Dublin, Ireland: Avebury Aviation.

Endsley, M. R. (1995b). Towards a theory of situation awareness. *Human Factors, 37*(1), 32–64.

Hamm, R. M. (1988). Moment-by-moment variation in expert's analytic and intuitive cognitive activity. *IEEE Transactions on Systems, Man and Cybernetics, 18*(5), 757–776.

Hammond, K. R. (1986). *A theoretically based review of theory and research in judgment and decision making* (CRJP 260). Boulder: Center for Research on Judgment and Policy, University of Colorado.

Hammond, K. R. (1988). *Judgment and decision making in dynamic tasks* (CRJP 282). Boulder: Center for Research on Judgment and Policy, University of Colorado.

Hartel, C. E., Smith, K., & Prince, C. (1991, April). *Defining aircrew coordination: Searching mishaps for meaning.* Paper presented at the Sixth International Symposium on Aviation Psychology, Columbus, OH.

Hinsley, D., Hayes, J. R., & Simon, H. A. (1977). From words to equations. In P. Carpenter & M. Just (Eds.), *Cognitive processes in comprehension.* Hillsdale, NJ: Lawrence Erlbaum Associates.

Jones, R. A. (1977). *Self-fulfilling prophesies: Social, psychological and physiological effects of expectancies.* Hillsdale, NJ: Lawrence Erlbaum Associates.

Klein, G. A. (1989). Recognition-primed decisions. In W. B. Rouse (Ed.), *Advances in man–machine systems research* (Vol. 5, pp. 47–92). Greenwich, CT: JAI.

Klein, G. A. (1993). A recognition primed decision (RPD) model of rapid decision making. In G. A. Klein, J. Orasanu, R. Calderwood, & C. E. Zsambok (Eds.), *Decision-making in action: Models and methods* (pp. 138–147). Norwood, NJ: Ablex.

Klein, G. A., Calderwood, R., & Clinton-Cirocco, A. (1986). Rapid decision making on the fire ground. In *Proceedings of the Human Factors Society 30th Annual Meeting* (pp. 576–580). Santa Monica, CA: Human Factors Society.

Klein, G. A., Orasanu, J., Calderwood, R., & Zsambok, C. E. (Eds.). (1993). *Decision making in action: Models and methods.* Norwood, NJ: Ablex.

Logan, G. D. (1988). Automaticity, resources and memory: Theoretical controversies and practical implications. *Human Factors, 30*(5), 583–598.

Manktelow, K., & Jones, J. (1987). Principles from the psychology of thinking and mental models. In M. M. Gardiner & B. Christie (Eds.), *Applying cognitive psychology to user-interface design* (pp. 83–117). Chichester, England: Wiley.

Shugan, S. M. (1980). The cost of thinking. *Journal of Consumer Research, 7,* 99–111.

Tversky, A., & Kahneman, D. (1981). The framing of decisions and the psychology of choice. *Science, 211,* 453–458.

Chapter 27

The Recognition-Primed Decision (RPD) Model: Looking Back, Looking Forward

◆

Gary Klein
Klein Associates Inc.

The RPD model was formulated 10 years ago (Klein, Calderwood, & Clinton-Cirocco, 1985) to explain how experienced fireground commanders could use their expertise to identify and carry out a course of action without having to generate analyses of options for purposes of comparison. We found that the fireground commanders rarely compared the merits of alternative actions. Rather, they were able to use their experience to identify a workable course of action as the first one they considered. If they needed to evaluate a course of action, they conducted a mental simulation to see if it would work. The current RPD model is presented in Fig. 27.1. Previously, the model contained two functions, labeled *Simple Match* and *Evaluate a Course of Action*. The Simple Match represented a straightforward case in which a decision maker identifies a situation (which means that the goals are obvious, the critical cues are being attended to, expectations about future states are formed, and a typical course of action is recognized) and reacts accordingly. The function of Evaluate a Course of Action, represented as Level 3 in Fig. 27.1, shows a more complex case in which the course of action is deliberately assessed by conducting a mental simulation to see if the course of action runs into any difficulties and whether these can be remedied, or whether a new course of action is needed. The function Diagnose the Situation, shown in the figure, has been added recently, and is discussed in a later section.

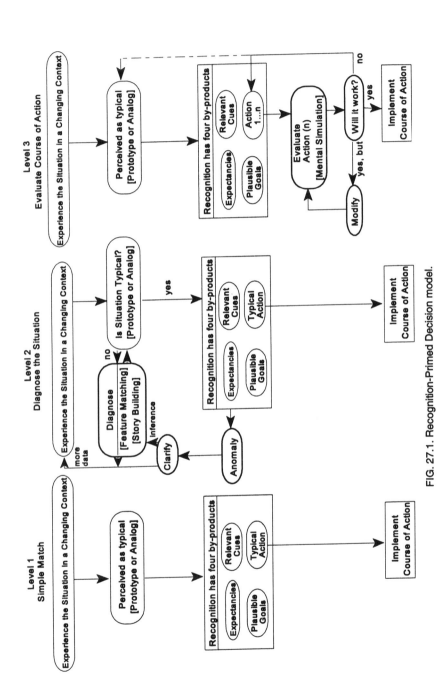

FIG. 27.1. Recognition-Primed Decision model.

The RPD model is an example of a naturalistic decision making model. It attempts to describe what people actually do under conditions of time pressure, ambiguous information, ill-defined goals, and changing conditions. It fits the four criteria of naturalistic decision making research presented by Zsambok (chapter 1, this volume): The model focuses on (a) experienced agents, working in complex, uncertain conditions, who face (b) personal consequences for their actions. The model (c) tries to describe rather than prescribe, and (d) it addresses situation awareness and problem solving as a part of the decision-making process.

There are other models of naturalistic decision making, as described in this volume, and the RPD model does not address all the concerns of naturalistic decision making, such as the influence of team and organizational constraints. Additionally, the RPD model does not reflect memory or attentional or metacognitive processes.

The function of the RPD model is to describe how people can use their experience to arrive at good decisions without having to compare the strengths and weaknesses of alternative courses of action. This is one of the hallmarks of naturalistic decision making, and the RPD model suggests that people can use experience to size up a situation, providing them with the sense of typicality shown in Fig. 27.1 (i.e., recognition of goals, cues, expectancies, and courses of action).

Prior to the RPD model and others like it, traditional decision researchers speculated that under certain task conditions people would not use a Rational Choice strategy, but no one presented a coherent idea of what the alternative might be. Most researchers assumed that it would be a defective version of the Rational Choice strategy, or else some sort of random process. The description of the RPD model provided a firm counterexample to Rational Choice, and made it easier for NDM researchers to take naturalistic decision strategies more seriously.

Additionally, the validity of the RPD model has been evaluated in a wide array of activities, including urban and rural firefighting, Army tank platoon operations, command and control in Navy AEGIS cruisers, hardware and software design, flight control in commercial airliners, chess tournament play, nursing in intensive care units, and so forth. All of these settings have provided evidence for the typicality of recognitional decision making.

Finally, the RPD model presents guidance for training people to make better decisions and for designing equipment that will support decision making. This guidance differs from alternative recommendations for improving decision making because it emphasizes the growth of experience in sizing up situations rather than the use of strategies for making rational choices among courses of action.

EMPIRICAL SUPPORT FOR THE RPD MODEL

The most critical assertion of the RPD model is that people can use experience to generate a plausible option as the first one they consider. If this assertion is invalid, the rationale for the RPD model disappears. Klein, Wolf, Militello, and Zsambok (1995) studied option generation in chess players. With the assistance of a chess master, we selected four reasonably challenging chess positions and presented each of these to players whose official ratings were at a "C" level (mediocre) and an "A" level (very strong). Each player was tested individually, and we asked each one to think aloud while trying to generate a move to play. The data consisted of the first option articulated by the player. When the player had repeated this process for all four boards, we asked for a rating of all the legal moves on each of the boards.

Results showed that the first move announced by the players was much stronger than would be expected if the players were randomly sampling from the pool of legal moves ($x^2 = 211.1, p < .0001$). Although most legal moves are rated as being of very low quality, the first moves generated by the players at both levels of skill were at least adequate, and the majority were judged as strong, using the players' own criteria. Therefore, according to the players' own standards, they were generating moves that were as strong as their skill permitted.

On an objective basis, how good were these moves? For each board position, we had available the move ratings assigned by a panel of grand-masters. For these four board positions only 16% (20/124) of the *legal* moves were judged acceptable by the grandmasters. However, two-thirds of the *first* moves generated by the chess players in our study received grandmaster points ($x^2 = 189.6, p < .0001$). Therefore, by objective standards, the players were generating strong moves as the first ones considered.

These data provide support for the RPD model. These findings are not a critical test of the model, and to my knowledge no decision theories explicitly predict random sampling from the set of available options. The significance of the results is that they constitute a new finding not previously hypothesized or demonstrated. One function of a model is to generate new hypotheses, such as this, in order to add to the storehouse of lawful relationships. If skilled decision makers are able to generate reasonable courses of action as the first ones considered, then the advice to generate large option sets (e.g., Janis & Mann, 1977) is less valuable. The findings show how important it is to take expertise into account in studying decision making, and in making prescriptive claims.

A second assertion of the RPD model is that time pressure need not cripple the performance of decision makers who have considerable exper-tise, because they use pattern matching. Analytical models would predict that time pressure would be likely to interfere with the analytical processes needed, and could result in degraded performance at all levels. An earlier study of chess playing provides support for this prediction from the model. Calderwood, Klein, and Crandall (1988) studied the quality of chess moves generated under tournament conditions, using either regulation time (approximately 2.6 minutes per move) or blitz conditions (approximately 6 seconds per move). We found that even under the extreme time pressure of blitz chess, move quality remained very high. The rate of blunders shown by class "B" players did increase, from 11% to 25% under time pressure, but the rate of blunders remained unchanged for chess experts, 8% versus 7%. This interaction was statistically significant ($F(1,10) = 6.58, p < .05$).

A third assertion of the model is that experienced decision makers can adopt a course of action without comparing and contrasting possible courses of action. Kaempf, Wolf, Thordsen, and Klein (1992) probed 78 instances of decision making in operational Navy antiair warfare incidents involving AEGIS cruisers. These were, for the most part, actual encounters with potentially hostile forces in which several courses of action theoretically existed. The 78 instances were probed retrospectively, during interviews, to determine the rationale for the decision. Kaempf et al. estimated that in 78% of the cases the decision maker adopted the course of action without any deliberate evaluation, and in 18% of the cases the evaluation was accomplished using mental simulation. In only 4% of the cases was there any introspective evidence for comparisons of the strengths and weaknesses of different options. Similarly, Randel, Pugh, Reed, Schuler, and Wyman (1994) probed electronic warfare technicians while they were performing a simulated task, and found that 93% of the decisions involved serial (i.e., noncomparative) deliberations, in accord with the RPD model. Only two of the 38 decisions studied were classified as showing comparisons between options. Mosier (1991) has obtained similar findings in a study of pilots who were videotaped during a simulated mission. Pascual and Henderson (chapter 21, this volume) have also obtained data supporting the prevalence of recognitional decision making in a study of Army command-and-control personnel.

Similarly, Driskell, Salas, and Hall (1994) found that training experienced Navy officers to follow vigilant procedures (e.g., systematically scanning all relevant items of evidence and reviewing the information prior to making a decision) resulted in worse performance than if the Navy officers were allowed to use "hypervigilant" procedures that were compatible with recognitional decision making (e.g., scanning only the information items needed to make a decision, in any sequence, and only reviewing items if necessary).

Taken together, these studies show a growing body of empirical support for the RPD model as a descriptive account of the way experienced people make decisions.

DIAGNOSIS

We have recently added an additional component to the RPD model—the diagnosis of a situation. Diagnosis is the attempt to link the observed events to causal factors; by establishing such a linkage the decision maker would obtain an explanation for the events. Diagnosis is important for the RPD model because the nature of the situation can largely determine the course of action adopted. Often, decision makers will spend more time and energy trying to determine what is happening, and distinguishing between different explanations, than comparing different courses of actions.

We have augmented the original RPD model by adding a reference to a diagnostic function. Diagnostic activity is initiated in response to uncertainty about the nature of the situation. The purpose of the diagnosis is either to evaluate an uncertain assessment of the situation or to compare alternative explanations of events. Two common diagnostic strategies are feature matching and story building. In their study of Navy antiair warfare incidents, Kaempf et al. (1992) examined 103 instances of diagnosis. Of these, the most common strategy was to use feature matching to assign a diagnosis. This occurred in 87% of the cases. In 12% of the cases the decision maker engaged in story building to accomplish the diagnosis (e.g., showing that the erratic course of an unidentified aircraft could be explained as a lost helicopter trying to locate the carrier from which it was launched). Despite this small proportion, the episodes of story building were, in several incidents, the key part of the decision-making activity. Feature matching consists of identifying the relevant features of a situation in order to categorize it. Story building (Klein & Crandall, 1995) often involves a type of mental simulation in which a person attempts to synthesize the features of a situation into a causal explanation that can be evaluated and used in a number of ways. Pennington and Hastie (1993) have also examined the various functions of stories, one of which is to build causal explanations. Mental simulation can be used to project a course of action forward in time (Level 2 in Fig. 27.1), and it also can be used to look backwards in time as a way of making sense of events and observations. Here, the decision maker is trying to find the most plausible story, or sequence of events, in order to understand what is going on—a process of diagnosis that is intended to result in situation awareness. The expansion of the RPD model to include diagnosis of situations should increase its ability to describe the key aspects of naturalistic decision making.

PROBLEM SOLVING

If the RPD model is expanded in the future, a possible direction will be the processes of option generation. Some of the most influential work in the area of problem solving rests on assumptions that we have challenged in the domain of decision making: that people need to generate large sets of options and "filter down" to obtain the best one. This traditional paradigm in Prescriptive Decision Analysis emphasizes the analytical processes of filtering, rather than the strengths of expertise for generating options in the first place. In the study of problem solving, the artificial intelligence paradigm has claimed that people begin by generating problem spaces of all solution paths, and that they then use algorithms or heuristics to locate a workable path from a problem state to a solution state (Greeno & Simon, 1988; Newell, 1990; Newell & Simon, 1972).

For purposes of running computer simulations, researchers have needed to configure the task of option generation into a search task, which is what computers can do well. Thus this approach claims that a problem space is generated. However, there is no evidence that people generate problem spaces except for context-limited combinatorial puzzles. The concept of searching through a problem space diverts us from looking more carefully at the process of option generation. We should be asking questions about how expertise comes into play during option generation, and how people recognize the problem situation. In so doing, we may be able to synthesize ideas about expertise and situation recognition in decision making and problem solving.

CONCLUSIONS

The Recognition-Primed Decision model has evolved during the past ten years, addressing the issue of situation diagnosis, and becoming more clear about the nature of mental simulation. At the same time, the model has received empirical support both from researchers working in other domains, and from testing hypotheses generated by the model. In addition the RPD model has been extended by other NDM researchers. Cohen, Adelman, Tolcott, Bresnick and Marvin (1993) presented a recognition/metacognition model that aims at incorporating metacognitive functions. Lipshitz and Shaul (chapter 28, this volume) have suggested elaborations of the situation awareness process. McLennan and Omodei (1996) showed that prepriming events that occur in preparation for a decision should be taken into account.

All of this work suggests that the RPD model is accomplishing the functions expected of scientific models—generating testable hypotheses, stimulating replications, and triggering related accounts as the NDM community gains a richer understanding of the way people use their experience to make decisions in field settings.

ACKNOWLEDGMENT

Early research was supported by funding from the U.S. Army Research Institute for the Behavioral and Social Sciences under Contract MDA903-89-C-0032.

REFERENCES

Calderwood, R., Klein, G. A., & Crandall, B. W. (1988). Time pressure, skill, and move quality in chess. *American Journal of Psychology, 101,* 481–493.

Cohen, M. S., Adelman, L., Tolcott, M. A., Bresnick, T. A., & Marvin, F. F. (1993). *A cognitive framework for battlefield commanders' situation assessment* (Technical Report 93–1). Arlington, VA: Cognitive Technologies, Inc.

Driskell, J. E., Salas, E., & Hall, J. K. (1994). *The effect of vigilant and hyper-vigilant decision training on performance.* Paper presented at the 1994 Annual Meeting of the Society of Industrial and Organization Psychology.

Greeno, J. G., & Simon, H. A. (1988). Problem solving and reasoning. In R. C. Atkinson, R. J. Herrnstein, G. Lindzey, & R. D. Luce (Eds.), *Steven's handbook of experimental psychology: Vol. 2. Learning and cognition* (pp. 589–672). New York: Wiley.

Janis, I. L., & Mann, L. (1977). *Decision making: A psychological analysis of conflict, choice, and commitment.* New York: The Free Press.

Kaempf, G. L., Wolf, S., Thordsen, M. L., & Klein, G. (1992). *Decision making in the AEGIS combat information center.* Fairborn, OH: Klein Associates Inc. (Prepared under contract N66001–90–C–6023 for the Naval Command, Control and Ocean Surveillance Center, San Diego, CA)

Klein, G. A., Calderwood, R., & Clinton-Cirocco, A. (1985). *Rapid decision making on the fire ground* (KA–TR–84–41–7). Yellow Springs, OH: Klein Associates Inc. (Prepared under contract MDA903-85-G-0099 for the U.S. Army Research Institute, Alexandria, VA)

Klein, G. A., & Crandall, B. W. (1995). The role of mental simulation in naturalistic decision making. In P. Hancock, J. Flach, J. Caird, & K. Vicente (Eds.), *Local applications of the ecological approach to human–machine systems* (Vol. 2, pp. 324–358). Hillsdale, NJ: Lawrence Erlbaum Associates.

Klein, G. A., Wolf, S., Militello, L., & Zsambok, C. (1995). Characteristics of skilled option generation in chess. *Organizational Behavior and Human Decision Processes, 62*(1), 63–69.

McLennan, J., & Omodei, M. M. (1996). The role of prepriming in recognition-primed decision making. *Perceptual and motor skills, 82,* 1059–1069.

Mosier, K. L. (1991). Expert decision-making strategies. In R. Jensen (Ed.), *Proceedings of the Sixth International Symposium on Aviation Psychology* (pp. 266–271). Columbus, OH.

Newell, A. (1990). *Unified theories of cognition.* Cambridge, MA: Harvard University Press.

Newell, A., & Simon, H. A. (1972). *Human problem solving.* Englewood Cliffs, NJ: Prentice-Hall.

Pennington, N., & Hastie, R. (1993). A theory of explanation-based decision making. In G. Klein, J. Orasanu, R. Calderwood, & C. E. Zsambok (Eds.), *Decision making in action: Models and methods* (pp. 188–201). Norwood, NJ: Ablex.

Randel, J. M., Pugh, H. L., Reed, S. K., Schuler, J. W., & Wyman, B. (1994). *Methods for analyzing cognitive skills for a technical task.* San Diego, CA: Navy Personnel Research and Development Center.

Chapter 28

Schemata and Mental Models in Recognition-Primed Decision Making

◆

Raanan Lipshitz
Orit Ben Shaul
University of Haifa

In this chapter we inquire into the role of mental models (specific situation representations) and schemata (abstract cognitive structures that guide the construction of mental models) in recognition-primed decision making, thereby relating three situation-assessment-based decision models (Endsley, chapter 26, this volume; Klein, chapter 27, this volume; Serfaty, MacMillan, Entin, & Entin, chapter 23, this volume) to one another, and to Neisser (1976) and Rouse and Morris (1986). The chapter consists of five sections. The first two sections present certain differences between expert and novice decision making in a sea-combat simulation that prompted the inquiry. The third section argues that the RPD model (Klein, chapter 27, this volume) cannot account for some of these differences satisfactorily. The fourth section introduces the distinction between schemata and mental models and discusses their role in recognition-primed decision making. The fifth section outlines the implications of conceptualizing recognition-primed decision making as schemata-driven mental modeling for theory, research, and training applications. We begin with a brief description of differences between the decision processes of expert and novice Israel Defense Force gunboat commanders in a high-fidelity simulator of sea combat.

THE SEA-COMBAT SIMULATION

The simulator is a computer-driven device that runs various scenarios of sea combat for training and planning purposes. It presents Israeli Defense Force trainees (who may be experienced commanders) with situations, tasks, and displays of information similar to those that are actually encountered at sea. Trainees sit in front of a radar-like screen that displays the sea and air situation in a radius of approximately 30 km. In addition to the screen, trainees receive (and send) information via simulated radio communications with their headquarters, a shore observation post, and the two other gunboats. The computer that drives the simulation updates the display on the screen in response to both the trainees' decisions and input from an instructor who oversees the simulation. The trainees whom we observed "commanded" three fast gunboats on coastal patrol with the mission of identifying and intercepting suspected targets. Once a target had been noted, they were required to make a series of decisions in regard to their own actions, the actions of the two other boats, and (potentially) the actions of other naval and air units.

We conducted our observations as a pilot to a project (which will not be reported here) of developing a training program in tactical decision making. In the pilot we observed the performance of two experts (gunboat commanders with considerable operational experience) and six novices (NCOs with no command experience) on three standard scenarios. The scenarios challenged even very experienced commanders. Data were collected from several sources including online probing, postsimulation debriefing, videotaping of radar screens, and audiotaping of radio communications. Because our sample size was small even for descriptive statistics, we analyzed our observations informally, looking for differences between the decision processes of experts and novices that made (or did not make) sense vis-à-vis existing models and studies. Five differences that we observed are worth noting.

EXPERT VERSUS NOVICE DECISION PROCESSES IN THE SEA-COMBAT SIMULATION

Experts Collected More Information on the Situation Before Making a Decision

Experts collected more information from more sources on more varied aspects of the situation before making a decision. For example, one expert asked, "On what alert is Air Force unit X and how long do I need to notify them in advance if I want their support?" On another occasion, this expert asked to be notified "immediately on any change in the target's course to let me make the necessary adjustments in our own courses. In addition, what is the location of missile boat Y and how long will it take to get missile boat Z's helicopter airborne?" The other expert inquired of the shore station: "What can you see with electronic device X? What else can you tell me about target Y? How long has target Z been in that location?" Novices never showed similar concerns on the whereabouts of friendly sea and air units, and were content to decide on the basis of the information provided by the screen (e.g., "This is a typical occurrence, no need to hurry" or, conversely, "This looks suspicious, better approach and investigate"). This observation is consistent with Cohen (in press), who found that experienced commanders wait longer than inexperienced commanders before making their decision; Amalberti and Deblon (1992) and Gott, Bennett, and Gillet (1986), who found that experts spend more time trying to understand the problem and construct a mental model of the task environment than novices; and Calderwood, Crandall, and Baynes (1988), who found that experts are more likely to engage in situation assessment, whereas novices are more likely to engage in option evaluation. Serfaty et al. (chapter 23, this volume) found that experts and novices did not differ in the number of questions which they asked, but that experts asked more pertinent questions. In addition to having empirical support, this observation is consistent with Neisser's concept of schema. According to Neisser (1976), schemata are situation or domain specific cognitive structures that (a) direct external information search, (b) specify which available information will be attended to and which information will be ignored, (c) organize information in memory, (d) direct the retrieval of information from memory, and (e) become more differentiated as a function of experience. Attributes (a), (b), and (e) explain why experts can conduct more detailed information search than novices during situation assessment.

Experts Engaged in More Efficient Information Search

Once given information on a given target from the shore station, experts would keep track of it on their own. In contrast, novices asked repeatedly for the same information on a given target. This observation is, again, consistent with Neisser (1976), as attributes (c), (d), and (e) imply that novices will not be able to encode, store, and retrieve information efficiently owing to underdeveloped relevant schemata (see also Kardash, Royer, & Greene, 1988).

Experts "Read" the Situation More Accurately

For example, experts were able to distinguish between legitimate and bogus targets, and estimate the correct distances between the various targets on their screens. Novices made many mistakes in both tasks. This observation is consistent with findings that experts consider more alternative interpretations of their situations (Calderwood, Crandall, & Klein, 1987), extract information that nonexperts either overlook or are unable to see (Shanteau, 1988), and therefore construct more complete and accurate mental models of the situation (Bostrom, Fischhoff, & Morgan, 1992; Calderwood et al., 1988; Howell & Cooke, 1989; Lesgold et al., 1988).

Experts Made Fewer Bad Decisions

The significance of this observation is that we could relate it directly to the previous observation. For example, experts were better able to distinguish between legitimate and bogus enemy targets and deploy their boats to deal effectively with enemy targets. In contrast, one novice mistook a blip left by a flock of birds for an enemy boat and literally chased the wind; another inexperienced trainee miscalculated the distances between two of the targets and ended up trying to handle all three single handedly. This observation is, of course, consistent with all models of situation-assessment-based decision making that relate the likelihood of making good decisions to the ability to construct more accurate mental models of the situation (Endsley, 1995; Klein, 1993; Lipshitz, 1993; Reason, 1987).

Experts Communicated More Frequently and Elaborately With Friendly Units

For example, one expert checked with the other two boats on their locations, asked them to keep him constantly informed, and communicated to them his detailed instructions: "Boat A—open to the westernmost point in sector X. Boat B—stay behind in sector X. Boat A takes on the western target in sector X, and I set course in order to intercept the target in sector Y in case that A fails. In any event, stay behind to cover Zone X and do not join the chase." In contrast, a novice at the same decision point simply instructed the two other boats that he was leaving the sector and ordered them to "coordinate your movements and change course to cover for me." This observation shows that expects are more likely to consider other players' perspectives in their decision making. Brezovic, Klein, and Thordsen (1987) and Calderwood et al. (1987) similarly reported that experts are more able to "decenter" and take into account unseen events such as friendly and enemy movements as well as the enemy's perspective.

AN RPD ANALYSIS OF EXPERT–NOVICE DIFFERENCES IN THE SEA-COMBAT SIMULATION

To make sense of our observations, we used Klein's (1993) RPD model. This model applies to the sea-combat simulation because trainees make decisions principally on the basis of what they see, and the model has a strong perceptual (i.e., recognition) component. The five observations discussed earlier are generally consistent with the model. For example, both experts and novices combined situation assessment with serial option evaluation, but experts conducted more thorough situation assessment and referred to imagined friendly and enemy actions, whereas novices focused on their own actions and reacted to the display on their screens. The model cannot account, however, for the observations that experts conduct more extensive and efficient information search prior to making a decision.

The RPD model specifies three precise conditions under which decision makers collect additional information prior to taking action (Klein, 1993): (a) The decision maker cannot recognize the situation, that is, identify plausible goals, relevant cues, expectancies, and a typical course of action; (b) some of the decision maker's expectancies are violated; and (c) the decision maker envisions that a course of action fails in a mental simulation. Given that experts tend to collect more information than novices these conditions imply, correspondingly, that: (a) Experts require more information (and time) than novices to identify goals, expectancies, relevant cues, and actions (the products of situation assessment in the model); (b) expectancies of experts are more likely to be violated than expectancies of novices; and (c) experts engage in mental simulation more often than novices. The latter implication has empirical support (Calderwood et al., 1987). The first two implications, however, contradict a well-established positive relationship between expertise and efficient and effective information processing (Lesgold et al., 1988). As we have seen, Neisser's schema theory can account for the observation that experts collect more information and for the observation (on which the RPD model is moot) that they search information more efficiently than novices. The question that suggests itself is to what extent are the RPD model and the schema construct compatible? Other situation-assessment-based NDM models include a schema component. Examination of these models shows that the question should concern the extent to which the RPD model is compatible not with one, but two hypothetical constructs—schemata and mental models.

SCHEMATA AND MENTAL MODELS
IN RECOGNITION-PRIMED DECISION MAKING

Several NDM models bear strong resemblance to the RPD model (Endsley, chapter 26, this volume; Reason, 1987; Serfaty, MacMillan, Entin, & Entin, chapter 23, this volume). Similar to the RPD, all these models conceptualize decision making as a recognition + reasoning process of serially matching situations with appropriate action. However, these models include two hypothetical constructs that do not appear in the RPD. The elements of the first construct, mental representations of specific situations, are variously labeled "mental models" (Serfaty et al., this volume; see also Bryant, Tversky, & Franklin, 1992; Whitefield & Jackson, 1982) or "situation awareness" (Endsley, this volume). The elements of the second construct, general cognitive structures that drive the construction of specific situation representations, are variously labeled "schemata" (Reason, 1987) and, unfortunately, "mental models" (Endsley, this volume; Rouse & Morris, 1986). Following Neisser (1976), we refer to the elements of this construct as schemata, reserving "mental models" uniquely to specific situation representations.

The terminology that one chooses is not essential. However, distinguishing conceptually between schemata and mental models is essential. In the final analysis, decisions are driven by mental models. In order to understand why experts and novices make different decisions in identical situations we have to inquire into differences in their mental models. For example, we claim that experts we observed made better decisions because they were able to construct more accurate and comprehensive mental models of the situation. Mental models, however, are labile entities that are constructed and discarded as decision makers move, in time and space, from one situation to another. Thus, in order to understand why our experts and novices constructed different mental models in identical situations, and in order to generally understand how novices carry experience from one situation to another until they become experts, we have to inquire into differences in their schemata and how schemata change as a function of experience, respectively.

Figure 28.1 delineates the role of schemata and mental models in recognition-primed decision making. The figure serves four functions: It explicates the relationship between several models of recognition-primed decision making models (Endsley, chapter 26, this volume; Klein, 1993; Serfaty et al., chapter 23, this volume) and schema theory (Neisser, 1976; Rouse & Morris, 1986); allows the RPD model to account for all our observations; and generates specific hypotheses for future research and applications.

Figure 28.1 posits three mediating and two moderating links in the process of matching situations with appropriate actions: A situation–display

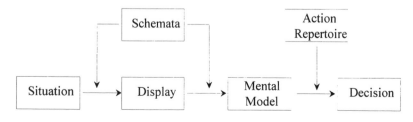

FIG. 28.1. Recognition-primed decision-making as schemata-driven mental modeling.

link and a display–mental-model link (moderated by the decision maker's schemata), followed by a mental-model–action link (moderated by the decision maker's action repertoire). These links are presented as unidirectional to simplify the discussion. Actual decision processes include complicated feedback loops such as the triggering of schemata by initial displays and the shaping of situations and schemata by the outcomes of actions. These complications are worked out in the models that this simplified scheme integrates.

The Situation–Display Link

Borrowing from Weick (1979), we use the term *display* to denote information about the situation that is available to the decision maker. In some situations, such as the sea-combat simulation or the control of production processes via instrument panels, "display" can be used literally. In most situations, however, the term is used figuratively. A fire-fighting commander's display consists of whatever he can see, hear, or smell on the scene, and a CEO's display for making a strategic decision consists of information received from his or her staff, or that the CEO collects on his or her own. The distinction between situation and display underscores a point that is essential to Endsley (this volume), Klein (this volume), and Serfaty et al. (this volume) as well as Weick (1979): Outside the laboratory, decision makers operate almost invariably on the basis of partial information about the situation.

The Schemata–Display Link

Displays are not given to decision makers but are constructed by them. The construction process is driven by schemata that specify which information is missing and where it can be obtained, and which information is irrelevant and should be ignored. This function of schemata is borrowed from Neisser (1976; Rouse & Morris, 1986) and is elaborated by Serfaty et al. (this volume), in the exploration phase of their model.

The Display–Mental-Model Link

Displays consist of equivocal sensory data. Transforming equivocal displays to mental models that are sufficiently unequivocal and meaningful to guide action is driven by schemata that moderate the display–mental-model link. This phase is elaborated by Serfaty et al. (this volume) as "exploration" and by Endsley (this volume) as the role of long range memory stores in the development of situation awareness. The three links described earlier expand the "experiencing the situation ⇨ recognition" node in the RPD model. "Recognition," which consists of critical cues, goals, expectations, and a typical course of action, is contingent on being "familiar" with the situation. Conversely, searching for information is contingent on not being familiar with the situation. Adopting Endsley's definition of situation awareness, we think that mental models cannot be reduced to a set of cue, goals and expectations. We also concur with Serfaty et al. (this volume) that to be nonrandom, information search requires sufficient familiarity to trigger an appropriate schema to guide this search. The difference is subtle enough to lead us to suspect that schemata and mental models are in fact implicit in the RPD model. Klein, however, clearly thinks that recognition-primed decision processes can be modeled without referring explicitly to these hypothetical constructs.

The Mental-Model–Action Link

This link appears as the situation-awareness-decision node in Endsley (this volume), the matching phase in Serfaty et al. (this volume), and identification of a typical course of action, development of a course of action and mental simulation in the RPD model.

In conclusion, Fig. 28.1 points to the similarities and differences between Klein's RPD model on the one hand, and Endsley's Situation Awareness and Serfaty et al.'s "Hourglass" models on the other hand. The three models are basically similar in that they all conceptualize decision processes as recognition primed, and as involving serial option generation and evaluation. In contrast to Klein, Endsley and Serfaty et al. use two hypothetical constructs—schemata and mental models. In the next section we argue that schemata and mental models are useful constructs rather than mere embellishments of the RPD model.

IMPLICATIONS FOR THEORY, RESEARCH, AND TRAINING

Adding elements to a model entails a loss in parsimony. Such loss is justified if the elaboration produces a more powerful model. We have already shown how Fig. 28.1 explains expert–novice differences that the RPD model cannot explain. The figure suggests additional hypotheses on why experts are more likely to make good decisions than novices. In addition to presenting these hypotheses, we outline how they can be tested, in principle:

Hypothesis 1: Experts are more likely to construct complete and accurate mental models. The proposition that "good" or "effective" decision making is contingent on a complete and accurate representation of the situation is basic in all situation-assessment-based decision-making models. Mistaking a flock of birds for a fast enemy boat or misreading the significance of an enemy maneuver are clearly likely to yield a bad decision. This hypothesis can be tested by deriving mental models of experts and novices using, for example, methods such as SAGAT (Endsley, in press), and comparing their accuracy vis-à-vis some objective standard. It may also be interesting to find if experts and novices, or effective and ineffective decisions, are more clearly differentiated by the accuracy of mental models or by some feature of option evaluation (e.g., use of compensatory vs. noncompensatory strategies).

Hypothesis 2: Experts are more likely to have superior schemata. Hypothesis 2 is entailed by Hypothesis 1. An incomplete or inaccurate mental model is a proximal cause of bad decision making. The more fundamental cause is inadequate schemata that guide the construction of mental models through information search and interpretation. In fact, mental models may not be interesting except as "windows" into underlying schemata.

The principal problem in testing this hypothesis is that there are no standard methods to operationalize schemata. The basic principle underlying all operationalizations is that schemata are reflected in the processes which they drive or the output that these processes produce. Following this principle, we tentatively suggest schemata can be derived by obtaining a sample of mental models (e.g., of experts and novices in identical situations) and constructing from them maps that specify (a) how specific cues are related to fewer higher order concepts, and (b) how these higher order concepts are interrelated in form of cause maps (Weick, 1979) or influence diagrams (Bostrom et al., 1992). Following these steps, Hypothesis 2 can be tested by identifying features that distinguish between schemata of experts and novices which are consistently related to making effective decisions.

Hypothesis 3: Experts are more likely to have richer action repertoires. This hypothesis is illustrated by the state of medicine prior to the advent of antibiotics. In that period, patients were often given inadequate treatment (by present standards) not because of mistaken diagnoses, but because effective treatment was simply unavailable. The hypothesis can be tested by ensuring that a group of experts and a group of novices have identical accurate mental models of a situation, and asking them to list all the relevant options in their order of appropriateness. According to the hypothesis, a panel of independent experts will judge the lists of experts to have a higher percentage of appropriate and a lower percentage of inappropriate options.

Hypothesis 4: Experts are more likely to use appropriate decision rules. This hypothesis pertains to situations in which decision makers construct an accurate mental model and have the appropriate option in their repertoire, but fail to connect one to the other. This hypothesis can be tested as a by-product of the procedure outlined for Hypothesis 4. A higher percentage of experts with the most appropriate option in their list will rank it as most appropriate than the percentage of novices in the corresponding subgroup.

Klein, Wolf, Militello, and Zsambok (in press) used a similar procedure to that outlined for Hypotheses 3 and 4 to compare expert and novice option generation in chess. Klein et al. did not control the accuracy of subjects' mental models. Consequently, it is not clear what factors in the RPD model were most important for experts' superior option generation.

The set of four hypotheses constitute a diagnostic tool for training in decision making. Basically, training consists of repeated cycles of performance, diagnosis, feedback, and repeated performance. To be effective, instructors should provide trainees with detailed and accurate feedback. The four hypotheses guide them to test if an inappropriate decision should be attributed to an inadequate mental model (Hypothesis 1); the trainee's mental modeling (Hypothesis 2); the trainee's ignorance of the appropriate option (Hypothesis 3); or inability to identify correctly when this option applies (Hypotheses 4). Effective coaches help practitioners reflect on their action this way (Schön, 1983). Identifying consistent expert–novice differences vis-à-vis the four hypotheses can help the process by alerting instructors to look for these differences online, and by guiding the design of exercises that let trainees experience them (Needham & Begg, 1991).

In conclusion, we have argued in this chapter that schemata and mental models should be explicitly considered in the study of recognition-primed decision making processes. Studying mental models and schemata "involves inferring the existence and nature of entities that cannot be empirically proven to exist" (Rouse, Cannon-Bowers, & Salas, 1992, p. 1304). Some encouraging progress has already been made in measuring mental models (Endsley, 1995). The more difficult, and important, challenge is still ahead. Until valid methods to describe domain-specific schemata are developed, doubts that schemata and mental models are unnecessary embellishments will remain.

APPENDIX

For example, in one scenario, which takes place at night, the screen suddenly displayed three "blips" signifying the appearance of three potential targets at three different locations. The trainee's basic mission was to identify these targets. His options at this point were to approach one of the targets himself, order another boat to approach one of the targets, order the other boats to change course without approaching the target, and ask a shore station to track a target and keep him informed of its speed, its course, etc.. This is not a trivial task. Because the three "blips" appear in rapid succession, the trainee has to decide when and how to engage which target, and divide his attention between tracking the targets and coordinating his own and the other two boats' actions. Data were collected from several sources: an observer sat next to the trainee, took notes, and, when possible, probed him on-line; all communications to and from the trainee were audio-taped; the screen and a second screen (available to the instructor) that displays a larger area were video-taped; and, finally, a debriefing was conducted following the simulation.

REFERENCES

Amalberti, R., & Deblon, F. (1992). Cognitive modeling of fighter aircraft process control: A step towards an intelligent on-board assistance system. *International Journal of Man–Machine Studies, 36*, 639–671.

Bostrom, A., Fischhoff, B., & Morgan, G. (1992). Characterizing mental models of hazardous process: A methodology and an application to Radon. *Journal of Social Issues, 48*, 85–100.

Brezovic, C. P., Klein, G. A., & Thordsen, M. (1987). *Decision-making in armored platoon command.* Yellow Springs, OH: Klein Associates Inc.

Bryant, D. J., Tversky, B., & Franklin, N. (1992). Internal and external spatial frameworks for representing described scenes. *Journal of Memory and Language, 31*, 74–98.

Calderwood, R., Crandall, B. W., & Baynes, T. H. (1988). *Protocol analysis of expert/novice command decision-making during simulated fire ground incidents.* Yellow Springs, OH: Klein Associates Inc.

Calderwood, R., Crandall, B. W., & Klein, G. A. (1987). *Expert and novice fire ground command decisions.* Yellow Springs, OH: Klein Associates Inc.

Cohen, M. V. (in press). Metacognitive strategies in support of recognition. In *Proceedings of the Fifth Workshop on Uncertainty and AI.*

Endsley, M. R. (1995). Measurement of situation awareness in dynamic systems. *Human Factors, 37*, 65–84.

Gott, S. P., Bennett, W., & Gillett, A. (1986). Models of technical competence for intelligent tutoring systems. *Journal of Computer-Based Instruction, 13*(2), 43–46.

Howell, W. C., & Cooke, N. J. (1989). Training the human information processor: A review of cognitive models. In I. L. Goldstein (Ed.), *Training and development in organizations.* San Francisco, CA: Jossey-Bass.

Kardash, C. A. M., Royer, J. M., & Greene, B. A. (1988). Effects of schemata on both encoding and retrieval of information from prose. *Journal of Educational Psychology, 80*, 324–329.

Klein, G. A. (1993). Recognition-primed decision (RPD) model of rapid decision making. In G. A. Klein, J. Orasanu, R. Calderwood, & C. Zsambok (Eds.), *Decision making in action: Models and methods,* (pp. 138–147). Norwood, NJ: Ablex.

Klein, G. A., Wolf, S., Militello, L., & Zsambok, C. (in press). Characteristics of skilled option generation in chess. *Organization Behavior and Human Decision Processes.*

Lesgold, A., Robinson, H., Feltovich, P., Glaser, R., Klopfer, D., & Wang, Y. (1988). Expertise in a complex skill: Diagnosing X ray pictures. In M. T. H. Chi, R. Glaser, & M. Farr (Eds.), *The nature of expertise,* (pp. 312–342). Hillsdale, NJ: Lawrence Erlbaum Associates.

Lipshitz, R. (1993). Converging themes in the study of decision making in realistic setting. In G. A. Klein, J. Orasanu, R. Calderwood, & C. Zsambok (Eds.), *Decision making in action: models and methods,* (pp. 103–137). Norwood, NJ: Ablex.

Needham, D. R., & Begg, I. M. (1991). Problem-oriented training promotes spontaneous analogical transfer. Memory-oriented training promotes memory for training. *Memory & Cognition, 19*, 543–557.

Neisser, U. (1976). *Cognition and reality.* San Francisco: Freeman.

Reason, J. (1987). A preliminary classification of mistakes. In J. Rasmussen, K. Duncan, & J. Leplat (Eds.), *New technology and human error* (pp. 15–20). New York: Wiley.

Rouse, W. B., Cannon-Bowers, J. A., & Salas, E. (1992). The role of mental models in team performance in complex systems. *IEEE Transactions on Systems, Man and Cybernetics, 22*, 1296–1308.

Rouse, W. B., & Morris, N. M. (1986). On looking into the black box: Prospects and limits in the search for mental models. *Psychological Bulletin, 100*, 349–363.

Schön, D. A. (1983). *The reflective practitioner.* New York: Basic Books.

Shanteau, J. (1988). Psychological characteristics and strategies of expert decision makers. *Acta Psychological, 68*, 203–215.

Weick, K. E. (1979). *The social psychology of organizing.* Reading, MA: Addison-Wesley.

Whitfield, D., & Jackson, A. (1982). The air traffic controller's picture as an example of mental model. In G. Johanssen & J. E. Rijnsdorp (Eds.), *Analysis, design and evaluation of man–machine systems* (pp. 37–42). London: Controller, HMSO 1982.

Chapter 29

Recognition-Primed Decision Making as a Technique to Support Reuse in Software Design

◆

Christine M. Mitchell
John G. Morris
Jennifer J. Ockerman
Center for Human–Machine Systems Research
Georgia Institute of Technology

William J. Potter
NASA Goddard Space Flight Center

As automated control systems proliferate in such applications as space, aviation, and manufacturing, the human role shifts from manual operator to supervisory controller, emphasizing monitoring, situation awareness, and failure management (Sheridan, 1992). With this change in emphasis has come a corresponding change in the role of the control system designer (Rasmussen, 1986). The designer is a critical determinant of the effectiveness of the overall control system, designing the symbiotic relationship between human and computer controllers. This chapter describes two research efforts directed at assisting the designer of command and control software—specifically: (a) how to facilitate reuse of existing and successful software designs; and (b) how to avoid repeating past mistakes.

This research is based on a set of assumptions. First, extensive exploration of software design in general, and command and control software in particular, suggests that the design process might be usefully characterized as a naturalistic decision making problem (Klein, Orasanu, Calderwood, & Zsambok, 1993). Second, to support more effective design, including reuse, the recognition-primed decision (RPD) making model (Klein, 1989) provides an organizing structure for the specification of computer-based decision-support for designers of large-scale software systems. Finally, complementing RPD, case-based reasoning (Kolodner, 1993) offers both an additional theoretical framework together with a computational methodology to specify and implement support architectures for designers.

The current research explores this collection of assumptions in two research projects: A Design Browser and a Designer's Associate. The former is a computer-based decision-support system that makes design characteristics of past software systems available to the designer(s) of future systems. The Browser serves as an institutional memory at the conceptual design level and is predominantly passive in its approach to facilitating the exploration of previous designs. The second project, the Designer's Associate, is more active; it serves as a vehicle for collecting as well as disseminating design knowledge. To this end, it handles an expanded range of design knowledge, retains much of the structural characteristics of such knowledge, and serves as a design editor.

THE PROBLEMS OF DESIGN AND REUSE IN LARGE-SCALE SOFTWARE DESIGN

The design of large-scale software systems in general, and command and control software in particular, is a complex process. It is a complicated by many factors, few of which are easily ameliorated. Designer expertise, in many cases, is the only solution to the many difficulties. Unfortunately, expertise is often highly specialized and thinly spread. Moreover, the very nature of software design makes it difficult to formalize expertise or capture it in expert systems. Nevertheless, domain knowledge is embedded in application-specific artifacts generated during the design process (e.g., concept definitions, domain analyses, requirements specifications, conceptual designs, detailed designs, source code). Reuse of such artifacts is widely believed to be a key to improving software development productivity and quality (Biggerstaff & Richter, 1987).

Typically the central difficulty of design is not a lack of problem-solving skills, nor an absence of relevant design knowledge *per se*, but the fact that much relevant knowledge has been forgotten, is located in someone else's head, or lies within some long-forgotten document. In short, the problem is not the existence of knowledge, but its inaccessibility.

The present research posits that one key to assisting the designers of command-and-control software for next-generation systems is the dissemination and diffusion of task-relevant knowledge as embodied in the artifacts produced during the design of similar systems. Thus, an effective decision support system for such designers is one that makes previous experience accessible, facilitates recognition of relevant past experience, and supports its adaptation to the needs of the current situation.

Because both the Design Browser and Designer's Associate are implemented in proof-of-concept form for the designers of spacecraft command management systems, a brief description of the domain of satellite ground control and the role of command management systems is in order. It is important to note that, although the implementations are embedded in the particular details of satellite ground control and command management, the description of the design task and associated problems, together with the proposed architectures to support the design process, are intended to generalize at least minimally to the design of command and control software, and potentially to the more general design process for large-scale software systems.

SPACECRAFT COMMAND MANAGEMENT SYSTEMS

Satellite ground control is an example of supervisory control in which human operators configure communications and computer equipment, monitor data, and diagnose and compensate for system problems. The particular facility studied for this research is NASA's Goddard Space Flight Center. Data from near-earth scientific spacecraft are periodically transmitted to the ground. Spacecraft are controlled by a combination of real-time and stored commands. Spacecraft commands configure instruments, control data recorders, change spacecraft orientation, initiate transmission of science data to the ground, and so on. Real-time commands are executed immediately on receipt by the spacecraft. Stored commands are transmitted to the spacecraft in "real time" but are stored in the spacecraft's memory for later execution. A spacecraft's command management system is the focal point of stored command operations.

Despite their fundamental similarity in purpose and function, each NASA spacecraft typically has a "new" command management system. Each design, formulated without systematic consideration of past experience, risks the repetition of previous mistakes, forgoes the productivity benefits of reusing successful features, and may squander creativity by reinventing past solutions. It is important to note that the literature suggests that the reuse problems at NASA are, in general, typical of software development for large-scale systems (e.g., Banker, Kauffman, & Zweig, 1993).

The current research was initiated in direct response to NASA's concerns about reuse and the failure of broad reuse mandates (e.g., reuse must exceed 80%) to bring about significant reuse, associated reductions in costs, or improvements in effectiveness. The Design Browser and the Designer's Associate offer two complementary support architectures to assist designers in software reuse.

THE DESIGN BROWSER

Focused on conceptual design, the Design Browser defines an architecture that functions as an institutional memory—a computational structure in which to store and examine previous design experience. The Design Browser is intended to support three constituencies: software developers, operational users, and management. The intended users of the CMS Design Browser are: (a) NASA management who are responsible for timely development and effective operation of the system; (b) designers who work under contract to NASA to develop the software; and (c) users, also under contract to NASA, but employees of a different company, who initially specify and, subsequently, operate the satellite ground control system, including its command management function.

The Browser addresses decision making and problem solving at the critical design-review level. At NASA, critical design specification and its associated review are the final technical and legal description and review of the proposed design. The designers present the critical design specification to an audience comprised of NASA management and users. This review is a major milestone in the design process: at its conclusion a proposed design is approved, approved with legally binding changes, or rejected and returned with requirements for substantial improvements in its functions or forms. Given the importance of this stage of NASA's design process, a critical design review offers an appropriate proof-of-concept application for the initial implementation of the Design Browser.

Two components comprise the Design Browser: a conceptual framework and a case base that archives previous designs. Each component is described in the following sections and illustrated with examples from the application of the generic architecture to the NASA command management system domain.

Conceptual Framework

To facilitate the accumulation and rapid assimilation of design experience, the Design Browser provides a conceptual framework to organize features and functions that are common across a class of design applications. The conceptual framework attempts to support the first stage of recognition-primed decision (RPD) making—situation awareness. The framework provides designers with a set of concepts and an associated vocabulary with which to specify a design. The conceptual framework is intended to emphasize similarities among designs.

The Design Browser's conceptual framework has two hierarchical decompositions: one part–whole and the other functional. The simultaneous decompositions allow for a multidimensional description of the artifact as both a collection of parts and as a purposeful entity. The Design Browser implements the decompositions with two complementary design descriptions: building blocks (i.e., a hierarchical decomposition of objects that comprise the system) and functional components (i.e., sets of system objects grouped together with a recognizable purpose). Building blocks are components of the software system that are manipulated by the functions. A third characteristic of the conceptual framework, system issues, allows designers to annotate individual designs with information concerning such features as cost and performance. Issues complement building blocks and functions by archiving system-level information such as hardware and software platforms.

The Design Browser as implemented for NASA's command management systems attempts to use its conceptual framework as a unifying framework for CMS software design. The framework was developed by reviewing seven command management systems. Its building blocks consist of CMS components common across CMSs: data inputs, commands, activities, and so forth (see the pull-down menu for Building Blocks of Fig. 29.1). Likewise its functions define the major functional requirements of command management, such as command scheduling, command preprocessing, and so on (see the pull-down menu for Functions of Fig. 29.1). Based on extensive conversations with NASA command management personnel (i.e., NASA management, designers, users), salient issues involved in CMS design and reuse were identified (see the pull-down menu for Issues of Fig. 29.1).

Case-Based Retrieval System

The Browser further attempts to facilitate recognition-primed decision making by storing design experience in a case-based reasoning system. The Design Browser uses its conceptual framework as a method to organize and display information about existing designs that is, its case or *stories*. Stories are interesting examples of previous designs that illustrate how existing systems implement various system components or functions, or describe how important design issues are resolved.

Stories about design features are represented hierarchically. As is appropriate, graphical explanations of the building blocks, functions, and issues are also available. The hierarchical explanations make the large quantity of

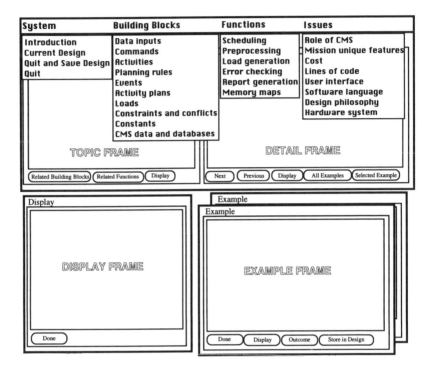

FIG. 29.1. Design browser interface.

data about previous designs more manageable and easily understood. The graphical view increases understanding of individual components or functions by augmenting text-based descriptions.

Stories have three parts: generic description, examples, and illustrations. The description is hierarchical and consists of a topic and associated details. The topic is a high-level description of the building block, function, or issue that is generalized from existing designs. The details provide lower level information about interesting aspects of the features common across designs (e.g., how particular instantiations differ). Details are direct links to the second part of the story—the examples. Examples consist of design-specific descriptions of features. The last part, illustrations, contains graphical depictions that complement the text describing a topic, detail, or example.

Fig. 29.2 illustrates a portion of the case base implemented for command management systems. In this example, the topic portion of the generic description characterizes the command component of a command management system. The associated detail describes, again in a somewhat generic manner, one of the features typically addressed in the design of the command—that is, how times are associated with time-delayed commands. Finally, the specific design of time-tagged commands for the SOHO space-

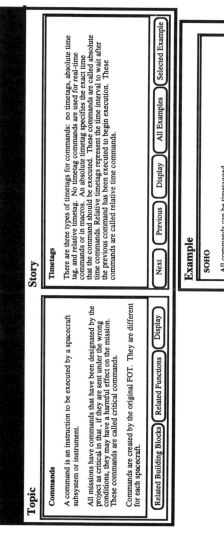

FIG. 29.2. Design browser example.

craft is described. The interface allows navigation throughout the case base to identify related building blocks and functions for individual topics and one or more examples, that is, cases for each detail. An individual design may also contain outcome information. For example, the outcome information may record an assessment of the extent to which an individual design was successful.

Empirical Evaluation

An empirical evaluation of the Design Browser for NASA's command management function was conducted. There were four subjects from each of the three user groups. The Design Browser was evaluated in two ways. First, a formal study assessed the extent to which users for specified tasks were able to retrieve useful information; the tasks were developed before hand in conjunction with the broader user communities. In addition, a survey was conducted to elicit subjects' perceptions regarding the utility of the Design Browser.

Two metrics were used in the formal analysis: product and process measures. A product measure attempts to ascertain if the subject arrived at the expected outcome; for example, object-oriented programming was used for three NASA systems. A process measure attempts to ascertain if the subject arrived at the expected answer—that is, the product—by the process expected by the designer and underlying the Browser's design, such as, movement from category to story to example. Two types of tasks were performed. The first type was used to demonstrate the Browser's ability to support subject navigation. The second type of task was used to demonstrate the Browser's ability to support representative design tasks. Over 2 weeks, 12 subjects—NASA managers, spacecraft operators, and developers—used the Design Browser to complete 10 tasks each. Observations were recorded manually, captured by the Browser in a computer log file, and recorded on video tape. An initial questionnaire was used to gauge subject expertise and a final questionnaire was used to ascertain subjects' perception of Browser usefulness.

Subjects had an average of 4.03 years of involvement in command management system design. Out of 120 tasks performed (12 subjects and 10 tasks) 84.17% resulted in the expected product and 85% resulted in the expected process. All subjects believed the Browser to be useful. One can infer from these data that the principles underlying the Design Browser can be effective in enhancing one's awareness of past conceptual designs.

BEYOND BROWSING: A DESIGNER'S ASSOCIATE

The Design Browser, combining a design experience case base and a conceptual framework, is a potentially useful first step to support software reuse. The Designer's Associate complements and extends the Design Browser by providing more extensive support for understanding previous designs and *creating* new designs based on reusing and adapting previous experience. The Designer's Associate differs primarily from the Design Browser in that it takes a more *active* role in the design process. It is not only a vehicle for preserving and disseminating design knowledge (i.e., the situation-awareness component of the RPD model), but also for collecting, synthesizing, and creating such knowledge as well. In this respect it provides a mechanism for supporting those facets of RPD concerned with formulating a course of action and evaluating that course of action through a process of mental simulation; that is, assembling the pieces of a plausible solution, based on past experience, and evaluating the result to see the extent to which it addresses the problem at hand.

Structures Comprising the Designer's Associate

The Designer's Associate is based on the assumption that a designer, in an unfamiliar situation, is unable to utilize a recognition-primed decision-making approach. There are two reasons. First, a designer may be unable to find useful cues in the sea of incomplete, ambiguous information that constitutes the statement of the problem. Second, even if such cues were obvious, to the less experienced designer (relative to the problem at hand) a satisfactory design strategy or solution may not be obvious.

There is nothing more basic to thought, perception, and action than categorization (Lakoff, 1987). As such, categorization plays a key role in recognition-primed decision making—particularly in the first RPD step of situation awareness. Thus, in order to effectively disseminate knowledge, the Designer's Associate must first assist the designer in categorizing the current problem relative to past problems. That is, it must assist the designer in: (a) drawing salient cues from the statement of the problem, and (b) in light of such cues, placing the current problem within the context of similar past problems.

The notion of categories plays a pivotal role in the Designer's Associate for two reasons. First, category systems can serve to organize large amounts of information in ways that require little cognitive effort. Second, category systems are made possible by the fact that, for a skilled decision maker, the perceived world comes as structured information rather than as arbitrary or unpredictable attributes (Rosch, 1978). The Designer's Associate uses taxonomies to convey the categorical structure of the domain. Utilizing and assimilating such structures may enable the less experienced designer to more rapidly acquire an expert's understanding of the problem domain.

The Designer's Associate architecture consists of four complementary knowledge structures that describe software systems and their components: part–whole decompositions, means–end hierarchies, taxonomies, and a case base. Design artifacts (e.g., descriptions of purpose and function, requirements specifications, detailed design documents, and source code) can be described in terms of part–whole decompositions. For example, a design problem description (i.e., a statement of requirements) consists of a number of somewhat independent subproblems. The relationships between these subproblems and the problem as a whole can be captured in a part–whole decomposition.

The artifacts of a given system are linked to one another via their components (i.e., the subelements of an artifact) in a means–end hierarchy (see Fig. 29.3). This representation allows one to determine and describe the relationships between artifacts and their components across the levels of the hierarchy. Research suggests that a hierarchy of this kind can facilitate understanding and evaluation by allowing the designer to consider the level of abstraction (i.e., detail) appropriate to the immediate needs (Rasmussen, 1986).

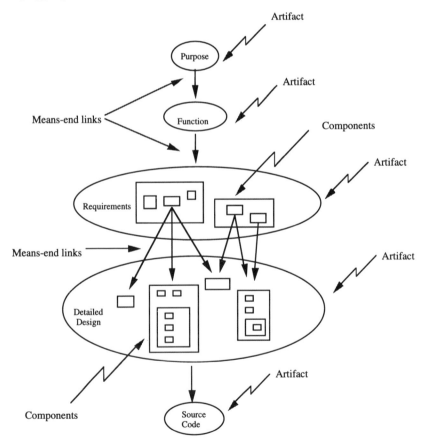

FIG. 29.3. Means–end hierarchy.

The Designer's Associate uses four taxonomies to link systems as a whole, as well as the problems and solutions they embody: system, problem, object, and function. The system taxonomy is used to categorize systems and is organized according to system purpose and function. For example, space-craft differ markedly based on the phenomenon they are designed to observe: Some are passive observatories, whereas others are continually redirected at targets of interest.

The problem taxonomy is used to categorize the problems a system is intended to solve and is organized with regard to both recurring and system unique subsets of requirements (i.e., problem statements). For example, load generation is a common design problem; however, there are many different types of loads one might construct (e.g., stored commands, spacecraft software updates, etc.).

The object and function taxonomies categorize the two primary facets of software solutions—data and behavior. The object taxonomy is used to categorize solution components that are implemented as data structures or classes and is organized according to the part–whole configuration of such data structures or classes. For example, in the parlance of object-oriented programming, a stored command load might be represented as one class and a database of permissible spacecraft instructions might be represented as another. Providing access to existing classes, particularly when linked with the problems they purport to solve, is one mechanism by which software reuse can be enhanced.

The function taxonomy is used to categorize solution components that are implemented in functional form—for example, organized according to input, output, and side effects. Stored command load generation, for instance, is a typical function of command management systems. Such functions take as input an operation plan and a database of spacecraft instructions and output a stored command load (i.e., a sequence of time-tagged spacecraft instructions).

This knowledge organization reflects to some degree the structure of codified knowledge within the command management system design domain. Systems are drawn from operations concept documents. Problems are drawn from requirements specifications. Objects and functions are drawn from both detailed design documents and source code.

The Designer's Associate Case Base

The taxonomies, means–end hierarchy, and part–whole decompositions utilized by the Designer's Associate provide a vehicle for the less experienced designer to locate and retrieve relevant cases from the case base. The case base contains two types of cases: instantiations of systems taken as a whole, and the individual components and subcomponents that comprise a specific instantiated system. Although one system may provide a template or architecture for a solution, no single system is likely to exhibit all the necessary capabilities or behaviors. Hence, the Designer's Associate enables a designer to utilize (a) a single large case (i.e., a specific system) to provide an initial, adaptable solution, and (b) smaller, minor cases (i.e., specific objects or functions from different systems) to guide the design of capabilities and behaviors absent from the case from which the initial solution is drawn.

The Designer's Associate as a Knowledge Source

The Designer's Associate supports two search strategies, categorical and means–end. The categorical strategy is predicated on the assumption that similar systems exhibit similar problems. Both strategies assume that similar problems are often solved by similar solutions. Locating and studying an existing system similar to that envisioned or the solution to a past problem similar to the one at hand is often a useful design strategy.

Categorical searches for design-relevant information begin at a basic-level category and proceed downward within the taxonomic structure. Such categories are intended to be maximally distinct—as are cats, dogs, chairs, and cars—and not easily confused—even by those with little experience. Entities at each level of the taxonomy are distinguished one from the other by some significant attribute and the Designer's Associate makes such distinguishing attributes obvious. Distinguishing attributes at higher levels within the taxonomy are more critical discriminators than those at lower levels. The Designer's Associate attempts to provide the less experienced designer with a more coherent view of the design domain via its taxonomic structure and basic-level categories.

Whereas the taxonomies are used to facilitate searches within a specific class of entities, the means–end hierarchy is intended to map one class of entities to another, namely problem to solution. In the Designer's Associate, searches within the means–end hierarchy (Fig. 29.3) can begin at the system level and proceed to the objects and functions via the problem level. However, because the means–end hierarchy, part–whole decomposition, and taxonomies intersect at each component, one is able to change the character of search at any point (i.e., shift from a means–end search to a categorical search). In other words, the Designer's Associate provides links from the means–end hierarchy, which defines a path tracing a *particular* solution space, to the various taxonomies, which at each level offer alternative solutions.

The Designer's Associate as an Active Aid

The knowledge structures that comprise the Designer's Associate are intended to facilitate situation awareness. A second facet of the Designer's Associate, a design editor, also assists the designer in the creation of new designs and in so doing supports the use of an RPD-based design strategy (i.e., it can facilitate the formulation and evaluation of a proposed course of action). The design editor has two functions: to elicit knowledge and to make the disseminated knowledge more readily usable. The design editor consists of two components: an informal designer's notebook and a formal design description editor. Formal design documents, such as detailed design specifications, do not capture all pertinent information. Designers typically take notes and draw sketches to record their progress, keep track of ideas that need to be considered at the appropriate time, record decisions, identify alternatives, and so on. It is just such information that the designer's notebook is intended to capture.

The Designer's Associate also enables a designer to create new, formal design descriptions from fragments of existing system descriptions. Because the process of design often involves the successive integration of components into a growing problem/solution statement, the Designer's Associate supports the design process by allowing the designer to create a new means–end hierarchy and "grow" it from the top down, reusing when appropriate components as well as the structures in which they exist (i.e., the means–end hierarchy and part–whole decompositions). Hence, the designer need not rebuild design descriptions from scratch, but can construct them in a cut-and-paste manner. In this way, Designer's Associate provides information in more readily usable form and the Designer's Associate knowledge grows and evolves in step with current problems.

CONCLUSION

This research offers two theories, associated architectures, and proof-of-concept implementations of decision support for designers of large-scale software systems such as command-and-control systems. The theories are engineering-oriented and rest on a combination of recognition-primed decision making as a model that describes expert decisions in naturalistic settings and case-based reasoning. The architectures are domain-general; (Clancey, 1987); that is, they specify both a conceptual structure (e.g., design features, decompositions, and taxonomies) and a computational structure (e.g., a case-based reasoning system) that characterize decision-support software for a class of applications, namely command and control system software. To demonstrate and empirically evaluate both the theory and proposed architectures, the Design Browser and the Designer's Associate are implemented as decision-support systems for participants in the design process for NASA satellite command management software. As noted previously, the Design Browser has been extensively evaluated with data that suggests that users found it both useful and usable. The Designer's

Associate is currently being implemented and will be formally evaluated using command management system software designers.

REFERENCES

Banker, R. D., Kauffman, R. J., & Zweig, D. (1993). Repository evaluation of software reuse. *IEEE Transactions on Software Engineering, 19*(4), 379–389.

Biggerstaff, T. J., & Richter, C. (1987). Reusability framework, assessment, and directions. *IEEE Software, 4*(2), 41–49.

Clancey, W. J. (1987). *Knowledge-based tutoring.* Cambridge, MA: The MIT Press.

Klein, G. A. (1989). Recognition-primed decisions. In W. B. Rouse (Ed.), *Advances in man–machine systems research.* (pp. 47–92). Greenwich, CT: JAI.

Klein, G. A., Orasanu, J., Calderwood, R. & Zsambok, C. E. (Eds.). (1993). *Decision making in action: Models and methods.* Norwood, NJ: Ablex.

Kolodner, J. L. (1993). *Case-based reasoning.* San Mateo, CA: Kaufmann.

Lakoff, G. (1987). *Women, fire, and dangerous things: What categories reveal about the mind.* Chicago, IL: University of Chicago Press.

Rasmussen, J. (1986). *Information processing and human–machine interaction: An approach to cognitive engineering.* New York: North-Holland.

Rosch, E. (1978). Principles of categorization. In A. Collins & E. E. Smith (Eds.), *Readings in cognitive science* (pp. 312–322). San Mateo, CA: Kaufmann.

Sheridan, T. B. (1992). *Telerobotics, automation, and human supervisory control.* Cambridge, MA: MIT Press.

Chapter 30

Myths of Expert Decision Making and Automated Decision Aids

◆

Kathleen L. Mosier

San Jose State University Foundation
NASA Ames Research Center

In the past decade, automated decision aids and expert systems have been implemented in domains such as medical diagnosis, nuclear energy plant operations, sales and scheduling support, and, in the aerospace domain, glass cockpits and Air Traffic Management. Expert decision makers are being exposed to and are required to utilize these aids as they do their jobs and make necessary decisions. In this chapter, I discuss several of the myths, or misconceptions, that are typically associated with human expertise, and with expert systems and automated decision aids. These have been derived from the literature on expert systems and naturalistic decision making, and from interviews with glass-cockpit pilots and instructors, system operators in the aviation domain. The intent of the discussion is not to diminish the importance of automated aids, but rather to address several areas in which common perceptions of these systems and of the users they are intended to serve do not coincide with reality, and to suggest directions for research on how automated systems might best enhance the expert decision-making process. Although most of the anecdotal examples come from aviation, the issues are relevant to any naturalistic domain in which human expertise is being supplemented or replaced by automated systems.

Consider the following scenario: You are the veteran captain of a major passenger airline, flying a new-generation, two-engine aircraft out of San Francisco. Prior to taxi, you were advised that birds have been sighted at the departure end of the runway. As you leave the ground and begin to climb, the fire handle for your left engine illuminates briefly, then goes out. At the same time, the message "ENGINE FIRE" appears on the Electronic Warning System, and the electronic "ENGINE FIRE CHECKLIST," triggered automatically, outlines the steps for #1 engine fire, including engine shutdown. The engine indications, displayed on the center panel, show the #1 engine deteriorating for a few moments, then recovering. Readings for the #2 engine, however, show that it is deteriorating rapidly, suggesting that it is actually the more severely damaged engine—but nothing about this engine shows up on your electronic aids. You need to go back and land at the same point from which you took off (Mosier, Palmer, & Degani, 1992).

Quite probably, you have ingested some of the birds from the runway into one or both of your engines. This is the dilemma that you face: All of the cues that you have traditionally used to diagnose engine problems—that is, engine parameter readings, engine fan and core vibration, and so forth—are telling you that the left engine is recovering strength and the right engine is deteriorating. However, your electronic decision aids, which have been implemented primarily to help you make quick, accurate decisions, are instructing you to shut down the "fire damaged" engine, leaving you with only one marginally operative engine. What do you do?

In order to ensure the most efficient and effective usage of automated decision aids, it is important to start out with accurate assumptions about the nature of these aids, and about precisely how they fit into theories and models of expert decision making in real-world environments (e.g., the Naturalistic Decision Making features presented by Orasanu & Connolly, 1993, or the Recognition-Primed Decision Making model described by Klein, 1993). These systems undeniably have an impact on expert decision making, if only because they introduce an additional "cognitive agent" into the process (Woods, 1994). The presence of automated decision aids also alters, subtly or blatantly, the role of the human expert.

The potential advantages of automated decision-aiding systems, in terms of speed, internal consistency, and ability to deal with large amounts of data, are obvious. However, potential negative consequences of their implementation, such as inappropriate over- or underreliance on automated aids (Riley, Lyall, & Wiener, 1993), or confusion over or inefficient use of systems (Sarter & Woods, 1993) have also been cited. Some of these negatives may be due to misrepresentation on both sides of the human–machine system. Many automated and expert systems have been designed to correspond with a model of expertise that may not be representative of expert decision making in a dynamic, naturalistic environment—that is, one characterized by factors such as ill-structured problems, high degrees of risk, ambiguous or incomplete information, or time constraints. This is not meant to be a criticism of such systems, but rather is discussed as a source of misunderstanding and misconstrued expectations.

Additionally, the users of automated decision aids, hindered by factors such as poor interface design, inadequate training, or lack of familiarity, may not understand the design intent or the workings of automated and expert systems. Many expert operators have misguided notions of what their automated system can and cannot be expected to do (Will, 1991), and system users, as well as those who design them or prescribe their use, may fall prey to misconceptions, or "myths" concerning the nature and function of automated aids.

MYTH: DECISION AIDS AND EXPERT SYSTEMS MAKE DECISIONS THE SAME WAY EXPERT HUMANS MAKE THEM

Automated systems are often advertised as functioning just like (or better than) expert decision makers:

> A goal in the design of an intelligent system is to encode, in a computer program, the facts an expert has and the methods of reasoning about them, together with formal methods of reasoning about unstructured situations. In that sense, an intelligent system may be viewed as a descriptive model of an expert reasoning process about a problem domain. (Stephanou & Sage, 1987, p. 780)

Although this may be true in isolated, structured, combinatorial domains, or in real-world systems for which our models are accurate and complete (Dreyfus, 1989), it is not necessarily true in dynamic, naturalistic domains. Sophisticated decision-aiding systems solve problems analytically, constructing solutions selectively and efficiently from a space of alternatives (Waterman, 1985). Their approach corresponds to the classical theories of decision making that focus on analytical problem solving (i.e., alternative selection) and pay only lip service to the process of how the problem was formulated in the first place (Winograd, 1989). This means that the critical nuances of situation assessment may be present only in the most generic, preprogrammable form, and assumes that the domain is structured and isolated from outside elements.

Human experts, in contrast, often make many decisions nonanalytically, often "intuitively" (Dreyfus & Dreyfus, 1986; Simon, 1983), and focus on situation assessment as the most critical aspect of decision making (Kaempf & Klein, 1994). When we observe experts making decisions in dynamic real-world settings, such as the operating room or the aircraft flight deck, we see that they act and react on the basis of their prior experience. They generate, monitor, and modify plans to meet the needs of the situation (Klein, 1993). In naturalistic settings, expert humans tend to deliberate more during the diagnosis phase of decision making than during response selection, and also tend to expend more effort on situation assessment than novices do. Knowledgeable, experienced decision makers are typically able to recognize and interpret situation types, and it is out of this process that the generation of a workable response option evolves. In most cases, once the situation is understood, the appropriate course of action is obvious (Kaempf & Klein, 1994).

In terms of emulating experts, there may be a vast difference between what automated systems can do and what they are sometimes advertised or perceived as being able to do. Although they are often referred to as "intelligent," the intelligence that even the most sophisticated systems possess consists of competent, analytical (rather than intuitive), rule-based reasoning, and is limited to if...then expert knowledge for which rules have been programmed. For example, the message on the electronic checklist in the bird-strike incident may have been generated by the code "if ... fire handle, then ... engine fire checklist." The system could not, however, go past this code to the understanding that the message was inappropriate in view of other factors.

The processes of expert systems and automated decision aids, then, do not always correspond to those of real-world experts in complex domains, but rather are limited to the context-limited, rule-based reasoning that Dreyfus and Dreyfus (1986) have described as the "novice" stage of expertise development. Within these confines, automated aids can be indeed faster than humans, and can be more accurate and internally consistent with respect to the application of the rules.

Unless competent decision making is sufficient, however, and important contingencies can be anticipated and rules for response can be programmed into a computer, the expert's ability to use intuition and nonanalytical reasoning is indispensable (Dreyfus & Dreyfus, 1986). Human expertise can be characterized by creativity, adaptability, the presence of a broad focus, and the ability to incorporate sensory experience, analogical reasoning, and "commonsense" knowledge. Because automated systems can lack these qualities, their advantages may be limited, unless supplemented by human expert capabilities.

MYTH: EXPERT SYSTEMS AND AUTOMATED DECISION AIDS CAN MAKE EXPERTS OUT OF NOVICES

Perhaps one of the most troublesome misconceptions of automated and expert systems is that they are able to make experts out of novices. "When a novice uses an expert system, the resulting expectation is expert performance" (Will, 1991, p. 176). The novice, armed with an expert system or decision aid, is expected to be able to solve problems beyond the realm of his or her knowledge or experience. "The expert system in effect amplifies his/her expertise in the given domain, and can sometimes be used in place of a specialist" (Basden, 1984, p. 64). In the domain of aviation, for example, decision aids are being proposed to augment the inexperience of general aviation pilots, and even to enable relatively inexperienced pilots to fly commercial glass cockpits in standard operations. The theory underlying this proposal is that the decision-aiding system can compensate for a lack of experiential knowledge and expertise in the human.

This proposal has several potentially dangerous pitfalls. The notion that inexperienced operator + automated aid = expert performance implies that the automated or expert systems can somehow replace the missing human expertise. This is a misleading notion. As mentioned, computers are limited to specific technical knowledge. They are not as versatile as human experts, and the artificial expertise they offer is different from human expertise. However, we often expect automated aids to provide precisely the capabilities they cannot, such as commonsense reasoning, interpretation of inconsistent information, or integration of contextual factors (Hart, 1992; Waterman, 1985). The electronic aids in the bird-strike cockpit, for example, could not help the crew deal with engine readings that were inconsistent with their advice, could not make the connection between the engine problems and the birds on the runway, and could not alert the crew to these inadequacies.

The notion of automated expertise may have dangerous implications when novices, aided by decision-aiding systems, believe they are as capable or skilled as experts, and enter into situations that are at the edge of or beyond their capacity to handle. The irony of this is that automated systems are best able to aid the novice when situations are routine, or covered by standard procedures. It is the complex, unanticipated, novel anomaly that most taxes the limits of both the inexperienced operator and the automated system. Moreover, most automated aids cannot recognize or convey to the user the limits of their abilities. "When pushed beyond their limits or given problems different from those for which they were designed, expert systems can fail in surprising ways" (Waterman, 1985, p. 182). Inexperienced operators will lack the knowledge base to recognize these limitations, and may be unable to recognize even blatant errors.

An additional danger associated with the combination of inexperienced operators and automated systems is that, in the long run, the novice may not get the experience or exposure requisite to the development of domain expertise. For example, the meanings associated with constellations of cues and changes in cue clusters are not available to novices (Kaempf & Klein, 1994). With continued reliance on automated aids, it is possible that novices may never develop the skills and learn the cue combinations and contextual interactions associated with situation assessment and response generation, making the human decision makers, in effect, less efficient and equipped to deal with events—especially those involving any kind of automated system failure—than they might have been without the aid.

MYTH: THE HUMAN OPERATOR CAN EASILY
IGNORE AUTOMATED INPUT
AND USE THE TRADITIONAL CUES

When an automated aid is introduced, it changes patterns of cue utilization and the way situations are assessed and dealt with. Operators learn (and are instructed) that the automated aid is the "best" cue for making a decision. If the system proves initially to be generally reliable, and reflective of the results of the operators' own diagnostic processes, what may (and does) happen over time is that operators check automated output *before* anything else, and may not look for traditional, meaningful cues even when they are available, especially if time is short. As described by Hopkin (1994), for example:

> The proposals made by the machine are usually highly efficient and safe, and do indeed offer the best practical option. [Air Traffic] Controllers are therefore trained to accept the solutions offered in most circumstances because they really are the best. Gradually as controllers learn to trust the computer assistance they may come to accept its proposals more wholeheartedly but also tend to examine them less critically and less rigorously. (p. 316)

Additionally, automated aids are designed to be highly salient, difficult to ignore, and easier to use than traditional methods. Their availability feeds into the general human tendency to "satisfice," or take the road of least cognitive effort (Fiske & Taylor, 1991; Simon, 1983). Results of research on salience effects on decision making indicate that in diagnostic situations, the brightest flashing light or the largest or most focally located gauge—that is, the most salient cue—will bias the operator toward processing its diagnostic content over that of other stimuli (Wickens, 1984). In some cases, automated cues may replace other information, making a cross-check with traditional cues impossible.

In the bird-strike flight, both the electronic warning system message and the electronic checklist appeared directly in front of the nonflying crew member, and required no interpretation or integration of data on the part of the pilot. The combination of salience and ease of use made the automated messages difficult to ignore. Diagnosing the significance of the engine parameter readings, in contrast, involved more effort, as the information was located in the center instrument display panel rather than directly in front of either pilot, and each engine's status had to be evaluated independently via an integration of several parameters, and compared with the other.

Automated aids, therefore, are not easily disregarded, and in fact may be accompanied by diminished access to traditional cues. Further, the tendency for automation to engulf the decision-making field is likely to be exacerbated by the extent to which organizations encourage or mandate the use of automated decision-aiding systems, and the extent to which automation is perceived to yield more accurate or reliable outcomes than operators' own judgment.

MYTH: AUTOMATED DECISION AIDS TAKE INTO ACCOUNT MORE FACTORS THAN HUMAN EXPERTS

Automation has greatly increased our capacity for measuring, differentiating, and integrating data (Bennett & Flach, 1994). The enormous data storage and computational capabilities of automated systems evoke the perception that they can, in theory, take into account far more information than a human expert. Although it is true that they perform prescriptive analyses, which humans usually have neither the time nor the ability to do, it is more accurate to say that automated systems can take some *different* factors into account than human experts. Automated systems can process only what they have been programmed to process. They can do this in a highly consistent manner, but they are by nature contextually blind (Winograd & Flores, 1986).

In naturalistic settings, external or situational factors can be critical to diagnosis and decision making. Human experts can incorporate contextual information, and also have access to information outside of their domain of expertise that may impact the problems being solved by them:

> Human experts can look at the big picture—examine all aspects of a problem and see how they relate to the central issue. Expert systems, on the other hand, tend to focus on the problem itself, ignoring issues relevant to, but separate from, the basic problem. (Waterman, 1985, pp. 14–15)

Automated flight systems, for example, can pick up electronic signals from the ground, and guide the aircraft to a safe landing in heavy fog. They cannot, however, know that a maintenance truck is stalled in the middle of the runway. Flight management computers, given a designated position and altitude, can determine the course and the precise spot at which the aircraft should start a standard descent in order to arrive at the prescribed point in vertical and lateral space. The computer does not consider, however, whether this descent path will put the aircraft on a direct collision course with another aircraft. The electronic checklist system described at the beginning of this chapter could sense the presence of fire in the #1 engine. Once the checklist was generated, however, the system could not rescind its directive if the fire went out, and did not have the "common sense" to warn the crew that shutting down the #1 engine would put them in the precarious position of trying to complete the return to the runway with only one rapidly deteriorating engine.

MYTH: HUMAN EXPERTS WILL BE ABLE TO TELL WHEN THE SYSTEM IS MAKING A MISTAKE

Vigilant human experts may be able to detect errors in automated systems, if the processes of the system are observable to the operators. Although experience and expertise would seem to afford human operators some advantage in detecting automated errors, this may not always be the case. Often, systems are represented as "intelligent," "knowledge-intensive," "learning," or "understanding," leading even the expert user to believe the systems are smarter or more capable than they are themselves. Will (1991) commented on this when, contrary to his predictions that domain experts would notice errors in a defective decision-aiding system, he found that experts were no more likely than novices to detect flaws. In fact, both experts and novices expressed confidence in their "wrong" answer to the problem, displaying reliance on a false technology.

One of the biggest obstacles to human recognition of system errors is the poor quality of feedback about the activities of automated systems (Woods, 1994). For example, it is often difficult or impossible to trace the processes or predict the activities of automated flight systems. In fact, three of the most common queries glass-cockpit pilots make about automated flight systems are: What is it doing? Why is it doing that? What is it going to do next? (Wiener, 1989). Automated systems also, as mentioned earlier, do not convey a sense of their limitations to the user, and may not make obvious what they cannot do. The pilots of the bird-strike flight had no clear information to tell them if the automated system "knew" whether or not the fire was still present in engine #1, whether it knew about the worsening #2 engine, or whether or not it knew that birds were the source of their engine problems.

Human inefficiency in vigilance tasks may also interfere in expert detection of automated errors. Many studies of vigilance have established the fact that the ability to monitor for the occurrence of infrequent, unpredictable events typically declines over time. Automated systems, especially those that are perceived to be highly reliable, may induce complacency in the operator, exacerbating the tendency toward nonvigilant monitoring (e.g., Parasuraman, Molloy, & Singh, 1993).

The combination of these factors may result in a human–computer system that is not conducive to automated error detection, even by domain experts. Additionally, a long-range hazard of extensive use of automated decision-aiding systems is what Hart (1992) has termed the *deskilling* of experts:

> If people come to rely on a computer system for advice and guidance then there is the chance that human expertise will become rare. This could result in deskilling and a "regression towards the mean" of knowledge ... [and a tendency to] trust the computer even when it is not authoritative (p. 30)

MYTH: IT IS OBVIOUS THAT ULTIMATE RESPONSIBILITY FOR DECISIONS REMAINS (AND SHOULD REMAIN) WITH THE HUMAN OPERATOR

Historically, automated and expert systems have functioned as *consultants* in the decision-making process. Through tradition or regulation, the user has maintained ultimate decision-making authority and borne responsibility for the outcomes of decisions. In aviation, for example, "The pilot in command of an aircraft is directly responsible for, and is the final authority as to, the operation of that aircraft" (U.S. Federal Aviation Regulation 91.3). As these systems become more capable of autonomous functioning, and are given more power in the guidance of human activity, however, the question of ultimate responsibility becomes more complicated. Human operators may endow automated aids with personae of their own, capable of independent perception and willful action (Sarter & Woods, 1994), and may see reliance on them as the most efficient way to get decisions made, particularly in times of high workload. In fact, because one explicit objective of automating systems is to reduce human error, automating a decision-making function may communicate to the operator that the automation *should* have primary responsibility for decisions.

The issue of responsibility in the use of expert systems and decision aids is complex and confusing. Who—or what—should be liable if the human–automation system fails? Although responsibility for actions still rests with the user (Mockler, 1989), the legal status of the user of expert system technology is not always clear (Will, 1991), especially in instances in which the user ignores or overrides the advice of the system. Lawsuits may, in the future, result if system aids are consulted and fail to perform correctly, give inaccurate or misleading indications, or are incorrectly used; or, conversely, operators may be liable for the nonuse of an available system (Zeide & Liebowitz, 1987). In fact, for "certain sensitive, delicate or hazardous tasks [such as aircraft requiring fast and accurate response beyond human capability], it may be unreasonable not to rely upon an expert system" (Gemignani, 1984, p. 1045).

This issue may seem to be somewhat peripheral to the interaction of experts and automated decision aids. The question of responsibility, however, is certain to take on more importance and relevance in the future. The presence of automated aids may subtly erode the decision maker's role, and foster almost an abdication of responsibility to the system. As automated systems become more prevalent and more autonomous, deskilled experts may no longer have the degree of expertise to judge when the system is performing correctly, or to carry out the necessary degree of supervision (Morgan, 1992). "There is also a problem of over-expectation that obscures responsibility ... there's a kind of mystique—an objectification—that if the expert system says so, it must be right" (Winograd, 1989, p. 62).

Five out of eight of the crews in the bird-strike simulation described earlier did follow the directive of the electronic systems. At least one of these captains, articulating the thought processes preceding his shutdown of the #1 engine despite contradictory indications on the engine parameter gauges, commented that the *checklist said they should* shut it down.

RESEARCH ON THE INTERACTION BETWEEN DECISION MAKERS AND AUTOMATED DECISION AIDS

The use of automated systems is likely to become more and more critical as naturalistic environments become more complex and data-intensive. As yet, we have only begun to explore the nature of the interaction between expert human decision makers and automated decision aids. It is clear from the foregoing discussion, however, that characteristics and capabilities of present generation automated aids may be misrepresented, and that they may conform to a notion of expertise that does not describe expert decision makers in real-world situations. Hypotheses about how expert decision making can best be supported and enhanced by these aids need to be formulated and tested in naturalistic settings.

One potential product of this research should be a set of guidelines for the design of automated decision aids. In order for a human–automation system to function most efficiently and effectively, the strengths of each should be capitalized on, and care must be taken to offset or compensate for each agent's limitations. Among the strengths of automated systems are their abilities to provide access to large databases of information, to sort through that information much faster and more comprehensively than the human, and to monitor the consistency of the experts' choices and behavior (Evans, 1987). These capabilities can be utilized by humans to reduce situational ambiguity (e.g., by calling up information that may aid diagnosis) and to guard against human error or bias (e.g., "Do you really want to delete this file?" or " ... this flight plan?"). Automated systems can also function as powerful learning tools, if their processes are transparent and observable to the learner.

Research guiding system design must also incorporate studies on the appropriate input and timing of information to be presented, as well as on the role of the decision aid. The capabilities of automated decision aids range across many levels. They can, for example, simply display data; monitor systems and alert the user to anomalies; provide trend information, advisories, or action directives; or provide wrong decision protection—that is, prevent the operator from implementing a "dangerous" decision. Billings (1996), in his discussion of aviation automation proposed a taxonomy of levels of "human-centered" automation. At one extreme is complete human manual control; at the other is autonomous automated operation. I would suggest that somewhere in the mid-range of this taxonomy, at the levels of assisted or shared control (i.e., the operator has authority and discretion both in the use of the aid and in the acceptance of its advice) or management by delegation (i.e., the operator delegates operational tasks to the system), are the automated capabilities which will be found to be most "expert-centered." That is, those levels of automation that assist the human operator but remain under human control will be most amenable to human expert intervention and error detection, and will supplement rather than attempt to replace human expertise.

Finally, given the character of automated aids that experts are and will be dealing with in the near future, the importance of comprehensive training for system users cannot be overemphasized. Operators must have sufficient knowledge of what an automated system can do, what it "knows," and how it functions within the context of other systems, as well as knowledge of its limitations, in order to utilize it efficiently and exploit its real capabilities. Only through careful instruction and comprehensive knowledge will the myths ascribed to automated systems be dispelled. Additionally, procedures can be implemented to foster the maintenance of expert skills along with the skillful use of automation.

The implementation of automated decision aids and expert systems presents challenges to system designers, to expert users, and to the research community. The requirement for designers is to create systems that are "expert-centered" by preserving and enhancing human expertise. Expert operators must develop techniques to utilize automated aids efficiently and effectively, neither over- nor underestimating their capabilities. The challenge to the research community is to enable expert users and system designers to meet their respective goals, through empirical exploration and the establishment of principles for the demythologized interaction of human experts and automated systems.

ACKNOWLEDGMENTS

Many thanks to the editors for their comments and suggestions. This chapter was written under the support of NASA grants NCC 2–798 and NCC 2–327.

REFERENCES

Basden, A. (1984). On the application of expert systems. In M. J. Coombs (Ed.), *Developments in expert systems* (pp. 59–75). New York: Academic Press.

Bennett, K. B., & Flach, J. M. (1994). When automation fails... In M. Mouloua & R. Parasuraman (Eds.), *Human performance in automated systems: Current research and trends* (pp. 229–234). Hillsdale, NJ: Lawrence Erlbaum Associates.

Billings, C. E. (1996). *Human-centered aviation automation: Principles and guidelines* (Tech. Mem. No. 110381). Moffett Field, CA: NASA Ames Research Center.

Dreyfus, S. (1989, Spring). In Davis, R. (Ed.), Expert systems: How far can they go? Part One. *AI Magazine* 61-67.

Dreyfus, H. L., & Dreyfus, S. E. (1986). *Mind over machine.* New York: The Free Press.

Evans, J. St. B. T. (1987). Human biases and computer decision-making: A discussion of Jacob et al. *Behavior and Information Technology, 6,* 483-487.

Fiske, S., & Taylor, S. (1991) *Social cognition.* Reading, MA: Addison-Wesley.

Gemignani, M. (1984). Laying down the law to robots. *San Diego Law Review, 21,* 1045.

Hart, A. (1992). *Knowledge acquisition for expert systems.* New York: McGraw Hill.

Hopkin, D. V. (1994). Human performance implications of air traffic control automation. In M. Mouloua & R. Parasuraman (Eds.), *Human performance in automated systems: Current research and trends* (pp. 314-319). Hillsdale, NJ: Lawrence Erlbaum Associates.

Kaempf, G. L., & Klein, G. (1994). Aeronautical decision making: The next generation. In N. Johnston, N. McDonald, & R. Fuller (Eds.), *Aviation psychology in practice* (pp. 223-254). Hants, England: Avebury Technical.

Klein, G. A. (1993). A recognition-primed decision (RPD) model of rapid decision making. In G. A. Klein, J. Orasanu, R. Calderwood, & C. E. Zsambok (Eds.), *Decision making in action: Models and methods* (pp. 138-147). Norwood, NJ: Ablex.

Mockler, R. J. (1989). *Knowledge-based systems for management decisions.* Englewood Cliffs, NJ: Prentice-Hall.

Morgan, T. (1992). Competence and responsibility in intelligent systems. *Artificial Intelligence Review, 6,* 217-226.

Mosier, K. L., Palmer, E. A., & Degani, A. (1992). Electronic checklists: Implications for decision making. In *Proceedings of the 36th annual meeting of the Human Factors Society* (pp. 7-11). Santa Monica, CA: Human Factors Society.

Orasanu, J., & Connolly, T. (1993). The reinvention of decision making. In G. A. Klein, J. Orasanu, R. Calderwood, & C. E. Zsambok (Eds.), *Decision making in action: Models and methods* (pp. 3-20). Norwood, NJ: Ablex.

Parasuraman, R., Molloy, R., & Singh, I. L. (1993). Performance consequences of automation-induced "complacency." *The International Journal of Aviation Psychology, 3*(1), 1-23.

Riley, V. A., Lyall, E., & Wiener, E. (1993). *Analytic methods for flight-deck automation design and evaluation, phase two report: Pilot use of automation* (FAA Contract Rep., Contract No. DTFA01-91-D-00039). Minneapolis, MN: Honeywell Technology Center.

Sarter, N. B., & Woods, D. D. (1993). *Cognitive engineering in aerospace application: Pilot interaction with cockpit automation* (NASA Contractor Rep. 177617). Moffett Field, CA: NASA Ames Research Center.

Sarter, N. B., & Woods, D. D. (1994). Decomposing automation: Autonomy, authority, observability and perceived animacy. In M. Mouloua & R. Parasuraman (Eds.), *Human performance in automated systems: Current research and trends* (pp. 22-27). Hillsdale, NJ: Lawrence Erlbaum Associates.

Simon, H. A. (1983). *Reason in human affairs.* Stanford, CA: Stanford University Press.

Stephanou, H. E., & Sage, A. P. (1987). Perspectives on imperfect information processing. *IEEE Transactions on Systems, Man, and Cybernetics, SMC-17,* 780-798.

Waterman, D. A. (1985). *A guide to expert systems.* Reading, MA: Addison-Wesley.

Wickens, C. D. (1984). *Engineering psychology and human performance.* Columbus, OH: Merrill.

Wiener, E. L. (1989). *Human factors of advanced technology ("glass cockpit") transport aircraft* (Tech. Rep. No. 117528). Moffett Field, CA: NASA Ames Research Center.

Will, R. P. (1991). True and false dependence on technology: Evaluation with an expert system. *Computers in Human Behavior, 7,* 171-183.

Winograd, T. (1989, Spring). In R. Davis (Ed.), Expert systems: How far can they go? Part One. *AI Magazine,* 61-67.

Winograd, T., & Flores, F. (1986). *Understanding computers and cognition: A new foundation for design.* Norwood, NJ: Ablex.

Woods, D. (1994). Automation: Apparent simplicity, real complexity. In M. Mouloua & R. Parasuraman (Eds.), *Human performance in automated systems: Current research and trends* (pp. 1-7). Hillsdale, NJ: Lawrence Erlbaum Associates.

Zeide, J. S., & Liebowitz, J. (1987, Spring). Using expert systems: The legal perspective. *IEEE Expert,* 19-22.

Chapter 31

Applying Hybrid Models of Cognition in Decision Aids

◆

David E. Smith
Naval Command, Control, and Ocean Surveillance Center[1]

Sandra Marshall
San Diego State University

Decision making in many domains is complicated and difficult, inspiring the development of decision aids intended to improve performance. Antiair warfare (AAW) is one such area. The decision space is complex, in that the number of variables is large and the interrelationships among them often convoluted. Absolute categorical analysis is often difficult, if not impossible, due to the absence of complete knowledge regarding the situation. Missing, imperfect, or ambiguous data, as well as incomplete knowledge of relevant prior probabilities makes it a near certainty that complete analysis of any given situation is impossible. Decision making in the absence of absolute knowledge, together with the complexity of the situations, often requires some form of approximate reasoning.

In addition, decision making in AAW is severely constrained by external time limits. The time constraints are applied by external forces; failure to take appropriate action in timely fashion may result in suffering damage due

---
[1]David E. Smith is currently at the University of California, San Diego, Department of Neurosciences.

to attacking forces. AAW provides an excellent opportunity to study decision making in time-constrained and ambiguous situations.

Systematic errors and biases have been demonstrated within the context of AAW (Barnett, Perrin, & Walrath, 1993). Other types of instance-specific errors have also been found in AAW scenarios (Hutchins & Kowalski, 1993). Attempts to provide decision aids to overcome these errors could, in theory, supply any of several different types of tools, possibly embracing formal methods such as Bayesian approaches or artificial intelligence methods such as heuristics. All of them, however, are oriented toward the same goal—providing expert-level assistance in making decisions within the domain of interest. As Orasanu and Connolly (1993) noted, experts rely on an ability to "use their knowledge and experience to size up the situation, determine if a problem exists, and, if so, whether and how to act upon it." (p. 12).

Klein and his associates (Kaempf, Wolf, Thordsen, & Klein, 1992; Klein & Zsambok, 1991; Klein, 1992) provided important groundwork concerning the nature of situation assessment decisions in dynamic AAW situations. They found that diagnosis of a situation was either the result of feature matching or story generation, with the former being far more prevalent. *Feature matching* is essentially pattern recognition, in that a pattern of activity is recognized as being sufficiently similar to a previous experience to motivate action on the part of the decision maker. *Story generation*, however, is occasioned when the current situation is not sufficiently similar to previous experiences, thereby failing to activate an existing template or schema. As a result, the decision maker constructs a causal explanation for the current circumstances.

Several questions concerning these strategies naturally arise. What might be the knowledge structures that support these strategies? What are the relationships between the different types of knowledge? What type of unitary structure could produce both feature matching and story generation?

The primary goal of our research has been to understand the nature and structure of knowledge used by experts to understand and make decisions about the situations they encounter. The point of origin for this ongoing research was the body of schema theory developed by Marshall (Marshall & Marshall, 1991). It is being extended from static problem solving to dynamic decision-making situations.

METHOD

We study the decisions made by the commanding officer (CO) and the tactical action officer (TAO), the two primary AAW decision makers in the shipboard environment. There are four other members of the team, but they are role players recruited and trained for the purpose.

Our approach is to present test subjects with realistic scenarios that place them in tactically demanding situations, gather data regarding their decision-making activities, analyze these data, and build models representing the observed activities. The team interacts with the scenarios in a laboratory, that is a mock-up of the combat information center (CIC) systems used on board ship. Each scenario occurs in a Persian Gulf geographic setting and provides a series of events that could be encountered there. There are six scenarios, which include two for training and four for experimental purposes. Each scenario has been designed to provide event data in such a manner as to cause decision making to be time constrained. Many of the event data are ambiguous and uncertain. However, these data have been analyzed by domain experts who have determined that there is sufficient content in the data to allow adequate decision making to proceed.

In order to gather data concerning decision-making behaviors, we videotape the computer screens used by the CO and TAO during the scenarios, we audio tape all verbal communications, we use a situation assessment questionnaire at one point in each scenario, and we conduct postexperiment interviews.

From these data, we construct a schema-theory model of decision-making behavior. The identification knowledge components have been completed and are being used to provide input to an experimental decision support system. Currently, we are completing the elaboration knowledge components, which we discuss in the following section.

SCHEMA THEORY

The structure and representation of knowledge required for intelligent action is one of the most problematic subjects confronting cognitive science (Brooks, 1991; Kirsh, 1991a; Kirsh, 1991b). We consider schemas to be the essential structural elements in human memory because they represent the organization of knowledge concerning how to interact with the environment.

There is an extensive literature on schema theory. It is not our intent to review it in this chapter. Rather, this research constitutes an extension of the schema theory developed by Marshall from static problem-solving situations to dynamic decision-making situations.

Marshall (1995) has proposed the following definition for a schema:

A schema is a vehicle of memory, allowing organization of an individual's similar experiences in such a way that the individual

- can recognize easily additional experiences that are also similar, discriminating between these and ones that are dissimilar;

- can access a generic framework that contains the essential elements of all these similar experiences, including verbal and nonverbal components;
- can draw inferences, make estimates, create goals, and develop plans using the framework; and
- can utilize skills, procedures, or rules as needed when faced with a problem for which this particular framework is relevant. (p. 39)

Marshall previously developed a schema theory regarding problem solving. In her earlier models, there were four basic components, each of which reflects a separate type of knowledge. The four types of required knowledge are *identification knowledge, elaboration knowledge, planning knowledge,* and *execution knowledge* (Marshall & Marshall, 1991). The relationships between these types of knowledge are depicted in Fig. 31.1.

Identification knowledge is required for recognition of patterns in the input data. The question that identification knowledge is required to answer is: "Does the stimulus ... contain a pattern of elements sufficient to activate an existing schema?" (Marshall & Marshall, 1991, p. 35). Identification knowledge is usually the mechanism for activation of a schema.

Elaboration knowledge enables the construction of an appropriate mental model of the situation. Here we find not only the general structure of the situation, but also descriptions of its components, that is, the details about what is allowable and how all of the necessary pieces accommodate each other.

Planning knowledge is instrumental in formulating goals and subgoals, and contains information about how to identify any unknown elements of the situation. This type of knowledge depends on having the appropriate mental model of the situation and on using that model comfortably.

Planning knowledge is required to determine which plan to follow for a given situation. Execution knowledge is used to establish how to enact the plan, what steps to follow, and in what order. As each piece of the plan is

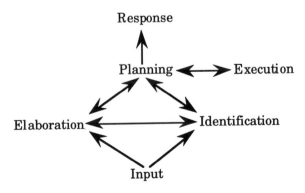

FIG. 31.1. The components of a schema.

completed, the execution knowledge is called on to address subsequent ones.

The complete schema model representing these types of knowledge is a *hybrid symbolic-connectionist model.* Neither symbolic nor connectionist models alone are capable of capturing all of human behavior. There are advantages and disadvantages to both types of models.

Symbolic (production) systems are well suited to sequential functions normally associated with rule-based behaviors. On the other hand, such systems tend to be brittle and inflexible, and the rules can be difficult to assemble.

Connectionist systems are particularly capable in those areas in which symbolic systems are weak, and relatively incapable in those areas where symbolic systems are strong. For instance, connectionist systems are adept at pattern recognition rather than logical sequences of actions. Processing in connectionist systems is holistic and depends on the activation of a number of nodes rather than on firing of individual rules. The result is that the behavior of neural network models is robust, even with incomplete input vectors.

This hybrid system employs both of these types of systems in a complementary fashion and takes advantage of the strengths of each. Fig. 31.2 depicts a hybrid schema model, with identification knowledge modeled as a connectionist network and elaboration knowledge and planning knowledge modeled as a production system (Marshall, 1995).

This outline formed the basis for two models. One was a performance model of problem solving with respect to arithmetic word problems. The second was a model of learning, in the sense that the model learned to behave in the same manner as human subjects when given appropriate feedback. The performance model was capable of solving arithmetic word problems with more than one unknown, and the learning model did, indeed, learn. (See Marshall & Marshall, 1991, for a complete description.)

MODEL OF DECISION MAKING

The incorporation of schemas into the TADMUS decision support system (DSS) offers a number of advantages. First, schemas provide a logical framework for analyzing situation assessment decisions. Both feature matching and story generation are easily derived and explained under schema theory. Each originates from a different aspect of schema knowledge. On the one hand, feature matching is a natural outcome of the application of identification knowledge. The co-occurrence of identifiable features occasions recognition of the situation. On the other hand, story construction depends primarily on elaboration knowledge and its associated mental

Output

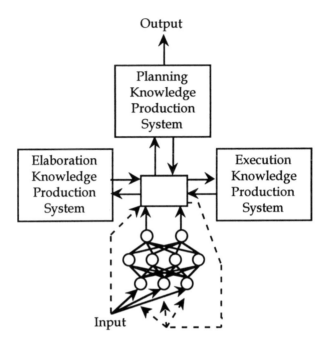

FIG. 31.2. Hybrid schema model. From *Problem-Solving Schemas: Hybrid Models of Cognition* (p. 46), by S. P. Marshall and J. P. Marshall, 1991, Center for Research in Mathematics and Science Education. Copyright 1991 by Center for Research in Mathematics and Science Education. Reprinted with permission.

model. When incoming features are insufficient to engender recognition using identification knowledge alone, elaboration knowledge may be called on to provide the underpinnings and the default characteristics for any missing data. This knowledge is then associated with the situation best approximated by the features extracted from the situation. The "best guess" from elaboration knowledge allows access to a mental model. Thus, feature matching allows all possible features to have influence. Story generation looks for particular features that match the story reflected by the mental model of the schema.

A second advantage of the schema approach is that it provides for explicit computer modeling of both strategies. The importance of modeling is that it allows us to observe the essential components of the decision and how they are intertwined. There are at least two benefits to this capability. One benefit is that it permits us to form hypotheses concerning the effects of the absence of one or more components, and then to test the hypotheses directly with the model. Another is that it provides a framework not just for deriving, but also for unifying the cognitive strategies of feature matching and story generation. These strategies have been previously described as distinct and

separable (Klein, 1992). Schema theory supplies a single cognitive framework, in which feature matching emerges primarily as instances of emphasis
on identification knowledge, and story generation emerges primarily as
instances of emphasis on elaboration knowledge.

Figure 31.3 depicts a hybrid schema model of decision-making behavior.
This represents an extension of Marshall's schema theory to include dynamic situations.

Identification knowledge is modeled as a collection of neural networks. Each
neural network functions with respect to a single contact (i.e., an airplane,
helicopter, or other ship). Elaboration, planning, and execution knowledge are
once again modeled as production systems. The architecture of the model is
blackboard-based, which means that the individual components of the model
operate independently and share results via a common blackboard.

The output from the neural networks in the first stage of the model is a
determination of whether a contact requires attention. Each contact is put

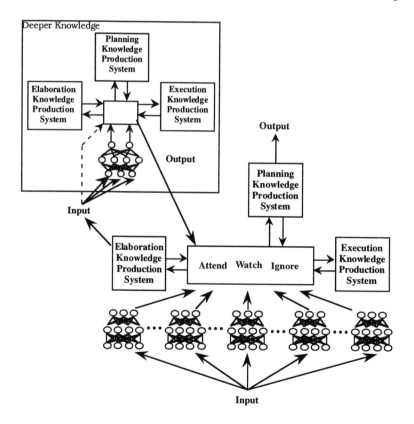

FIG. 31.3. Schema model of decision making.

into one of three categories: "attend," "watch," and "ignore." Putting a contact into the "attend" category indicates that contact requires immediate attention from the decision maker. More than one contact may be in this category, meaning that the multiple contacts must compete for his attention. "Watch" denotes contacts that do not require immediate attention but should be monitored for changes which might indicate that they do require attention. These contacts should not be ignored, but do not merit special attention at the moment. "Ignore" indicates those contacts that can be safely disregarded.

Data relevant to each contact are posted to the blackboard, where they are accessed by the other types of knowledge. In the case of elaboration knowledge, these data are passed on to later stages of the model. Elaboration knowledge is required to prioritize the "attend" contacts, as well as to determine whether additional information is required for any of them. Planning knowledge builds on the results of elaboration knowledge and requires knowledge of how to sequence any actions that need to be taken. This is provided by execution knowledge, which consists largely of knowing what actions to take, who should take them, and how they should be taken.

The structure of these later stages is similar to that of the first. The principle difference is that there is only one neural network, and the stage is processing data for only one contact. One of the stages computes the identification of the contact, and the other computes the intent. This is analogous to the process of progressive deepening.

A third advantage of the schema approach is that it will enable us to model the schemas of many different experts. Next, we combine separate models of individual experts to form a single "super expert" model, representing the synthesis of the individual expertise of each decision maker. Such a model, when completed in the domain of AAW, will be incorporated into a decision support system (DSS) and evaluated experimentally for its ability to provide assistance in time-constrained and ambiguous situations.

APPLICATION IN A DECISION AID

The model just described can readily serve as the central processing element for a DSS. Display elements will be driven by individual components of the model. The neural network outputs posted to the blackboard from the first stage will function to prioritize contacts, generate alerts, and trigger both planning and elaboration knowledge. Planning knowledge produces a plan of action, which, in turn, activates specific actions from execution knowledge. These can be displayed in a response manager, hopefully assisting the decision maker in knowing what actions to take, when to take them, and which actions have already been taken.

However, the emphasis in this environment is on understanding the situation. Expertise, according to Donald Norman (1993), is rapid and efficient recall of correct procedure. We expand this statement: It is not sufficient for the decision maker to know what to do; the decision maker must first *understand* the situation. This understanding is facilitated by identification knowledge and elaboration knowledge. Identification is produced by one of the second-stage elements, operating as directed by elaboration knowledge. Intent is produced similarly.

A naturalistic decision aid, such as the one described in this chapter, is likely to provide better assistance than one based on a normative model of the decision process. Although the intention may be to support rather than replace the user, "*the user's approach to the problem*, if not the user himself or herself, is replaced" (Cohen, 1993, p. 266) in tools predicated on decision analysis or mathematical optimization. This occurs, in part, because of the focus on outcomes in normative decision aids. A naturalistic decision aid, on the other hand, supports processes familiar to the decision maker. The result may be that the processes "may produce outcomes that are not merely just as good as, but better than Bayesian procedures—because they more effectively exploit the decision maker's knowledge and capacity" (Cohen, 1993, p. 266).

A decision aid based on schema theory provides a unique opportunity to support processes familiar to decision makers because the processes represented in the tool are models of the decision makers themselves. The model we presented here is certainly being built to generate the same outcome as we observe human decision makers producing. However, the focus has been on reproducing the internal and intermediate processes, rather than only the outcome. One result of this emphasis is that the first level of the model represents the allocation of attentional resource. It determines which contacts require the decision maker's attention. Subsequent levels determine the identification and intent of the contact.

It is not sufficient, however, for the processes (in the decision aid) to be familiar to the decision maker. The processes must also be appropriate to the situation at hand. Norman (1993) distinguished between two broad categories of cognition: *experiential cognition* and *reflective cognition*. He defined experiential cognition as an effortless process of perception and reaction to events in the environment surrounding us. Reflective cognition, on the other hand, is an analytical mode of operation. He cautions that these two categories do not capture all of human behavior, but he uses them to illustrate how an inappropriately designed tool can have serious consequences. One way for this to happen is to provide decision aids designed to support experiential behavior that require reflection. A second way is to provide decision aids intended to support reflection that do not support comparisons, exploration, and problem solving.

It is not the case that the tool should guide the decision maker to a specific mode of operation or attempt to limit the user to one or another cognitive strategy. The decision aid must provide assistance for the strategy employed by the decision maker, rather than providing assistance at the expense of one of the strategies. This is another way of saying that a decision aid should not constrain decision makers to using a particular cognitive strategy.

For both Klein (1992) and Norman (1993), a pattern-matching strategy is the typical method for expert decision making. For the expert, it is quicker, easier, and facilitates a rapid and nearly automatic response. However, performance of such a strategy presumes the existence of a template, some preexisting structure against which the current system can be compared in a holistic sense. Without such a structure, pattern matching is impossible.

In the absence of this structure, stories provide significant expressive power: they encapsulate, into one compact package, information, knowledge, context, and emotion (Norman, 1993). An appropriate task for a decision aid is, therefore, to assist with story construction, in addition to aiding pattern matching. By this, we mean that the DSS should assist with causal reasoning. We adopted a method in which available evidence that causally supports an identified hypothesis concerning the current situation is presented to the decision maker, along with counterevidence and assumptions that must be made if the hypothesis is to be accepted.

Our schema-based decision aid is not limited to a single type of knowledge, strategy, or performance. Multiple knowledge forms are incorporated into the model. Furthermore, one type of knowledge is not given greater weight or importance over any other. Instead, the tool provides simultaneous access to the different types of knowledge.

The role of the decision aid is to provide assistance by offering intelligent advice in a nonthreatening manner. It can review the available evidence, provide insight into possible alternatives relating to the situation, and critique different courses of action. There must be a rich interaction between the user and the decision aid, such that there is a type of natural discourse between the two. This can only occur if the decision maker is able to relate to the aid and the processes it represents.

KEY POINTS

- The schema is the essential structural element in human memory.
- Schema theory supplies a framework for understanding cognitive behavior regarding decision making.
- A schema model of decision-making behavior affords the possibility of unifying otherwise separate cognitive strategies.
- The same schema model will be the central processing component of a new decision support system, which we are currently building.

REFERENCES

Barnett, B. J., Perrin, B. M., & Walrath, L. C. (1993). *Bias in human decision making for tactical decision making under stress* (Tech. Rep., Task 1). Naval Command, Control, and Ocean Surveillance Center.

Brooks, R. A. (1991). Intelligence Without Representation. *Artificial Intelligence, 47,* 139–160.

Cohen, M. S. (1993). The bottom line: Naturalistic decision aiding. In G. Klein, J. Orasanu, R. Calderwood, & C. E. Zsambok (Eds.), *Decision making in action: Models and methods* (pp. 265–269). Norwood, NJ: Ablex.

Hutchins, S. G., & Kowalski, J. T. (1993). Tactical decision making under stress: Preliminary results and lessons learned. *1993 Symposium on Command and Control Research.* Washington, DC: The Technology Panel for C3 of the Joint Directors of Laboratories.

Kaempf, G. L., Wolf, S., Thordsen, M. L., & Klein, G. A. (1992). *Decisionmaking in the AEGIS Combat Information Center* (Tech. Rep., Task 1). Fairborn, OH: Klein Associates Inc. Prepared under contract N66001–90–C–6023 from NCCOSC, San Diego, CA.

Kirsh, D. (1991a). Foundations of AI: The big issues. *Artificial Intelligence, 47,* 3–30.

Kirsh, D. (1991b). Today the earwig, tomorrow man? *Artificial Intelligence, 47,* 161–184.

Klein, G. A. (1992). *Decisionmaking in complex military environments* (Tech. Rep., Task 4). Fairborn, OH: Klein Associates Inc. Prepared under contract N66001–90–C–6023 from NCCOSC, San Diego, CA.

Klein, G., & Zsambok, C. E. (1991). Models of skilled decision making. *Human Factors Society 35th annual meeting.* San Francisco, CA: HFES.

Marshall, S. P. (1995). *Schemas in problem solving.* Cambridge, England: Cambridge University Press.

Marshall, S. P., & Marshall, J. P. (1991). *Problem-solving schemas: Hybrid models of cognition.* Center for Research in Mathematics and Science Education.

Norman, D. A. (1993). *Things that make us smart.* New York: Addison-Wesley.

Orasanu, J., & Connolly, T. (1993). The reinvention of decision making. In G. Klein, J. Orasanu, R. Calderwood, & C. E. Zsambok (Eds.), *Decision making in action: Models and methods* (pp. 3–20). Norwood, NJ: Ablex.

Chapter 32

Finding Decisions
in Natural Environments:
The View From the Cockpit

◆

Judith Orasanu
NASA–Ames Research Center

Ute Fischer
Georgia Institute of Technology

In keeping with the naturalistic decision making (NDM) tradition of study-ing "real people making real decisions" in their everyday contexts, our mission was to understand flight-related decision making by commercial airline pilots: What constitutes effective flight crew decision making? What conditions pose problems for crews and lead to poor decisions?

Our initial examination of decision strategies that distinguished more effective from less effective crews in simulated flight showed striking vari-ability in the decision behaviors of the most effective crews. Sometimes the crews were very quick and sometimes they were slow and methodical. In retrospect we should not have been surprised, but as psychologists we were looking for simple patterns, such as, "Good crews always make the fastest decisions."

These observations suggested that the most effective crews tailored their decision strategies to the situation. Thus, to understand what constitutes an effective decision strategy we need to understand the problem situations that crews encounter. Our research question was expanded to: "How can we assess the sensitivity and appropriateness of decision strategies in light of situational features?"

We adopted an approach used by ethnographers (e.g., Hutchins & Klausen, 1991) and cognitive engineers (Woods, 1993): close examination of a phenomenon of interest in its everyday context, seeking natural variations in critical features. Our approach builds on Klein's (1993) Recognition-Primed Decision (RPD) model and on Hammond's Cognitive Continuum Theory (Hammond, Hamm, Grassia, & Pearson, 1987). Our work also echoes the theme of Hart's work on "strategic behavior" (Hart & Wickens, 1990); namely, operators make decisions that serve overall task goals, capitalizing on their strengths and minimizing work.

A SEARCH FOR DECISION EVENTS IN CONTEXT: DATA SOURCES

As our starting point shifted from strategies to situations, we began a search for decision events in context. Our initial observations were based on crews "flying" a mission in a high-fidelity flight simulator, which yielded three distinct types of decisions. However, we realized that our opportunity to observe decisions was restricted by the particular scenarios used in those studies, so we sought a broader set of situations that might present other types of decision events.

The Aviation Safety Reporting System (ASRS) data base satisfied this need. The ASRS is a confidential reporting system maintained by NASA (with funding from the FAA). Pilots (and others) can submit a report describing an incident that may have involved a risky situation that was problematic in some way. Key words used to search the database were *Problem Solving and Decision Making*. The resulting set of incident reports describes diverse events that required crew decision making. However, because of the self-report nature of the descriptions, what we know about the actual decision strategies used by crews is what they chose to tell us. Likewise, information about conditions that may have led to poor decisions is limited. To address these limitations, we pursued a third data source.

The National Transportation Safety Board's (NTSB) accident investigations offer deep analysis of actual crashes, based on crew conversations documented by the cockpit voice recorder, physical evidence, aircraft systems, and interviews with survivors or observers. We chose reports in which crew actions were identified by NTSB analysts as contributing or causal factors in the accidents. These case studies provide a detailed picture of what happened immediately prior to each accident, what the crew focused on, how they managed the situation, what decisions were made, and what actions were taken. The analyses are good sources of hypotheses about contextual factors that make decisions difficult, types and sources of error, and effective strategies.

What we learned about decision situations and decision strategies from these three data sources is described in the remainder of this chapter. First we describe six types of decision events that were identified. Then we describe decision strategies that are associated with each type of decision, and differences in strategies used by more and less effective crews. Finally, we describe a decision process model and a model of decision effort derived from the first two activities.

DECISION EVENTS

Simulator Data

Our analyses were based on two full-mission simulator studies conducted at NASA–Ames Research Center. The first (Foushee, Lauber, Baetge & Acomb, 1986) was designed to study the effect of fatigue on the performance of two-member crews. The second (Chidester, Kanki, Foushee, Dickinson, & Bowles, 1990) investigated leader personality effects in three-member crews. All crews were exposed to the same events, which allowed between-crew comparisons. Crew performance in the simulator was videotaped and all communications were transcribed.

The scenario flown by all crews included a missed approach at their original destination due to bad weather (excessive cross-winds) and diversion to an alternate landing site. During climbout following the missed approach, the main hydraulic system failed. As a result, the gear and flaps had to be extended by alternate means. Moreover, the flaps could only be set to 15 degrees, resulting in a faster than normal landing speed, and the gear could not be retracted once extended, meaning that further diversion was not desirable because of fuel constraints.

Three major decisions were present in this scenario: (a) at the original destination, crews had to decide whether to continue with the final approach or to perform a missed approach; (b) once the crew realized that the weather at their destination was not improving, they had to select an alternate airport; (c) the hydraulic failure required crews to coordinate the flap- and gear-extension procedures during final approach, an already high-workload period. How to manage this coordination was the third decision. These problems imposed different cognitive demands on the crews: The situations differed in the number of constraints a solution had to satisfy and in the extent to which a solution was prescribed.

Problem (a) calls for a *Go–No Go decision*. A course of action is prescribed: If all facilitating conditions are normal, then Go. If the "go" conditions are not met, an alternate action is prescribed (No Go condition). Conditions for Go and No Go are clearly defined and the actions to be taken in both cases are also clearly prescribed. Selecting an alternate landing site as in problem (b) is an example of a *Choice problem*. Several legitimate options or courses of action exist from which one must be selected. No rule prescribes a single appropriate response. Options must be evaluated in light of goals, possible consequences, and situational constraints (such as fuel, runway length, or weather). *Scheduling problems* like problem (c) require the crews to decide on what is most important to do, when to do it, and who will do it. Several tasks must be accomplished within a restricted window of time with limited resources.

Incident Reports From the Aviation Safety Reporting System

Ninety-four ASRS reports were analyzed in depth and classified in terms of their precipitating events, phase of flight during which the event and subsequent decisions occurred, and focus of the decisions. Some 234 decisions were discerned in these cases, because a single precipitating event often set the stage for a series of decisions. For example, an engine problem may first require the crew to decide what to do with the engine (shut it down, reduce power to idle, or continue operation), then to decide whether or not to divert, where to divert, and any specific considerations about landing configuration as a consequence of the engine problem. Our analyses of the ASRS reports yielded three additional types of decision events.

Condition–Action Rules

The situation requires recognition of a predefined condition and retrieval of the associated response. These decisions mirror Klein's (1993) RPD, but are prescriptive in the aviation domain. They do not depend primarily on the pilot's personal experience with similar cases, but on responses dictated by the industry, company or FAA. Neither conditions nor options are bifurcated, as in Go/No Go cases, though both types rely on underlying rules. Examples include decisions to pull the fire handle in case of an engine fire or to descend to a lower altitude in case of cabin decompression. Thus, the pilot must know the rule and then decide whether conditions warrant applying it.

Procedural Management

The essence of this class of decisions is the presence of an ambiguous situation that is judged to be of high risk. The crew does not know precisely what is wrong, but recognizes that conditions are out of normal bounds. Standard procedures are employed to make the situation safe, often followed by landing at the nearest suitable airport. These decisions look like condition-action rules but lack prespecified eliciting conditions. The response also is generalized, such as "get down fast." One case studied was a decision to reduce cruise speed when an airframe vibration was experienced (which turned out to be due to a loose aileron trim tab). The defining features of this type of problem are ambiguous high-risk conditions and a standard procedural response that satisfies the conditions. No specific rules in manuals or checklists guide this type of decision; pilot domain knowledge and experience are the source of the action.

Creative Problem Solving

These are ill-defined problems and are probably the least frequent types of decision events crews ever encounter. No specific guidance is available in standard procedures, manuals, or checklists to guide the crew to a course of action. The nature of the problem may or may not be clear. The important distinction from procedural management situations is that standard procedures will not satisfy the demands of the situation. New solutions must be invented. Perhaps the most famous case is the DC-10 (UA flight 232) that lost all flight controls when the hydraulic cables were severed following a catastrophic engine failure (NTSB, 1990). The crew had to figure out how to control the plane. They invented the solution of using alternate thrust on the two remaining engines to "steer" it.

National Transportation Safety Board Accident Analyses

The six types of decision events just described could account for all problem situations analyzed in a dozen NTSB accident reports. Because the NTSB seeks to understand causal and contributing factors in accidents, we used their reports primarily as a source of hypotheses about decision processes and causes of poor decisions, rather than to expand the set of decision types.

Decision Event Taxonomy

The six types of decisions were identified using simulator performance and ASRS data bases. They fall into two subgroups that differ primarily in whether a prescriptive rule exists that defines a situationally appropriate response or whether the decision primarily relies on the pilot's knowledge and experience. These are referred to as "rule-based" and "knowledge-based" decisions.[1]

---

[1]The concepts are taken from Rasmussen (1983), but are used somewhat differently here because they apply primarily to decision situations, not to responses. Skill-based decisions, Rasmussen's third category, were not included in our analysis because of their automatic psychomotor nature.

Rule-based decisions include two subtypes: Go/No Go and Condition-Action decisions. They differ in whether a binary option exists or whether a simple condition-action rule prevails. The crucial aspect of the decision process for rule-based decisions is accurate situation assessment. The major impediment is ambiguity. Such decisions are often made under high time pressure and risk; thus, the industry has prescribed appropriate responses to match predictable high-risk conditions. Once the situation is recognized, a fast response may be required for safety. An example is deciding whether to abort a takeoff when an engine fails some time during the takeoff roll.

Knowledge-based decisions vary in how well-structured the problems are and in the availability of response options. "Well-structured" problems are those in which the problem situation and available response options are unambiguous and should be known to experienced decision makers. In one case ("choice" problems), the decision maker must choose one option after evaluating constraints and outcomes associated with various options. In the second case ("scheduling" problems), effective performance depends on good judgment about relative priorities of various tasks and accurate assessment of resources and limitations.

"Ill-structured" problems entail ambiguity, either in the cues that signal the problem or in the available response options. Cues may be sufficiently vague or confusing that the crew cannot identify the problem ("situational management" decisions), or crews do not know what to do even if the problem is understood ("creative problem solving" required).

Analysis of the 94 ASRS reports indicates that rule-based decisions were slightly more frequent in our sample (54%) than knowledge-based decisions (46%; Orasanu, Fischer, & Tarell, 1993). Three out of four rule-based decisions were Condition-Action decisions, the rest being Go/No Go decisions. This distribution is not surprising, because Go/No Go decisions occur in narrowly specified situations during takeoff and landing, whereas Condition-Action decisions can occur anytime. About a third of the decisions (36%) required choices, and the remainder were other types of knowledge-based decisions (4% Scheduling, 3% Procedural Management, and 2% Creative Problem Solving).

DECISION STRATEGIES

The earlier description of decision types was based on properties of the situation. Now we turn to crew strategies. We describe how crews responded to the various types of decision events and differences in behaviors associated with more and less effective crew performance. Crews flying full-mission simulations provided the richest source of strategy data. Little reliable strategy data could be obtained from ASRS reports due to the self-report nature of these descriptions. Corroborating strategy data were obtained from the NTSB accident reports.

Simulator Data

Videotapes of crew performance in simulators allowed us to observe decision making in action rather than relying on post-hoc accounts, as in the other databases. How decision making evolves over time in response to dynamic situations could be analyzed. These data provided not only records of *behavior* but also of crew *communication* as a "window" into the crew's thinking. Within-crew comparisons can be made as each crew faces several decision events, thus yielding the greatest generality of findings between and within crews.

Crew performance in the simulator was evaluated by two independent expert observers both online and from videotapes. Operational and procedural errors (not decision behaviors) were assessed. Crews were rank ordered by error scores and divided into higher and lower performance groups using a median split. Decision-relevant behaviors of the two groups were compared, based on their videotaped performance. Time-stamped transcripts of cockpit conversation and action timelines permitted detailed analyses of communication and decision behaviors. Our analyses of decision strategies were independent of the initial error assessments by check pilots.

The decision taxonomy guided our examination of decision behaviors, providing a structure that directed our focus. Working with aviation experts, we defined behaviors appropriate to each decision, cues that signaled the problems, available options, temporal parameters, relevant constraints, and standard procedures. For detailed descriptions of these analyses see Fischer, Orasanu, and Montalvo (1993) and Orasanu (1994).

We found differences between groups in two types of behaviors: (a) strategies specific to each decision type, and (b) differences in generalized strategies that cut across decision types.

Decision-Specific Strategies

Consider first the Go–No Go decision (the missed approach). Higher performing crews made the decision significantly earlier than the less effective crews, which provided a greater safety margin. One reason they could make this decision early was because they had attended to cues signaling the possibility of deteriorating weather. They sought weather updates as the approach progressed, and planned for the possibility of a missed approach.

The second decision was a knowledge-based choice decision. After the missed approach and the hydraulic failure, crews faced the problem of choosing a landing site. An alternate was listed on their flight plan, but the unexpected hydraulic failure raised constraints that made the designated alternate a poor choice (short runway with bad weather, mountainous terrain). Recognizing these constraints, realizing that the designated alternate was not a good option, retrieving other options, and evaluating them in light of the constraints were all required to make a good decision. The more effective crews in fact verbalized concern with the constraints, gathered more information about several options, and took longer to make their decision than the less effective crews. No differences were found in the number of options considered by the two groups despite differences in amount of information used to evaluate them. Relatively little attention (beyond standard checklist procedures) was devoted to defining the problem. The emphasis was on assessing potential solutions.

In the third type of decision, which required scheduling the manual gear deployment and alternate flap extension, both the nature of the problem and the actions to be taken were clear. What had to be decided was how these tasks were to be accomplished. What differentiated the more and less effective crews was the manner in which the tasks were planned and carried out. These abnormal procedures were unfamiliar to many crews (being relatively infrequent events) and required additional work during the normally busy final approach phase of flight. Preparation included review of the procedures in the checklists and manuals, becoming familiar with the location of the gear handle, assessing how long the tasks would take, determining when the tasks would be initiated and their sequencing, and assigning tasks to the crew members. Higher performing crews reviewed the written guidance in advance, during a low workload period. They rehearsed what would be done and how (e.g., use the alternate procedure to extend the flaps to 10 degrees, manually lower the gear, then continue extending the flaps to 15 degrees). Because they had planned for these tasks, the higher performing crews began the tasks earlier and completed them faster than the lower performing crews, thereby giving themselves a cushion of time to accomplish other essential tasks and maintaining better control of the aircraft during the final approach and landing.

Generalized Strategies

Strategies that cut across various decisions and characterized higher performing crews include the following: (a) they monitored the environment closely and appreciated the significance of cues that signaled a problem; (b) they used more information in making decisions and if necessary manipulated the situation to obtain additional information in order to make a decision; (c) they adapted their strategies to the requirements of the situation, demonstrating a flexible repertoire; (d) they planned for contingencies and kept their options open when possible; (e) they did not overestimate their own capabilities or the resources available to them; (f) they appreciated the complexity of decision situations and managed their workload to cope with it. Less effective crews showed significantly lower levels of all these behaviors and generally failed to modify their behaviors in response to different types of situational demands.[2]

INTEGRATION OF DECISION EVENT AND DECISION STRATEGY DATA

Our examination of crew decision making from the perspective of the three different data sources has led to several converging observations about cockpit decision making. We used the taxonomy and strategy data to develop a simplified decision-process model appropriate to the aviation environment, and a model of factors that determine the amount of cognitive work that must be done to make a decision (a surrogate for decision difficulty, because we presently have no empirical difficulty data).

A Simplified Decision Process Model

The decision process model we adopted is conceptually a simple one (see Fig. 32.1). It draws on Klein's (1993) RPD model and on Wickens and Flach's (1988) information-processing model. Our model is tailored to the structure of the decision taxonomy and includes only components that were visible in crew performance in the simulator.

The model consists of two major components: situation assessment and choosing a course of action. Situation assessment requires definition of the problem and assessment of risk level and time available to make the decision. Available time appears to be a major determinant of subsequent strategies. If the situation is not understood, diagnostic actions may be taken, but only if sufficient time is available. External time pressures may be modified by crews to mitigate their effects (Orasanu & Strauch, 1994). If risk is high and time is limited, action may be taken without thorough understanding of the problem.

Selecting an appropriate course of action depends on the affordances of the situation. Sometimes a single response is prescribed in company manuals or procedures. At other times, multiple options may exist from which one must be selected, or multiple actions must all be accomplished within a limited time period. On some rare occasions, no response may be available and the crew must invent a course of action. In order to deal appropriately with the situation, the decision maker must be aware of what response options are available and what constitutes a situationally appropriate process (retrieving and evaluating an option, choosing, scheduling, inventing).

ASRS reports revealed the importance of situation assessment. In many cases extensive diagnostic episodes occurred. These were not minor efforts

[2]It should be noted that our description of more and less effective strategies is limited by the flight scenarios used in these studies. Other effective strategies might be observed in situations differing in features not included here.

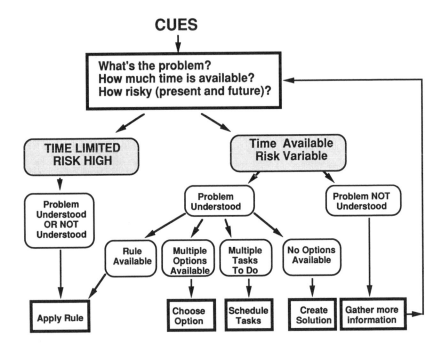

FIG. 32.1. Decision process model. The upper rectangle represents the Situation Assessment component. The lower rectangles represent the Course of Action component. The rounded squares in the center represent conditions and affordances.

but decisions in and of themselves, such as deciding that insufficient information was available to make a good decision and arranging conditions to get the needed information (e.g., to fly by the tower to allow inspection of the landing gear; send crew member to the cabin to examine engine, aileron, etc.). Certain diagnostic actions served a dual purpose: The actions could solve the problem as well as provide diagnostic information about the nature of the problem. The idea seemed to be, "If this action fixes the problem, we will know what the problem was."

Efforts are currently under way to validate the components of the process model. In one set of studies, pilots were asked to sort decision events into piles of scenarios that required similar decisions (Fischer, Orasanu, & Wich, 1995). Multidimensional scaling analyses suggest that pilots identified risk, time pressure, situational ambiguity, and response determinacy as decision-relevant dimensions. Although these aspects verify components of the process model, further studies are required to shed light on how they contribute to the process for different types of decisions. The decision-process model can now serve as a frame for analyzing crew performance in NTSB accident reports and in full-mission simulation.

Decision-Effort Model

Although we do not yet have experimental data on the cognitive demand level or difficulty of various decision events, we have a model that allows us to predict which decisions might involve the greatest amount of cognitive work, and where decision errors might be most likely. The model is based on the two components of the decision-process model. Its two dimensions are *situational ambiguity* and *response availability*, paralleling the processes of situation assessment and choosing a course of action.

Situation Ambiguity

If a situation is ambiguous, more effort will be required to define the nature of the problem than if cues clearly specify what is wrong. Three types of ambiguity have been identified that may differ in their demands on the crew.

Vague Cues.   These cues are inherently ambiguous and nondiagnostic. They consist of vibrations, noises, smells, thumps, and other nonengineered cues. Pilot knowledge and experience are critical to their interpretation. ASRS reports include cases of a ramp vehicle bumping into parked aircraft, a vibration during flight due to a loose aileron-trim tab, and the sound of rushing air in the cockpit.

Conflicting Cues.   Cues of this type are clear and interpretable, often engineered diagnostic indicators. The ambiguity lies in the simultaneous presence of more than one cue that signal conflicting situations and imply opposing courses of action. For example, the presence of a stall warning on takeoff and engine indicators of sufficient power for climb are conflicting cues.

Uninterpretable cues.   Again, these cues in themselves are clear, but in context are uninterpretable. As a result, the crew may disregard them or suspect that the indicator is faulty. A case of uninterpretable cues was the rapid loss of engine oil from both engines in synchrony during an overwater flight. The crew could not imagine a plausible scenario to explain these indicators, and continued the flight. Only on landing did they discover that caps had been left off both engine oil reservoirs.

Response Availability

The second dimension determining problem demand level is response availability. The least work is required if a single response is prescribed to a particular set of cues (rule-based decisions). More work is required if multiple responses must be evaluated and either one must be chosen (choice decision) or multiple actions must be prioritized (scheduling decision; Payne, Bettman, & Johnson, 1993). The greatest effort will be required if no response options are available and one or more candidates must be created (ill-defined creative problem solving).

Two other factors enter into the equation, but probably operate in different ways—time pressure and risk. When time pressure is high, little time is available for either diagnosing a problem or generating and evaluating multiple options, so greater error might be expected than when time pressure is low (Wright, 1974). The second factor, risk, may induce caution or increased attention to a problem at moderate levels. At high levels, dysfunctional stress responses may be expected, such as narrowing of perceptual scan, fixation on inappropriate solutions, and reduction of working memory capacity (see Stokes, Kemper, & Kitey, chapter 18, this volume).

At this point the decision-effort model serves as a framework for examining the relations among the various elements. We do not yet know whether situation ambiguity and response availability carry equal weight in terms of cognitive work, but the NTSB accident reports suggest that situation assessment may be the more vulnerable component.

NTSB Accident Analyses

Our examination of NTSB reports in which crew factors contributed to accidents found that in most cases crews exhibited poor situation assessment rather than faulty selection of a course of action based on adequate situation assessment (Orasanu, Dismukes, & Fischer, 1993). This conclusion is based primarily on crew communications captured by the cockpit voice recorder. Crews that had accidents tended to interpret cues inappropriately, often underestimating the risk associated with a problem. For example, several crews have flown into bad weather on final approach and crashed, rather than removing themselves from a dangerous situation. A second major factor was that they overestimated their ability to handle difficult situations or were overly optimistic about the capability of their aircraft. One crew decided to fly on to their destination on battery power after losing both generators shortly after takeoff. Unfortunately, the batteries failed before they reached their destination, resulting in loss of flight displays (NTSB, 1983).

The NTSB recently analyzed flightcrew-involved accidents from 1978 to 1990 (NTSB, 1994). Of the 37 accidents in which crew errors were identified as contributing factors, 25 involved what the authors called "tactical decision errors." Examples included deciding to continue the flight in the face of a system malfunction, unstable approach, or deteriorating weather.

Using our decision taxonomy as a frame to classify the tactical decision errors, we found that a large proportion of them (66%) were Go–No Go decisions, which should have been the simplest decisions in terms of response availability. These included rejected takeoffs, descent below decision height, go-arounds, and diversions. In all but one case, the crew decided to continue with the current plan in the face of cues that suggested discontinuation. However, in many of these cases the cues were ambiguous and it was difficult to assess with great confidence the level of risk inherent in the situation. Most significantly, most of the Go–No Go decisions were made during the most critical phases of flight, namely takeoff and landing, when time to make a decision was limited and the cost of an error was highest. Little room was available for maneuvering or for gathering more information. In contrast, decisions made during cruise, even very difficult decisions, usually are not burdened with the double factors of time pressure and high risk. (There are a few notable exceptions like a cockpit fire or rapid decompression.)

Data from our simulator studies provided an additional perspective on this issue. When the cognitive demands were great, the higher performing crews managed their effort by buying time (e.g., requesting vectors or holding) or by reducing the load on the captain by shifting responsibilities to the first officer (e.g., flying the plane). They also used contingency planning and task structuring to reduce the load. In contrast, lower performing crews apparently tried to reduce effort by oversimplifying situational complexity. They often acted on the first solution they generated, even though it was not very satisfactory. They also allowed themselves to be driven by time pressures and situational demands, rather than managing their "windows of opportunity."

CONCLUSIONS

Different perspectives on crew decision making were obtained from each of the data sources we examined. The ASRS reports provided insights into the many different types of decision events that crews encounter. The simulator data were most useful for providing evidence on more and less effective decision strategies because of their controlled nature and the opportunity they afforded to observe multiple crews facing the same situations. The NTSB analyses were a source of hypotheses about decision difficulty and where crews go wrong in making decisions. Analysis of different types of decision events allowed us to identify some of the differences in their underlying requirements and affordances, as well as the strategies most appropriate to each. Crew performance in a controlled simulator environment revealed some generic strategies that are beneficial in all decision contexts. These include good situation assessment, contingency planning, and task management to allow time to make a good decision. Other strategies are decision-specific and vary considerably, primarily in their temporal aspects. Effective crew performance consists of flexible application of a varied repertoire of strategies. Less effective crews did not appear to distinguish among the various types of decisions, applying the same strategies in all cases regardless of variations in their demands.

Decision difficulty may hinge on situational ambiguity and absence of planned response options. Time pressure clearly increases the likelihood of poor decisions and has a major impact on decision strategies. The effect of risk is not yet well understood, but our sorting study (Fischer, Orasanu, & Wich, 1995) indicates that it is a salient dimension to pilots, especially to captains. We have not directly examined the effects of high workload on decision error, but we imagine it might operate like time pressure. The best antidote for both appears to be appropriate task and situation management behaviors that serve to buy more time or to shed tasks from the decision maker.

Our findings have several implications for crew training: Programs should emphasize the importance of identifying the temporal demands, risks, affordances, and constraints inherent in a problem situation and the development of skill at adapting strategies to match situations. A theory of naturalistic decision making must be sensitive to significant situational variations and broad enough to account for a range of effective decision strategies.

ACKNOWLEDGMENTS

We wish to express our appreciation to NASA, Code UL, and to the FAA–ARD for their support of the research on which this chapter was based. Special thanks go to Eleana Edens, our project manager at the FAA, for her continued support.

REFERENCES

Chidester, T. R., Kanki, B. G., Foushee, H. C., Dickinson, C. L., & Bowles, S. V. (1990). *Personality factors in flight operations: Volume I. Leadership characteristics and crew performance in a full-mission air transport simulation* (NASA Tech. Mem. No. 102259). Moffett Field, CA: NASA–Ames Research Center.

Fischer, U., Orasanu, J., & Montalvo, M. (1993). Efficient decision strategies on the flight deck. In R. S. Jensen & D. Neumeister (Eds.), *Proceeding of the Seventh International Symposium on Aviation Psychology* (pp. 238–243). Columbus, OH: Ohio State University Press.

Fischer, U., Orasanu, J., & Wich, M. (1995). Expert pilots' perceptions of problem situations. In *Proceedings of the Eighth International Symposium on Aviation Psychology* (pp. 777–782). Columbus, OH: Ohio State University Press.

Foushee, H. C., Lauber, J. K., Baetge, M. M., & Acomb, D. B. (1986). *Crew factors in flight operations: III. The operational significance of exposure to short-haul air transport operations* (Tech. Mem. No. 88322). Moffett Field, CA: NASA–Ames Research Center.

Hammond, K. R., Hamm, R. M., Grassia, J., & Pearson, T. (1987). Direct comparison of the efficacy of intuitive and analytical cognition in expert judgement. *IEEE Transactions on Systems, Man, and Cybernetics, 17*(5), 753–770.

Hart, S. G., & Wickens, C. D. (1990). Workload assessment and prediction. In H. R. Booher (Ed.), *MANPRINT: An approach to system integration* (pp. 257–296). New York: Van Nostrand Reinhold.

Hutchins, E., & Klausen, T. (1991). *Distributed cognition in an airline cockpit.* Unpublished manuscript, University of California, San Diego.

Klein, G. A. (1993). A recognition-primed decision (RPD) model of rapid decision making. In G. Klein, J. Orasanu, R. Calderwood, & C. Zsambok (Eds.), *Decision making in action: Models and methods* (pp. 138–147). Norwood, NJ: Ablex.

National Transportation Safety Board. (1983). *Aircraft accident report: Hawker Siddley 748, Pinckneyville, IL.* Washington, DC: Author.

National Transportation Safety Board. (1994). *A review of flightcrew-involved, major accidents of U.S. Air Carriers, 1978 through 1990* (PB94–917001, NTSB/SS–94/01). Washington, DC: Author.

Orasanu, J. (1994). Shared problem models and flight crew performance. In N. Johnston, N. McDonald, & R. Fuller (Eds.), *Aviation psychology in practice* (pp. 255–285). Hants, England: Avebury Technical.

Orasanu, J., Dismukes, R. K., & Fischer, U. (1993). Decision errors in the cockpit. In L. Smith (Ed.), *Proceedings of the Human Factors and Ergonomics Society 37th annual meeting* (Vol. 1, pp. 363–367). Santa Monica, CA: Human Factors and Ergonomics Society.

Orasanu, J., Fischer, U., & Tarrel, R. (1993). A taxonomy of decision problems on the flight deck. In R. Jensen (Ed.), *Proceedings of the Seventh International Symposium on Aviation Psychology* (pp. 226–232). Columbus, OH: Ohio State University Press.

Orasanu, J., & Strauch, B. (1994). Temporal factors in aviation decision making. In L. Smith (Ed.), *Proceedings of the Human Factors and Ergonomics Society 38th annual meeting* (Vol. 2, pp. 935–939). Santa Monica, CA: Human Factors and Ergonomics Society.

Payne, J. W., Bettman, J. R., & Johnson, E. J. (1993). *The adaptive decision maker.* New York: Cambridge University Press.

Rasmussen, J. (1983). Skill, rules, and knowledge: Signals, signs and symbols, and other distinctions in human performance models. *IEEE Transactions on Systems, Man and Cybernetics, 13*(3), 257–267.

Wickens, C. D., & Flach, J. M. (1988). Information processing. In E. L. Wiener & D. C. Nagel (Eds.), *Human factors in aviation* (pp. 111–155). San Diego, CA: Academic Press.

Woods, D. D. (1993). Process-tracing methods for the study of cognition outside of the experimental psychology laboratory. In G. Klein, J. Orasanu, R. Calderwood, & C. Zsambok (Eds.), *Decision making in action: Models and methods* (pp. 228–251). Norwood, NJ: Ablex.

Wright, P. L. (1974). The harassed decision maker: Time pressures, distractions, and the use of evidence. *Journal of Applied Psychology, 59,* 555–561.

Chapter 33

How Can You Turn a Team of Experts Into an Expert Team?: Emerging Training Strategies

◆

Eduardo Salas
Janis A. Cannon-Bowers
Joan Hall Johnston
Naval Air Warfare Center Training Systems Division[1]

In modern society, teams are routinely called on to perform complex and critical tasks. Examples of such teams are cockpit crews, fire-fighting teams, military command and control teams, and trauma surgical teams. All of these teams make decisions in environments characterized by time pressure, heavy workload, and potentially conflicting and ambiguous information, and where the consequences for error are high. In these kinds of environments multiple experts come together as a "collective" to pool multiple sources of information in support of task accomplishment. Unfortunately, in the last decade we have witnessed a number of accidents and mishaps where innocent lives have been lost through the propagation of human error in decision making, coordination, and communication. One wonders why. Why does a highly trained set of individuals not always operate effectively as a coordinated team? Clearly, the answer to this question requires a fuller understanding of the processes and mechanisms that transform a team of experts into an expert team.

---

[1]The views expressed herein are those of the authors and do not reflect the official positions of the organization with which they are affiliated.

In 1993, Orasanu and Salas examined naturalistic team decision making (TDM) and provided an overview of the theories and research that emerged in the late 1980s. They pointed out that although some conceptual issues were gaining attention, much work still needed to be done. For example, they stated that the shared mental model construct appeared to be a viable framework for conceptionalizing team performance, because effective teams under stress appear to rely on implicit coordination. However, they concluded that few studies had examined TDM empirically, and even fewer had identified what constitutes effective (or ineffective) TDM. Without such information, training strategies could not be designed, developed, nor tested. Means, Salas, Crandall, and Jacobs (1993) also reviewed the literature on training decision makers. They concluded that not much conceptual work had been developed to guide training research, nor had empirical work been conducted on which to draw principles of training for TDM. This chapter was motivated, in part, by the desire to revisit, update and expand these two previous reviews.

The naturalistic decision making (NDM) paradigm has helped further our understanding of TDM performance and, more importantly, it has directed attention toward the development of a number of training strategies to enhance TDM. A primary question then becomes, how does the NDM paradigm help us to conceptualize and design training for TDM? First, NDM has focused our attention on *real* teams performing *real* tasks in their *real* environments. This is not to say that NDM cannot be studied in a laboratory setting. That is, NDM does not specify a place in which to conduct research; rather, it demands an approach or a "way of thinking" about the problem. What is necessary is that lab conditions mimic the environment to which we would like to generalize. Second, NDM highlights the requirement for focusing on the *process* by which decisions are made, and information is communicated and coordinated among team members. Finally, the NDM paradigm highlights the view that traditional methods of data recording and collection will not suffice. This means that if we are going to have a richer understanding of naturalistic TDM performance, we need to design and develop (or adapt) new protocols for data recording collection and interpretation.

In light of the aforementioned, the purpose of this chapter is twofold. First, we briefly discuss one of the theoretical underpinnings that has emerged as potential guidance for TDM training research. Second, we describe three emerging training strategies on which empirical work has been initiated. Finally, we delineate the lessons learned from research conducted in the last few years.

In doing this, we focus on a comprehensive program that was initiated in the late 1980s to enhance tactical decision making in complex environments by applying recent advances in decision theory and training (Cannon-Bowers, Salas, & Grossman, 1991). This program, called Tactical

Decision Making Under Stress (TADMUS), has focused on testing interventions for the purpose of developing theoretically based and practical principles and guidelines for a variety of team training strategies that are necessary in order to minimize the adverse effects of stress.

We begin with a description of a theory for enhancing team decision making in naturalistic environments, along with findings from three TADMUS-supported empirical studies regarding the development of team training strategies. It should be noted that there is ongoing research in other domains dealing with NDM and they are beginning to draw implications for training. However, it is beyond the scope of this chapter to fully cover those programs of research (Orasanu, 1993, 1994).

SHARED MENTAL MODELS

It has been argued that the nature of how teams cope and adapt during stressful conditions can best be understood in terms of shared mental model theory (Cannon-Bowers, Salas, & Converse, 1990). The notion of "mental models" has been invoked as an explanatory mechanism by those studying skilled performance for a number of years (Gentner & Stevens, 1983; Johnson-Laird, 1983). With respect to team training to reduce the harmful effects of stress, several researchers have suggested that the goal of instruction should be to foster shared mental representations of the team role structure, task structure, and the process by which the two interact (Cannon-Bowers, Salas & Converse, 1993; Rouse & Morris, 1986; Volpe, Cannon-Bowers, Salas, & Spector, 1996).

Despite the fact that work teams have been the focus of countless research efforts over the past few decades, relatively little is known about the attributes of teamwork, or how to train teams to perform in an optimal manner (Salas, Dickinson, Converse, & Tannenbaum, 1992). What is recognized is that intermember communication can be severely restricted during periods of high stress, and that team members must, therefore, possess the ability to adapt to stressful situations without the use of explicit communications (Kleinman & Serfaty, 1989). This skill appears to involve the ability of team members to predict the needs and information requirements of their teammates, and to anticipate the actions of other team members in order to adjust their behavior accordingly (Cannon-Bowers, Salas, & Baker, 1991). This ability has been explained by hypothesizing that the team members exercise shared or common knowledge bases.

The shared mental model framework allows for a better, and maybe more comprehensive description of *how* team members perform. For example, it allows for the full repertoire of the behavioral, attitudinal, and cognitive aspects of TDM to be examined. Second, this theoretical perspective specifies what team decision making *is* and *how* it may be trained, thus creating an opportunity for designing training strategies that go beyond practice and feedback (Salas & Cannon-Bowers, 1994). The following sections describe how the shared mental model construct is particularly useful in the development of training strategies aimed at fostering effective team decision making in complex environments.

Shared Mental Models and Training

In keeping with what has been stated thus far, we can hypothesize that members of effective teams are likely to need a number of accurate mental models (or knowledge structures) in order to generate predictions and expectations about team tasks and demands without employing overt communications. Cannon-Bowers et al. (1993) proposed that multiple mental models allow each member to understand a system at several levels. They include: an understanding of the equipment with which the member is interacting, an understanding of the task and how to accomplish it, and an understanding of the role the team member plays in the task, as well as the roles of other team members. Therefore, a reasonable proposition regarding mental models is that teams with shared mental models of the task and team member roles are more likely to have *accurate expectations* regarding the needs of the team, allowing them to adjust their behavior accordingly during stressful situations.

Therefore, the basic objective of training is to foster in the team members an accurate and sufficient mental representation of the team task structure, team role structure, and the process by which the two interact (Baker, Salas, Cannon-Bowers, & Spector, 1992). The learner must possess sufficiently clear mental models of these structures before he or she can perform the necessary teamwork, taskwork, and decision-making behaviors in an efficient manner. Furthermore, if team members share the same mental models, they presumably would be working under the same assumptions, have similar expectations regarding the roles and responsibilities of their fellow teammates, and would use similar decision-making strategies. This in turn would facilitate overall communication, coordination, and performance (Cannon-Bowers et al., 1993).

The individual approach and format for fostering mental models may vary considerably, depending on the task, the learner characteristics, the context, and a host of other variables; these factors are even more critical at the team level. Until recently, there have been no guidelines or principles for guiding team training strategies to foster shared mental models. In the last several years, enough information has been developed in the team and mental model literatures to formulate a set of training strategies that have considerable potential for fostering shared mental models. Under the TADMUS program, three team training strategies (cross-training, team model training, and team adaptation and coordination training) were developed based on the shared mental model approach. In all studies, the testbed was a simulation of a military task (e.g., fighter plane, ship's combat information center). Typical Navy simulation team training consists of having teams respond to a scenario generated by the simulation. The interventions developed under TADMUS were design to enhance simulation-based training. Therefore, although the research design was to compare performance of "trained" teams to "no training" teams, the "no training" teams actually received exposure to the typical military simulation training approach. Following is a description of each training strategy and the initial empirical findings for each.

CROSS-TRAINING

Many organizations face the challenge of maintaining high levels of performance in their teams despite conditions of frequent personnel turnover. Frequent turnover leads to an ongoing need to train novices to perform well within an experienced team. The objective of cross-training is to mitigate team performance decline due to turnover. The cross-training should be designed to develop commonly shared expectations (called interpositional knowledge) regarding specific team-member functioning. This can be accomplished by providing all parties with "hands-on" experience regarding the other job functions within the team. The general hypothesis is that exposure to, and practice on, *other* team member tasks should result in enhanced team-member knowledge regarding other team members' task responsibilities, resource needs, and coordination requirements (Baker et al., 1992; Cannon-Bowers & Salas, 1990; Salas et al., 1992).

Volpe et al. (1996) tested the effects of cross-training and workload on team processes, communications, and performance of 40 two-person teams operating a PC-based aircraft simulator. In the cross-training condition, each position was assigned specific tasks, but neither party was able to accomplish the objective without the task input of the other position. However, the objective could be reached individually in the no cross-training condition. Verbal communications were via intercom headsets and microphones. Teams in the cross-training condition were informed about and received practice on operationally relevant tasks pertaining to both parties' functional responsibilities. Workload consisted of increased informational demands within the primary task. Volpe et al. (1996) found that cross-training was an important determinant of effective task coordination, communication, and performance. These findings suggest that cross-training a new member on the team's role structure and each member's task responsibilities may circumvent degraded coordinated task activity, team interactions, and communication effectiveness.

Although the Volpe et al. (1996) study was conducted in a lab setting, the task had fairly high "real-world" validity in that it actually mimics the real environment quite well, thereby increasing the generalizability. Described in the next section is an empirical study of the team model trainer, which successfully adapted the cross-training concept to a higher fidelity Navy Antiair Warfare (AAW) tactical decision making trainer (Naval Underwater Systems Center, 1991).

TEAM MODEL TRAINING

The Team Model Trainer (TMT) was developed for the purpose of enhancing team performance through the cross-training of individual combat information center (CIC) team members. The TMT concept resulted from observations and interviews with Navy CIC teams, and from the hypothesized knowledge representations needed to enhance shared identity of the task (see Rouse, Cannon-Bowers, & Salas, 1992). The observations identified five team training issues in support of TMT development. Specifically, current training: (a) lacked standard measures of individual and team performance; (b) was unsystematic; (c) overloaded the trainees with many details, but often lacked a review of important scenario events; (d) provided unguided practice with little or no moment-to-moment feedback; and (e) provided little feedback about alternative options that might have been taken.

Results of the interviews suggested that CIC team members may have multiple mental models regarding: (a) the equipment; (b) individual tasks; (c) ship's systems; (d) CIC team performance; (e) the ship; and (f) situations (e.g., how the ship fits within a battlegroup context). As a result, a low-cost commercial, off-the-shelf PC-based training device was developed in order to train individual CIC team members in expert mental models that would, in turn, enhance team performance.

By way of description, a training session with the TMT begins with learning about any of the six task-related, team-related, or both, knowledge domains, and then participating in an event-based scenario simulation where the various knowledge domains can be applied and practiced. The knowledge domain information is presented in a text- and graphics-based format.

Participation in the scenario simulation allows the individual team member to engage in a simulation in one of two ways. For any one of the five watch stations represented, the trainee could choose to *observe* a 30-minute antiair warfare (AAW) scenario, or the trainee could *perform* a team-member role.

In the *observation* mode, the trainee hears the audio communications of an expert team as they engage in identifying air traffic and taking actions in line with the rules of engagement. In the *perform* mode, the trainee is responsible for carrying out the duties and responsibilities of a watch station. This is accomplished by listening to teammates, making reports, and entering data into the system. The TMT records participant responses from a point-and-click mouse function so that immediate performance feedback can be accessed by the trainee any time during the scenario. As feedback, a record of trainee performance is presented along with expert performance for the selected scenario event.

Duncan, Cannon-Bowers, Johnston, and Salas (1994) tested the impact of the TMT on the communications and performance of three teams on a PC-based simulation of five CIC positions referred to as DEFTT (Decision Making Evaluation Facility for Tactical Teams). Four additional teams serving as a control group for comparison did not receive the TMT. Performance and communications were recorded and evaluated for all teams performing two 30-minute scenarios.

Following the first practice scenario, the training condition consisted of individuals participating in a one-on-one 2-hour TMT session. A session began with task familiarization regarding the five AAW positions. Next, trainees observed a 30-minute AAW scenario, and heard simulated audio communications of an expert team as they engaged in identifying air traffic and took actions in accordance with rules of engagement. Then the trainees participated in the TMT "perform" mode, in which they were responsible for carrying out the duties and responsibilities of their own watch station. Trainee performance was compared to expert performance for the selected scenario event. Finally, each trainee "played" a position different from the one they had been assigned. Following TMT, subjects participated in another AAW DEFTT scenario.

Results showed that only the experimental group had significant improvement in AAW watchstation performance behavior measures that had been established for specific scenario events. A follow-on study was conducted with an additional four teams, using the same experimental procedures, to test the hypothesis that individual member's mental models of the team would improve with training on the TMT. Duncan et al. (1994) developed a team communications questionnaire to measure how the team believed they should communicate. Therefore, two types of questionnaires were developed: one indicating what the team members thought they should do, and one which recorded what they actually did. Items consisted of probes that tapped the member's model of the task and team, including: with whom he or she communicated, what he or she expected to communicate, and what information was expected to be received from team members. Responses from the first questionnaire were compared with the second questionnaire following training. Duncan et al. concluded that the trained group significantly improved their interpositional knowledge as a result of team model training. Specifically, they found a 10% improvement in team members' knowledge about appropriate team communications.

The TMT approach has illustrated that complex team interactions can be demonstrated, practiced, and learned using a low-cost, off-the-shelf PC-based system. Typically simulations focus only on the equipment that a person is learning to use. In addition, practice opportunities in highly dangerous and uncertain conditions are relatively infrequent, thus making the maintenance of complex cognitive decision making skills very difficult. Results from TMT suggests that simulated realism in PC-based training could be made readily available to a diverse audience, and that the scope of the simulation can be enlarged to include training for such teamwork skills as situation assessment, communication, team leadership, and compensatory behavior.

TEAM ADAPTATION AND COORDINATION TRAINING

Another training approach based on the shared mental model theory that has implications for team decision making is Team Adaptation and Coordination Training (TACT; Entin, Serfaty, & Deckert, 1994). TACT was developed as a result of findings by Serfaty, Entin, and Volpe (1993) that successful teams were able to adapt to increased stress levels by altering their coordination strategies. TACT was designed to reduce communication and coordination overhead within the team.

Entin et al. (1994) tested TACT using an experimental protocol, and tasks similar to those in the study by Duncan et al. (1994). Twelve 5-person teams of Navy students participated, with six teams receiving the TACT intervention. TACT consisted of the following four training modules. Following pretraining practice on two DEFTT scenarios, team members in the TACT condition were instructed on how to identify signs and symptoms of stress. Second, they received instruction on five adaptive strategies (preplanning, use of idle periods, information transmission, information anticipation, and redistribution of workload) to deal with increases in workload. Third, the team leader was instructed to transmit periodic situation-assessment updates to all team members in order to elicit anticipatory information from them. Finally, teams practiced the situation-update strategy while performing on three training scenarios. Following training, teams performed on two DEFTT scenarios.

Results showed that teams receiving TACT committed few errors and coordinated teamwork significantly better in response to the stressful scenarios than untrained teams. Improved performance of the TACT team was attributed to their ability to: (a) adjust individual taskwork to prevent overload; (b) increase congruence of team mission between team members and the team leader; and (c) increase communication of crucial information needed by the team leader. Controlling for stress, TACT teams performed significantly better on posttraining scenarios than teams assigned to control conditions. Changes in communication patterns were dominated by a stronger push of information to the team leader, and more anticipatory behavior.

Implications of these findings for team decision making training complement those described in the TMT section. Whereas the TMT enhances team performance through cross-training, TACT enhances team performance through changes in communication dynamics of team members under stress. TACT demonstrates the importance of practicing teamwork strategies in order to change behaviors to adapt to increasing task demands.

LESSONS LEARNED . . . SO FAR

So what have we learned (so far) about naturalistic TDM? Since 1993, progress has been made. First, at the conceptual level, shared mental model theory has emerged as a framework to guide TDM research and to focus training interventions. Second, training derived from shared mental model theory can be an effective means to improve TDM. More specifically, we have tested and successfully demonstrated effective interventions to improve communication and teamwork skills as well as decision-making performance. Third, there is more to training than simply practice and feedback. That is, these training strategies highlight the importance of knowledge structures in TDM and begin to tell us how to impart, reinforce, and evaluate team performance. Furthermore, these strategies demonstrate the importance of linking knowledge structures to performance strategies. The following section provides more specific lessons with regard to team training design.

Team Training Design

A number of important lessons with respect to team training design are worth discussing. First, we have learned that task stressors influence team communications and team performance strategies. Therefore, a training needs analysis to enhance TDM should identify and incorporate appropriated stressors into training simulations. Second, training needs analysis for TDM should emphasize major tasks and skills that are necessary conditions for effective team coordination and decision making. These lessons seem rather obvious, but are seldom followed. One problem is the lack of a taxonomy that could be used as a point of reference to identify skills required for effective team decision making. Recently, Cannon-Bowers, Tannenbaum, Salas, and Volpe (1995) synthesized the teamwork literature and offered a set of competencies and conditions for selecting team training strategies. Essentially, this framework links the requirements of the task and environment to the competencies required in team members. It may be useful as a means to highlight knowledge, skills, and attitudes required for TDM under various conditions.

Third, the studies reported here have shown that training that fosters anticipatory communication of team members can enhance TDM. Therefore, training designed to improve TDM should specify mechanisms and procedures (e.g., communication) that will enhance anticipatory behaviors. Fourth, we learned that training that fosters communication of the team leaders' mental picture of the situation can enhance the team's anticipatory behaviors. Therefore, practice with provisions for periodic situation updates by team leaders could be and perhaps, should be, included in training interventions.

In sum, team mental models of taskwork and teamwork strategies should be identified and fostered by: (a) providing relevant and meaningful *information* (knowledge structures) about other team members, the task, the equipment, and the situation; (b) *demonstrating* effective and ineffective TDM performance; (c) allowing for *practice* (via simulation, role play, behavioral modeling) with the relevant characteristics of the tasks and the context; and (d) providing *feedback* regarding crucial aspects of team performance.

HOW CAN YOU TURN A TEAM OF EXPERTS INTO AN EXPERT TEAM?

The answer to this question is not simple, because much more research needs to be done. However, a few guidelines are beginning to emerge from the research described here. We turn a team of experts into an expert team by: (a) fostering shared (or compatible) mental models of the task and of other team member roles; (b) training team members on teamwork skills such as situation awareness, communications, team leadership, adaptability, and compensatory behavior; (c) providing team members with guided practice on skills needed to perform under naturalistic conditions; (d) developing simulations that allow team members to experience different course of actions; (e) linking cue patterns to response strategies not only with regard to the tasks, but other team members; (f) training (via demonstration, practice, and feedback) team members on each others' roles and on building realistic expectations about the task requirements; and (g) training team leaders to maintain shared situational awareness by providing periodic updates to team members. Clearly, the NDM perspective helped us to get this far by focusing attention on the *process* of TDM in real teams. We feel assured that additional principles of TDM training will be forthcoming as further research is conducted.

CLOSING REMARKS

We stated at the onset of this chapter that progress had been made in the area of TDM training in the past 5 years or so. Unfortunately, besides the few efforts we have described here, few, if any, empirical investigations of training interventions directed at TDM have been conducted. This may be due in part to the difficulties inherent in conducting "action" research—especially at the team level. However, without solid, rigorous tests of the theoretical propositions generated from NDM theories, we will never have solid guidelines from training practitioners (Cannon-Bowers, Tannenbaum, Salas, & Converse, 1991). Therefore, our recommendation for future work is simple: Conduct more empirical tests of TDM training strategies, and make the results of those experiments available to both the research and practitioner communities.

REFERENCES

Baker, C. V., Salas, E., Cannon-Bowers, J. A., & Spector, P. (1992, April). *The effects of inter-positional uncertainty and workload on teamwork and task performance.* Paper presented at the annual meeting of the Society for Industrial and Organizational Psychology, Montreal, Canada.

Cannon-Bowers, J. A., & Salas, E. (1990, April). *Cognitive psychology and team training: Shared mental models in complex systems.* Symposium presented at the Fifth Annual conference of the Society for Industrial and Organizational Psychology, Miami, FL.

Cannon-Bowers, J. A., Salas, E., & Baker, C. V. (1991). Do you see what I see? Instructional strategies for tactical decision making teams. In *Proceedings of the 13th Annual Interservice/Industry Training Systems Conference* (pp. 214–220). Washington, DC: National Security Industrial Association.

Cannon-Bowers, J. A., Salas, E., & Converse, S. A. (1990). Cognitive psychology and team training: Training shared mental models and complex systems. *Human Factors Society Bulletin, 33,* 1–4.

Cannon-Bowers, J. A., Salas, E., & Converse, S. A. (1993). Shared mental models in expert team decision making. In N. J. Castellan, Jr. (Ed.), *Current issues in individual and group decision making* (pp. 221–246). Hillsdale, NJ: Lawrence Erlbaum Associates.

Cannon-Bowers, J. A., Salas, E., & Grossman, J. D. (1991, June). *Improving tactical decision making under stress: Research directions and applied implications.* Paper presented at the International Applied Military Psychology Symposium, Stockholm, Sweden.

Cannon-Bowers, J. A., Tannenbaum, S. I., Salas, E., & Converse, S. A. (1991). Toward an integration of training theory and practice. *Human Factors, 33,* 281–292.

Cannon-Bowers, J. A., Tannenbaum, S. I., Salas, E., & Volpe, C. E. (1995). Defining team competencies: Implications for training requirements and strategies. In R. Guzzo & E. Salas (Eds.), *Team effectiveness and decision making in organizations* (pp. 333–380). San Francisco: Jossey-Bass.

Duncan, P. C., Cannon-Bowers, J. A., Johnston, J., & Salas, E. (1994). *Using a simulated team to model teamwork skills: The team model trainer.* Unpublished manuscript.

Entin, E. E., Serfaty, D., & Deckert, J. C. (1994). *Team adaptation and coordination training* (Tech. Rep. No. 648–1). Burlington, MA: ALPHATECH, Inc.

Genter, D., & Stevens, A. L. (Eds.). (1983). *Mental models.* Hillsdale, NJ: Lawrence Erlbaum Associates.

Johnson-Laird, P. N. (Ed.). (1983). *Mental models.* Cambridge, England: Cambridge University Press.

Kleinman, D. L., & Serfaty, D. (1989). Team performance assessment in distributed decision making. In R. Gilson, J. P. Kincaid, & B. Goldiez (Eds.), *Proceedings of the Interservice Networked Simulation for Training Conference* (pp. 22–27). Orlando, FL: University of Central Florida.

Means, B., Salas, E., Crandall, B., & Jacobs, T. O. (1993). Training decision makers for the real world. In G. Klein, J. Orasanu, R. Calderwood, & C. Zsambok (Eds.), *Decision making in action: Models and methods* (pp. 306–326). Norwood, NJ: Ablex.

Naval Underwater Systems Center. (1991). *Decision-making evaluation facility for tactical teams (DEFTT) design analysis report.* New London, CT: Naval Underwater Systems Center.

Orasanu, J. (1993). Decision-making in the cockpit. In E. Wiener, B. Kanki, & R. Helmreich (Eds.), *Cockpit resource management* (pp. 137–170). San Diego, CA: Academic Press.

Oransanu, J. (1994). Shared problem models and flight crew performance. In N. Johnston, N. McDonald, & R. Fuller (Eds.), *Aviation psychology in practice* (pp. 255–285). Brookfield, VT: Ashgate.

Orasanu, J., & Salas, E. (1993). Team decision making in complex environments. In G. Klein, J. Orasanu, R. Calderwood, & C. Zsambok (Eds.), *Decision making in action: Models and methods* (pp. 327–345). Norwood, NJ: Ablex.

Rouse, W. B., Cannon-Bowers, J. A., & Salas, E. (1992). The role of mental models in team performance in complex systems. *IEEE Transactions on Systems, Man, and Cybernetics, 22,* 1296–1308.

Rouse, W. B., & Morris, N. M. (1986). On looking into the black box: Prospects and limits on the search for mental models. *Psychological Bulletin, 100,* 349–363.

Salas, E., & Cannon-Bowers, J. A. (1994, October). *Beyond practice and feedback: What shared mental model theory offers.* Panel presented at the 38th Annual Meeting of Human Factors Society, Nashville, TN.

Salas, E., Dickinson, T. L., Converse, S. A., & Tannenbaum, S. I. (1992). Toward an understanding of team performance and training. In R. W. Swezey & E. Salas (Eds.), *Teams: Their training and performance* (pp. 3–29). Norwood, NJ: Ablex.

Serfaty, D., Entin, E. E., & Volpe, C. (1993). Adaptation to stress in the team decision-making and coordination. In *Proceedings of the 37th Annual Human Factors and Ergonomics Society Annual Meeting* (pp. 1228–1232). Seattle, WA: The Human Factors Society.

Volpe, C. E., Cannon-Bowers, J. A., Salas, E., & Spector, P. (1996). The impact of cross-training on team functioning: An Empirical Examination. *Human Factors, 38*(1), 87–100.

Chapter 34

Managerial Problem Solving: A Problem-Centered Approach

◆

Gerald F. Smith
University of Minnesota

Like any human behavior, effective thinking is adapted to the task. Having been argued by Tolman, Brunswik, Simon, and others, this claim has the status of a truism. However, it has been overlooked by research on managerial thinking, research that often ignores problem-side variables and makes few provisions for analyzing tasks. This chapter describes a problem-centered strategy for improving managerial thinking. The chapter begins with a discussion of decision theory. It then assesses process-oriented theories of decision making and problem solving. Against this background, the chapter argues for a problem-centered approach to improving managerial thinking. Such an approach is proposed and is illustrated through a case analysis. Its prospects are assessed in the chapter's conclusion.

DECISION THEORY

One justification for a new approach to managerial thinking is the inadequacy of existing theory. Although battered by its critics, decision theory is still the reigning paradigm, offering descriptions of and prescriptions for action-oriented thought. Deriving from economic theories of rationality, this paradigm encompasses expected utility theory, multiattribute utility theory, decision analysis, behavioral decision theory, and similar models of preference-driven choice.

The major objection to decision theory has been its failure to admit knowledge constraints, Simon's (1976) claim that decisions are only "boundedly rational." This objection is valid, but should go without saying. Decision theory has three other inadequacies when viewed as an account of managerial thought.

First, it is theoretically vacuous, almost a tautology. Decision theory proposes that people select utility-maximizing courses of action. However, because the theory does not say what constitutes utility—what people will value in a situation—it is virtually nonfalsifiable (Tversky, 1975): One can always add regret, computational ease, or whatever to the outcomes of a choice in order to preserve the theory from violations. Were someone to claim that his decisions maximized, not utility, but x-ness, a decision theorist would respond that x-ness had utility for that person. The claim that humans pursue what they value is true by definition, the conceptual consequence of what it means to value something.

Second, decision theory suffers from an inadequate, judgment-centered, psychology. Decisions are conceived as compositions of predictive and evaluative judgments. Although judgment is an important part of our cognitive repertoire, so too are reasoning, memory recall, imagination, and other mental activities barely mentioned in decision theoretic research. The cause of its meager psychology is decision theory's behavioristic origins (Thagard, 1982). The effect is an inability to account for the richness of goal-directed thought in real-world situations.

Third, decision theory offers an overly narrow view of managerial tasks. Managers are assumed to face decision situations in which they make preference-driven choices from among alternatives. If management was like grocery shopping, this might be a good model. However, management is more like chess, choices being driven by assessments of outcome effectiveness rather than of preference. Then too, not all managerial tasks culminate in a decisive act of choice. Managers often face design problems in which an artifact—a new incentive plan, for instance—must be created. If an existing system is not performing acceptably, the key challenge is diagnosis, determining the cause of the performance shortcoming. Many decisions are made while solving such problems, but few could be usefully addressed with decision-analytic tools.

PROCESS ACCOUNTS OF THINKING

Dissatisfaction with decision theory (Grayson, 1973; March, 1981) and the opportunity created by cognitivism have increased interest in the study of decision-making/problem-solving processes. Process research is based on two premises: that there are understandable regularities in action-oriented thought and that an explication of the thinking process will enable useful prescriptions. Models that represent thinking as a sequence of functionally-defined stages are a familiar product of such research (Lang, Dittrich, & White, 1978). Some models depict problem solving (problem formulation, alternative generation) as the front end and decision making (evaluation, choice) as the back end of a unitary process. Many process researchers focus on particular elements of thinking: for instance, judgmental heuristics, fallacies of informal logic, problem categories, recall of information from memory, and the generation of alternatives (Sternberg & Smith, 1988). Others study the thought processes of experts in different fields (e.g., medicine, computer programming) and of people in special settings (e.g., juror decision making).

Much has been and will continue to be learned from process studies. However, there are reasons to question their ultimate effectiveness for understanding and improving managerial problem solving. These reasons derive from the two premises of process research.

First, the thinking process is more variable than has been assumed. This variability derives in part from individuals, but primarily from thinking tasks. Goal-oriented thought is "situated" (Suchman, 1987), shaped by the evolving task context. An empirical study of managerial problem solving led Isenberg (1986) to conclude that such thinking was "opportunistic." Schon (1983) characterized professional thinking as a "conversation with the situation" (p. 76). There is no standard problem-solving process, nor even a manageable set of process variations. Traditional functional models cannot accomodate, for instance, the fact that diagnosis is the key problem-solving task in some situations while being absent in others. Some problem-solving strategies have a meaningful measure of generality—for instance, "weak methods" like means–ends analysis and situation recognition as discussed by Clancey (1985) and Klein (1993). However, these elements of order in managerial problem solving are swamped by process variability. As a result, attempts to explicate the process and thereby improve it are unlikely to succeed.

Second, process understanding may have little prescriptive value. Scientists often assume that prescriptions follow directly from theoretical understanding of a phenomenon. However, that is not always true. One's understanding can be at too high a level to afford prescription. Thus, a theory which claims that problem solving involves the activation of problem schemas may be on the mark, but have scant prescriptive value, absent an account of the schemas themselves.

Most of the mental processes implicated in managerial problem solving are content-driven. Thus, memory recall relies on retrieval cues. General prompts—"What are the goals?"—are weak, compared to specific cues based on characteristics of the situation—"What goals does labor usually pursue in contract negotiations?" Perception, attention, learning, imagination, and other cognitive processes can be aided by prescriptions that tell one what to perceive, attend to, learn, and imagine in situations of a given kind. Lacking an account of problem types or characteristics, process theories are unlikely to provide strong, situation-specific prescriptions.

In summary, the prescriptive potential of process research on managerial problem solving seems limited. Owing to process variability, it is doubtful that a useful general theory of real-world problem-solving processes can be devised. Accounts of processes used to solve particular problems can be substantive, although wanting generality they will seem less impressive as theories. Researchers should be able to determine our basic repertoire of problem-solving behaviors, common failings, and useful heuristics. However, they are unlikely to offer powerful prescriptions from process knowledge alone, because effective problem solving is tailored to the problem at hand. In the remainder of this chapter, a more promising prescriptive approach is proposed.

TAKING PROBLEMS INTO ACCOUNT

If problem-solving behavior is to be adapted to the task, we must have a way of characterizing problems. Scientists describe real world problems as being complex and unstructured, marked by uncertainty and conflict (MacCrimmon & Taylor, 1976). General attributes like these are of little value in customizing solution efforts to particular situations. Some real-world problems pose severe time pressures, so that certain solution strategies are not workable, whereas others allow feedback on actions taken over time, so that performance can be improved through learning (Orasanu & Connolly, 1993). Although useful, these characterizations are still too broad.

To develop an adequate framework for problem analysis, we must recognize that experiential knowledge is critical for solving managerial problems, which typically involve behavioral and social phenomena. Also, although managerial problems are infinitely varied, they contain significant commonalities. They lie between the poles of "Everything is the same" and "Everything is different." For instance, organizations constantly confront the challenge of reducing costs, yet each such challenge has individuating features, any of which could be decisive for problem solving.

Real-world problems are often said to be unstructured. This can be taken to mean that there is no obvious, reliable solution method (Smith, 1988), in which case the claim is true. However, if structure is construed as the parts and relationships that constitute some thing, then all problems have structures. Each is a situation comprised of certain agents, goals, constraints, means, causes, knowledge, and other physical and conceptual entities in a complex set of relationships. A problem can exhibit multiple structures, each an identifiable pattern that comprises part of the overall situation. The sales shortfall experienced by a company might include such structures or subproblems as inadequate coordination within the distribution channel and resistance to change by salespeople asked to implement a new marketing program.

One characteristic of a problem structure is its breadth, how much of the overall situation it encompasses. Problem structures are also of varying depth, the degree to which they are obvious aspects of the situation or relatively deep and hidden. Finally, problem structures vary in generality, some being rare or one-of-a-kind, whereas others occur so frequently as to be generic. All companies experience sales shortfalls, a generic revenue problem that can affect any agent in an exchange economy.

Good solutions to problems are not good by accident; they exploit characteristics of the situation. For instance, to get a program implemented, one might persuade a manager that its implementation will promote acceptance of a pet project. Problem solvers must identify and take advantage of a problem's key characteristics. This can happen in either of two ways. First, by recalling similar situations from the past, one can recollect what did or did not work before. Unless the current situation is insightfully characterized, however, recall efforts will either be barren or will yield volumes of irrelevancies. By characterizing problems in structural terms, problem analysis can promote effective recall. One may be helped to think of situations similar to the current problem in critical ways. The strategy of using past experience to solve current problems has been implemented in case-based reasoning systems (Kolodner, 1993). These AI systems incorporate indexing schemes that guide memory retrieval. Problem structures may be able to serve the same function for managerial problems.

Even when a situation is structurally unique so that similar past experiences are not available, problem solving may be improvable through analysis. By thinking about the situation, one may be able to identify the logical or conceptual possibilities it contains. Consider the famous "Elevator Problem" in which the management of an office building was receiving complaints from users about slow elevator service. Defining the situation as one of inadequate elevator capacity, management considered ways of increasing that capacity. However, the notion of inadequacy implies a mismatch between supply and demand. Problem analysis suggests that management also consider strategies for reducing the demand for elevator service. In addition, reflecting on the distinction between causes and conditions suggests that if the cause of user complaints—slow elevator service—cannot be remedied, perhaps the condition for complaints—people do not like waiting—can be addressed by interventions (e.g., putting mirrors, TVs, or a snack bar in the lobby) that make the wait less onerous (Smith, 1990). Good thinking identifies a situation's structure and the solution possibilities implicit therein.

A problem-solving method is powerful by virtue of reliably delivering high-quality solutions to problems in its domain. A method is general for having a large domain, being applicable to many problems. Newell (1969) argued that methods face an inevitable trade-off, power being achieved at a loss in generality and vice versa. This trade-off derives from problem solving's dependence on characteristics of the problem. Powerful methods exploit certain features of a situation and are inapplicable when those features do not pertain. General methods are not so dependent on a problem's defining features, but as a result, are not very powerful. To be both strong and general, a method must encompass problem-specific activities within a general problem-solving framework. Such an approach to problem solving is outlined in the next section.

A PROBLEM-CENTERED APPROACH

Consider the following case, adapted from Smith (1994), which will be used to illustrate how problems can be analyzed:

> A publishing company has contracted with a husband–wife author team who have completed their first book. To secure their commitment, the acquisitions editor promised these authors more-than-normal input into production decisions. Now that the book has entered the editorial phase, difficulties have arisen for the production editor, responsible for finalizing the book design and having it printed. The authors have been making many revisions to the text, often on the basis of seemingly whimsical preferences. The production editor has tried unsuccessfully to limit the changes, which risk delaying the book's introduction past the prime marketing season. As a result, the relationship between the production editor and the authors has deteriorated. The authors are threatening to back out of the contract, alleging that the company has not done an acceptable job of publishing their book.

The case can be viewed as a negotiation problem, a conflict or disagreement between two parties. Whereas many situations of conflict involve incompatible goals, here the parties share the overriding goal of having a successful book. Their disagreement is about means, what is needed for the book to be successful.

Difficulties often arise when people have invalid beliefs. The authors may overestimate the impact of aesthetics and underestimate the importance of timely publication as regards the book's prospects. Because the company presumably knows best, it should help the authors understand this. And, as we know from experience, rancorous personal relationships can keep parties from reaching a mutually beneficial agreement. Someone other than the production editor should handle discussions with these authors.

A useful heuristic is to ask how a problem came about: How did we get into this mess? This situation occurred because the acquisitions editor gave the authors more-than-normal control over production decisions, creating difficulties for the production editor. This larger problem has a familiar structure: One part of the organization does something that makes its job easier, while making life more difficult for others. Sales promises delivery dates that Manufacturing must satisfy; R & D designs a new product and "throws it over the wall" to Manufacturing, which has to build it. Recognizing this structure, people can recall common responses. In this case, one might require the production editor's prior approval of special arrangements with authors, institute project teams that get the production editor involved from the start, or base individual incentives on overall project success.

There is not yet, and may never be, a comprehensive, step-by-step method for analyzing managerial problems. Such situations may be too variable to allow a formal procedure. Nonetheless, an informal analytic framework has been developed. It begins by attempting to characterize a situation in terms of problem types (e.g., negotiation problem) and structures. More than one structure may be pertinent—for instance, the larger, systemic weakness in the publishing company case. It is important to recognize what is distinctive about the situation. How is it atypical of problems of its kind? What are its defining features, things that may be responsible for the situation's being problematic in the first place? In the publishing company case, it was important to exploit the fact that all parties will benefit from the book's success, and to respond to the authors' apparent lack of knowledge about the importance of timely publication.

Having characterized the problem, two other analytic steps naturally follow: The identification of pitfalls, mistakes to avoid in situations of this kind; and positive prescriptions, heuristics that can be usefully applied in such situations. For example, an internal auditor claimed that her department suffered from a "perceptual problem": Audit recommendations were ignored, arguably because management misperceived the quality and relevance of the department's work. However, could there be an element of truth in these perceptions? A pitfall associated with perceptual problems is rationalization, defining a situation as a perceptual problem when other inadequacies are truly at fault. A related heuristic: Identify the alleged misperceptions and consider whether they have a basis in fact.

Empirical research indicates that managers are sometimes able to recognize structural features of the problems they face (Smith, 1993). Problem structures, pitfalls, and heuristics have been cited in various literatures, including economics and systems theory. Senge (1990) discussed *eroding goals*, the error of solving a problem by lowering aspiration levels without good reason. People can fail to identify problems due to *scapegoating*, blaming others for trouble. *The tail wagging the dog* occurs when proposed solutions require larger changes than the problem's importance would justify. *How did we get into this mess?* was cited earlier as a useful heuristic. Another, *good cop, bad cop*, allows an agent (the good cop) to take an unpopular action by having someone else (the bad cop) bear responsibility.

Problem types and structures are the cornerstone of analysis. Many structures involve trade-offs: *ethical dilemma, quality–quantity trade-off, speed–accuracy trade-off, long term–short term trade-off,* and *over/under costs,* as of making too many or not enough copies for a meeting. Problems of *delayed feedback* occur due to time lags between inputs and a system's response. There are several *resource-allocation problems,* including *success to the successful* (Senge, 1990), situations in which a claimant's initial successes allow it to capture even more resources at the expense of competitors. Some noteworthy structures are listed in Table 34.1. Others remain to be discovered.

ASSESSMENT AND PROSPECTS

What has been proposed is not a theory of problem solving, unless the notion of theory is used very loosely, as it often is. Rather, it is a claim that effective problem solving must proceed from an analysis of the problem. The chapter presented a framework for this analysis, a partially filled-in method that can direct thinking about managerial problems. Development of the method would be simplified if there existed an overarching theory of real-world problems, a principled account from which problem structures, pitfalls, and

TABLE 34.1

Problem Types and Structures

Cannibalization: A new activity initiated by an agent is successful at the expense of the agent's initial set of activities.

Chicken and egg: Situations in which one must have X to get Y, but also needs Y to get X.

Cost–benefit visibility: An activity is difficult to evaluate because its costs are more easily measured than its benefits.

Cost–benefit displacement: The benefits of an activity accrue to one party, whereas its costs are borne by another.

Escalation: Two or more competitors successively "up the ante" in an attempt to outdo the others.

Falling between two stools: An entity or activity, trying to satisfy multiple goals, fails for not doing anything well enough.

Frontrunner's burden: Any competitive disadvantage that falls on frontrunners.

Parasitism: An activity that is not cost-justified is perpetuated by being lumped in with worthwhile endeavors.

Responsibility–authority split: Situations in which an agent is held responsible for a task but lacks the authority needed to get it done.

Slippery slope: Making an exception to a rule leads to requests for ever more dramatic exceptions.

Start-up difficulties: Problems that occur when a new system is initiated or an existing system is significantly changed.

Thrashing: A system spends so much time and effort on task management that it has little left for task performance.

Unrewarded activities: An activity that is not included in performance evaluations tends to be done carelessly or ignored.

heuristics could be derived. Alas, no such theory seems possible. This pessimism is grounded in history: Decades of intensive research have yielded nothing that could be called a theory of managerial problems. Pessimism is also implied by philosophical analysis: Dupre (1993) argued that many phenomena lack the crisp identities and principled origins required for the development of scientific theories. Managerial problems reflect fundamental mismatches or incidental aspects of our natural and social worlds. Those worlds are too complex and arbitrary to support a substantive theory of problems.

Thus, the method's further development rests on inductive analysis of cases, a less glamorous strategy than top-down deduction, but one that has been successfully employed in the applied behavioral sciences (e.g., research on expertise). The approach lends itself to training. However, rather than a set of rules or steps, what must be learned are large numbers of problem structures, pitfalls, and heuristics, embedded in an analytic way of thinking about problems. The approach may be amenable to expert systems implementation, as through case-based reasoning (Kolodner, 1993). Indeed, it shares the principle that underlies case-based reasoning: When actions cannot be based on deep explanatory principles, they can be guided by a network of experience-derived generalizations, having varying degrees of breadth and power, that capture the essence of good thinking in particular situations.

REFERENCES

Clancey, W. J. (1985). Heuristic classification. *Artificial Intelligence, 27*, 289–350.
Dupre, J. (1993). *The disorder of things.* Cambridge, MA: Harvard University Press.
Grayson, C. J. (1973). Management science and business practice. *Harvard Business Review, 51*(3), 41–48.
Isenberg, D. J. (1986). Thinking and managing: A verbal protocol analysis of managerial problem solving. *Academy of Management Journal, 29*, 775–788.
Klein, G. A. (1993). A recognition-primed decision (RPD) model of rapid decision making. In G. A. Klein, J. Orasanu, R. Calderwood, & C. E. Zsambok (Eds.), *Decision making in action: Models and methods* (pp. 138–147). Norwood, NJ: Ablex.
Kolodner, J. (1993). *Case-based reasoning.* San Mateo, CA: Kaufmann.
Lang, J. R., Dittrich, J. E., & White, S. E. (1978). Managerial problem solving models: A review and a proposal. *Academy of Management Review, 3*, 854–866.
MacCrimmon, K. R., & Taylor, R. N. (1976). Decision making and problem solving. In M. D. Dunnette (Ed.), *Handbook of industrial and organizational psychology* (pp. 1397–1453). Chicago: Rand McNally.
March, J. G. (1981). Decisions in organizations and theories of choice. In A. H. Van de Ven & W. F. Joyce (Eds.), *Perspectives on organizational design and behavior* (pp. 205–244). New York: Wiley.
Newell, A. (1969). Heuristic programming: Ill structured problems. In J. Aronofsky (Ed.), *Progress in operations research* (Vol. 3, pp. 361–414). New York: Wiley.
Orasanu, J., & Connolly, T. (1993). The reinvention of decision making. In G. A. Klein, J. Orasanu, R. Calderwood, & C. E. Zsambok (Eds.), *Decision making in action: Models and methods* (pp. 3–20). Norwood, NJ: Ablex.
Schön, D. A. (1983). *The reflective practitioner.* New York: Basic Books.
Senge, P. M. (1990). *The fifth discipline.* Garden City, NY: Doubleday.
Simon, H. A. (1976). *Administrative behavior* (3rd ed.). New York: The Free Press.

Smith, G. F. (1988). Towards a heuristic theory of problem structuring. *Management Science, 34,* 1489–1506.

Smith, G. F. (1990). Heuristic methods for the analysis of managerial problems. *Omega, 18,* 625–635.

Smith, G. F. (1993). Defining real world problems: A conceptual language. *IEEE Transactions on Systems, Man, & Cybernetics, 23,* 1220–1234.

Smith, G. F. (1994). Managerial problem solving. In A. Kent (Ed.), *Encyclopedia of library and information science* (Vol. 53, Sup. 16, pp. 210–236). New York: Marcel Dekker.

Sternberg, R. J., & Smith, E. E. (1988). *The psychology of human thought.* Cambridge, UK: Cambridge University Press.

Suchman, L. A. (1987). *Plans and situated action: The problem of human-machine communication.* Cambridge, UK: Cambridge University Press.

Thagard, P. (1982). Beyond utility theory. In M. Bradie & K. Sayre (Eds.), *Reason and decision* (pp. 42–49). Bowling Green, OH: Bowling Green State University.

Tversky, A. (1975). A critique of expected utility theory: Descriptive and normative considerations. *Erkenntnis, 9,* 163–173.

Part V

Naturalistic Decision
Making:
Where Are We Going?

Chapter 35

Naturalistic Decision Making: Where Are We Going?

◆

Gary Klein
Klein Associates Inc.

This book describes a great deal of research and theorizing about naturalistic decision making (NDM). At the time of the 1994 conference, enthusiasm had been growing as more researchers were using NDM paradigms, and as researchers from neighboring fields were discovering bridges to NDM work. In this chapter I try to describe some of the major challenges that will need to be faced and try to anticipate where the field might go.

The main reason for the interest in NDM is that more and more researchers are growing impatient with artificial paradigms, and want to learn about ecologically valid tasks. Along with that objective is the eagerness to study topics that are relevant to sponsors of applied research. It is easier and more satisfying to search for ideas and solutions to high priority questions than to try to find ways to convince sponsors that an artificial task can somehow provide meaningful insights. (See Hoffman & Deffenbacher, 1993, for a useful discussion of ecological and epistemological validity with regard to applied cognitive psychology.) Yet another reason for the enthusiasm is the shift in focus of the research, from the traditional effort of setting up controlled conditions, to the opportunity to study subject matter experts and learn something useful from them.

A number of people have skeptical reactions to NDM. We have included some examples in this volume (see Howell, chapter 4, this volume), as we did in the earlier volume (e.g., Doherty, 1993). Much of the skepticism centers around whether NDM has anything really new to offer, anything

383

that has not already been said, and also whether NDM has any valuable messages that might allow it to succeed where earlier attempts have failed.

The argument that there is nothing new here has several different themes. One theme, that NDM is just a rehash of previous critiques of analytical decision research, is valid but not compelling. The previous critiques did not offer alternative lines of investigation, the way NDM has. Certainly we can trace the evolution of ideas backward in time, and we can raise questions about when counterreactions actually began, but these historical hindsights do not change the fact that large numbers of researchers are studying how people in field settings make decisions, with an intensity that has not previously occurred. Figure 35.1 shows a wide sampling of NDM researchers. The number and range of investigators suggests that the NDM movement has tapped into a widespread area of interest (see Fig. 35.1). The range of domains studied includes:

- Commercial and military aviation.
- Structural and wildland firefighting.
- Command and control operations (including Army, Navy and Air Force activities).
- Anesthesiology.
- Nuclear power plant operations.
- Software design.
- Off-shore drilling management.
- Corporate planning.
- Jury deliberations.
- Highway design.

Another theme to this argument is that many of the topics addressed by NDM are already being investigated by researchers in fields such as problem solving, ecological psychology, and situated cognition. This is also valid, but I prefer to see it as a challenge rather than a critique. The challenge is to see if NDM researchers have anything to offer to these existing and emerging approaches. We must try to build on existing work wherever possible, and one of the objectives of the 1994 conference was to help NDM researchers become more familiar with current work going on in related fields.

Another argument is that NDM research has not yet amassed a large number of tangible and successful outcomes. We agree with this concern. Without a body of accomplishments, it is not possible to evaluate the usefulness of the NDM approach. Ours is a new approach, and the clock is ticking. If we still have this concern in 5 or 10 years, then our enthusiasm for NDM research will have to be questioned. We are seeing a great deal of action, but action sometimes masquerades as progress. One goal of this

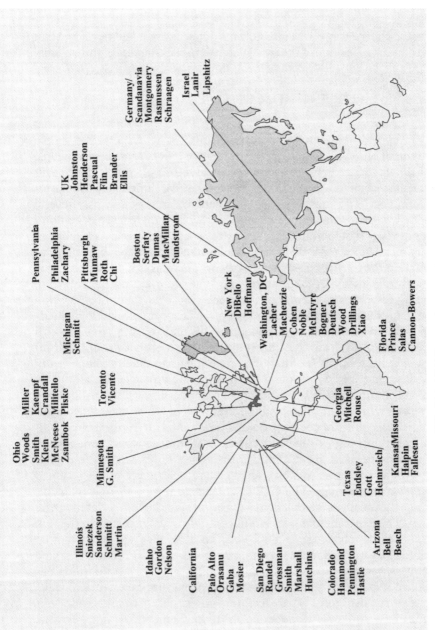

FIG. 35.1. NDM Research community.

volume is to provide a progress check of current work, and to describe efforts to apply NDM concepts.

To describe the way NDM research and applications might develop, I first need to summarize some of the issues covered in the chapters of this volume.

PUTTING NDM INTO PERSPECTIVE

One of the most common questions we hear is, "What *is* NDM?" A simple definition of NDM is the one offered in chapter 1: "NDM is the way people use their experience to make decisions in field settings." At first glance, this may seem so obvious as to be trivial. However, it is a surprisingly controversial definition. The reason is that decision research of the past several decades has centered around normative theory—analytical accounts of decision tasks that permit researchers to assess the quality of subjects' performance. Normative theory provides a standard for the research. Unfortunately, the reliance on normative theory also makes it difficult to study decision tasks that are ill-structured, and therefore are not amenable to analyzing responses as correct or incorrect. For some traditional decision researchers, the idea of studying how people make decisions within ill-structured tasks and complex environments seems bizarre. At a conference in Washington, DC during the summer of 1995, a well-known decision researcher heard about NDM for the first time and had difficulty comprehending that anyone would seriously try to study how people actually made decisions, in the absence of any normative standard.

The emphasis on experience in the definition of NDM is also important. Analytical accounts of decision making have relied on inexperienced subjects, for several reasons. One is that an experimenter can exert more control over a study by using naive subjects performing novel tasks. The experimenter does not have to worry about counterbalancing for experience, because no one has any. The experimenter does not have to worry about learning the nuances of the domain, because she or he is defining the domain. The experimenter does not have to worry about designing the task well enough to challenge the subjects, because they are starting from scratch. The experimenter does not have to worry about recruiting skilled performers, because unskilled subjects are readily available. Unfortunately, we have not learned a great deal about how people use experience to make judgments and decisions. Where researchers were able to develop paradigms for studying experienced decision makers (e.g., Hammond, Hamm, Grassia, & Pearson, 1987), their work was embraced by the NDM community. For the field of NDM, expertise is an important concept, and developments in our understanding of expertise will help to enrich NDM research and theory, just as NDM applications should help to expand our understanding of expertise. (The topic of expertise is discussed in greater detail in the next section.)

The term *naturalistic decision making* was coined during the 1989 NDM conference, and has since come into more general use. Nevertheless, several participants in the 1994 NDM conference raised objections to the term. These were in the minority, yet it is instructive to trace their criticisms. If they had all converged on the same criticism, we would have some reservations. However, the criticisms take opposing lines. Howell (chapter 4, this volume) argues that NDM is still decision-making research, and that the "naturalistic" adds nothing. According to Howell, decision researchers have been studying naturalistic settings from the beginning, so it is pretentious to assert otherwise. On the other hand, Vicente (see Beach et al., chapter 3, this volume) argues with the "decision making" part of the term "naturalistic decision making." He does not see any connection to traditional decision research, and instead he sees linkages to problem solving and ecological psychology. His argument is that we are misleading people when we refer to decision making, a term that is retained more as an historical accident than anything else. Our position is that the term naturalistic decision making is achieving the status of a common referent, and the fact that it is criticized from two opposite directions at the same time does not shake our confidence in the utility of the term.

By trying to understand how people make decisions, solve problems, and assess situations, the NDM movement is broadening its lines of inquiry. Ten years ago several researchers were excited to realize that the study of decision making could be expanded to cover ill-structured tasks, and could go beyond the moment of choice. This may have led some to view NDM solely as a reaction against traditional decision research, even though it was acknowledged that strategies such as multiattribute utility analysis also were useful in naturalistic settings. Many of the chapters in the Klein, Orasanu, Calderwood, and Zsambok (1993) volume do contain an "antiestablishment" tone. I believe that the negative reactions were against two features of the behavioral decision-making work: the implication that formal analytical methods could have strong benefits in most naturalistic settings, and the implication that failure to use formal, analytical methods would generally result in poor decision making. Neither of these implications seems justified. However, no one in the NDM community asserts that people *never* compare courses of action. No one claims that analytical methods are *never* useful. In order to place these methods in a proper perspective, we need to appreciate the role of domain expertise in decision making.

By now, the NDM community is spending less time defining differences with analytical decision paradigms, and more time assimilating research about problem solving, situation awareness, and expertise, and trying to see what can be learned from neighboring perspectives such as ecological psychology and the "situated cognition" view. These are the sorts of linkages to which Vicente was referring. As these linkages evolve, the focus of NDM research will change. The description of NDM offered by Zsambok (chapter 1, this volume) shows that NDM has a core set of objectives (to study how people make decisions under all sorts of adverse and complicating conditions) and methods (typical reliance on Cognitive Task Analysis techniques to examine proficient decision strategies). If analytical decision research ceased, many NDM researchers would not notice, which suggests that the NDM movement is not merely a reaction against the earlier tradition. We NDM researchers have our own agenda, and although we are prepared to learn from traditional, normative paradigms, we are not depending on those paradigms for inspiration, methodology, or identity.

For NDM researchers, an important challenge is to demonstrate applications and to collect data to document the "value added" of an NDM perspective. In order to accomplish those goals, we will need to clarify the procedures by which such a demonstration and evaluation is conducted.

EXPERTISE AND NDM

The theme of expertise is repeated throughout the chapters of this volume. Methods of Cognitive Task Analysis are continuing to be developed to elicit and represent aspects of expertise. In the absence of normative theory, training is geared toward helping people achieve expertise, relying on the strategies of the experts as a "gold standard." Models of NDM center around how people use expertise to make decisions.

For all this, we have not yet seen a true synthesis of the models and research in NDM with the models and research on expertise (e.g., Chi, Glaser, & Farr, 1988; Ericsson & Smith, 1991). This raises another challenge for NDM researchers, to take advantage of work that has already been done on the development of expertise, as a platform for generating hypotheses about decision-making performance. We anticipate that in the years to come, this will be a fruitful line of work, as NDM researchers elaborate their models and explore the implications of different accounts of the nature and development of expertise.

A potentially useful line of investigation concerns individual differences as they affect decision strategies. The differences may involve the nature and extent of expertise, or they may involve other types of factors such as need for cognition, and reflective-impulsive styles. These should impact the types of strategies employed and the way they are carried out. A related question is whether there are cross-cultural differences in decision strategies. I am not aware of any NDM studies of this issue, whereas Yates and Lee (1996) have examined cross-cultural differences in analytical decision strategies; this provides a useful lead for future NDM research projects.

Because expertise is so central to NDM work, we anticipate that there will be greater interest in understanding the natural development of expertise for different types of tasks, along with more careful criteria for distinguishing experts and novices in different domains. These accomplishments would have a significant effect on our appreciation for the way expertise is brought to bear in making difficult judgments and decisions.

THE NATURE OF ERROR IN NDM

Woods, Johannesen, Cook, and Sarter (1995) have claimed that errors can be expressions of expertise. The use of expectancies that permit skilled performance can on occasion result in misidentification of situations. Once errors are made, we have an indication that something went wrong. However, Woods et al. show that this is just the beginning of the story, not the ending. The error can be traced to the design of the equipment, the training provided, the culture that resulted in inadequate training or equipment, and so forth. Reason (1990) claimed that the person committing the error is merely the sharp end of the wedge, the unfortunate who is closest in time and space to the error, but is only a part of the influence chain. The rest of the chain, the "latent pathogens" as Reason described them, are too often ignored. The conclusion that is drawn by NDM researchers is that the concept of error has little value as an end judgment. Its function is to trigger the investigation of the entire chain of causal factors.

This argument contrasts with the position of normative theory, which is to define optimal choice strategies so that deviations can be detected and errors measured. Normative theory depends on having well-structured tasks and well-defined goals. For ill-structured tasks and ill-defined goals, researchers will have difficulty in identifying errors.

If error analysis is fundamentally limited, NDM research will often not be able to specify what counts as a success independent of the goals and situation awareness of the individual decision maker (Lipshitz, chapter 15, this volume). One strategy for finessing this difficulty is to use Cognitive Task Analysis to describe the processes used by subject matter experts, and use those descriptions for evaluating performance. However, experts may differ in their strategies, and in many tasks expertise does not develop to a very high level. One of the major challenges facing NDM research is to develop methods of performance evaluation.

THE ROLE OF THEORY IN NDM

Lipshitz (1993) compared alternative models of NDM. The present volume shows continual development of theory. Klein (chapter 27, this volume) has expanded the RPD model to cover diagnostic activities. Lipshitz (chapter

15, this volume) has suggested his own expansion of the RPD model. Cohen et al. (chapter 25, this volume) have presented yet another expansion, to incorporate metacognitive processes into a recognitional model. Endsley (chapter 26, this volume) has presented a model of situation awareness. Orasanu and Fisher (chapter 32, this volume) have differentiated different types of decisions in aviation. G. Smith (chapter 34, this volume) has synthesized a wide variety of studies of problem solving into some common frameworks. Klein and Crandall (1995) have presented a model of mental simulation. Klein (in preparation) has derived a nonlinear model of problem solving.

These models take a different form than normative theories. Shanteau (1995) has pointed out that the normative models of decision making center around mathematical equations, whereas the NDM models center around flow-charts and diagrams. Because the NDM models lack the precision of normative theory, we do not expect to see the same type of hypothesis testing. Nevertheless, there are some efforts in this direction. Lipshitz (chapter 15, this volume) has offered some suggestions about ways to falsify NDM models. Klein, Wolf, Militello and Zsambok (1995) generated a hypothesis from the RPD model, to the extent that for skilled decision makers, the first option they consider should be feasible. They tested this prediction, and obtained results supporting the model. The study allowed for the possibility of not supporting the RPD model. Beach (1990) conducted several studies that were tests of Image Theory.

Another source of concern about NDM theories is that the models resemble "trading card" collections, in which the effort is to begin with a limited set of cards (processes), then to add new cards (processes) to fill in the gaps. The result is an ever-growing diagram that has greater completeness, without increasing explanatory power or precision.

P. Smith (see Beach et al., chapter 3, this volume) has suggested that NDM researchers should concentrate more on domain-specific models that are basically descriptions of strategies, rather than global models. That is, as investigators study skilled decision making in new tasks and new domains, the greatest contribution may be to model what is happening within a domain. Smith, Griffin, Rockwell, and Thomas (1986) have provided a domain-specific description of how skilled pilots use scripts to troubleshoot malfunctions. Kaempf, Klinger, and Wolf (1994) have provided a domain-specific description of how baggage screeners decide whether to scrutinize an item of luggage. Kaempf, Klein, Thordsen, and Wolf (1996) developed a domain-specific description of how experienced naval officers commanding AEGIS cruisers infer the intent of potentially threatening aircraft. Militello and Lim (1995) described how experienced neonatal intensive care unit nurses identified early signs of necrotizing enterocolitis. Xiao et al. (chapter 19, this volume) described a problem-solving strategy used by anesthesiologists.

These descriptions of domain-specific strategies fall at a useful level of precision, and permit contrasts between experts and novices. I anticipate that progress will be made first by increasing the number and variety of these domain-specific models, and then perhaps by synthesizing them to provide global models that are better grounded in observation. A second type of progress in theory may result from efforts to incorporate advances in research concerning expertise, problem solving, and situation awareness into existing NDM models.

PARADIGMS OF NDM RESEARCH

Most NDM research projects involve case studies within a domain. Instead of comparing experimental and control groups using dependent variables such as performance accuracy, NDM research usually compares people with greater and lesser experience, using dependent variables such as the cues and strategies used. Some NDM experiments do rely on factorial designs, control conditions, and inferential statistics. However, many investigations take the form of case studies for the purpose of exploration and theory building, rather than theory testing.

One of the important criteria for evaluating research paradigms is the extent to which they permit useful generalization. A limitation of hypothetico-deductive studies is that they can constrain the task context so much that it can be difficult to gauge how much the results will generalize outside of those conditions. A limitation of case studies is that they apply so little control of contextual features that it is unclear to what extent the findings generalize to other domains.

A growing number of NDM researchers are taking the effort to develop simulations, which can permit controlled experimentation. These simulations (e.g., of AEGIS cruiser Combat Information Centers, of airline cockpits, of aviation dispatch centers, or nuclear power plant control rooms) are sufficiently rich and challenging that subject matter experts can be recruited as participants in studies. Therefore, it is possible to study how expertise is used to make decisions within controlled conditions.

We anticipate a more widespread use of simulations, perhaps framed around common tasks. Some NDM investigators have been able to gain access to skilled pilots and navy commanders, but few people have access to those types of specialized domain experts. If we were to find simulations for studying more common tasks, such as driver decisions rather than pilot decisions, then we might find a greater interest in simulation studies. More important, such a development would facilitate simulation studies in university settings. Currently, the labor-intensive nature of case studies serves as a barrier to university research projects. A single case-study project can take several years, including preparation and domain familiarization. Anything that increases the efficiency of conducting the research would be welcomed. In addition to accessible simulations that would engage skilled decision makers, we expect to see a codification of research paradigms so that new researchers have some guidance on how to structure their projects.

Further, NDM needs to clarify the dependent variables used. For tasks that are ill-structured and strongly context-dependent, it is difficult to measure accuracy of performance. Some other methods for assessing performance will need to be derived. One possibility is to use the strategies of subject matter experts as the standard, but Shanteau (1995) has noted that experts differ in their strategies. This does not indicate that there is not expertise, merely that fine points of strategy permit variability. Shanteau claimed that experts tend to converge on the same actions even while some of their strategies may diverge. This hypothesis requires investigation because it has important implications for a number of fields as well as for NDM research design.

One possible way of structuring NDM studies is to concentrate more on the way situation awareness shifts during an incident, than on the content of situation awareness. It can be a daunting task to characterize what a subject matter expert knows, as in sizing up a situation. It may be easier to study how situation awareness shifts when the conditions and context change. In addition to trying to capture the strategies that experts use in performing a decision task, we can look at what conditions lead to shifts in strategies. Payne, Bettman and Johnson (1988) effectively studied decision strategy shifts using tasks that required little expertise. The same type of paradigm could be very useful for NDM researchers as well.

COGNITIVE TASK ANALYSIS

Because Cognitive Task Analysis is used so frequently in NDM studies, and because the methods are themselves so new, there is currently a great deal of interest in this methodology. In addition to the chapter by Gordon and Gill (chapter 13) on CTA, several of the other chapters discussed CTA explicitly.

One of the most important accomplishments in the past 5 years has been this development and elaboration of different methods of CTA. The NDM community now has a much wider array of techniques to use, and a much greater appreciation of the boundary conditions of these methods. That could translate into greater efficiency and fewer wasted hours struggling with techniques that are not suited to a given task. It could also translate into improvements in the representation of the results from knowledge elicitation sessions, so that the findings can be more clearly described and communicated. Moreover, a number of studies have reported very good reliability for CTA methods (e.g., Kaempf, Klein, Thorsden, & Wolf, 1996).

Howard (1994) has also addressed the issue of validity, citing data showing that self-reports can have a higher validity than behavioral measures of performance.

The challenge is to develop more efficient methods of Cognitive Task Analysis than are currently available. The existing techniques are often difficult to master and to apply. Once we derive more streamlined techniques, then we can expect to see an even greater interest in carrying out CTA-based projects.

A second challenge is to develop paradigms for contrasting different CTA methods and evaluating the validity of each. Only when this happens could we be able to have strong confidence in any particular methods, along with clear guidelines about the boundary conditions for using alternative methods.

APPLICATIONS OF NDM

I believe that the field of NDM is just at the point of establishing procedures for different types of applications. I anticipate that NDM could have a strong impact on training, by enabling the training community to identify difficult judgments and decisions as training objectives, and by demonstrating the worth of different interventions. Team training should benefit from a more direct analysis of how to prepare teams to make better and faster decisions, rather than just emphasizing ways of improving harmony among team members.

I also anticipate that NDM could have a great deal to say about design, by enabling the design community to identify decision requirements as objectives to be supported by systems and by human–computer interfaces (Klein, chapter 5, this volume).

A challenge facing the NDM community is to show that the NDM perspective results in applications that are measurably superior to existing practice, and to show that it was the NDM perspective, rather than the skill of individuals, that made the difference. The recommendations will have to be traced, via some sort of audit trail, to the tenets of naturalistic decision making.

Even more is needed. Practitioners seeking to apply NDM concepts will require clear guidelines. These will include guidelines for performance measurement, showing how to achieve it, what to measure, and when to measure it. For example, Salas (1995) pointed out that for training, NDM efforts will have to do more than encourage feedback—they will have to specify the type of feedback, how it is to be presented, and when. It may be too much to ask for NDM researchers to generate entirely new training practices, but it is imperative to ask how the NDM framework leads to something different and superior to "good practices."

Another type of application is to support users who are trying to take advantage of new technologies, but are fearful that the technologies might turn out to be inappropriate or disruptive. NDM researchers can provide a valuable service by helping the users anticipate how the technologies might support or interfere with decision-making expertise.

ANTICIPATING THE FUTURE DEVELOPMENT OF THE NDM FRAMEWORK

It may be instructive to consider a scenario in which the NDM movement develops into the form of a guild, in which a set of researchers and practitioners derives and shares methods for addressing critical challenges for improving decision making through training, system design, and other means. The field of NDM might be called upon to improve decision making, problem solving, planning, and situation awareness.

One of the requirements of such a guild is that the members of the community be able to articulate and reliably implement a common set of applied methods. To some extent, this is occurring with regard to Cognitive Task Analysis methods, particularly as we learn the boundary conditions for each method. We should be able to study a problem and categorize it according to the Cognitive Task Analysis methods that are best suited.

A second requirement of a guild is to have vehicles for disseminating new techniques. This is also occurring, in the form of conferences on NDM and situation awareness research, the establishment of a new technical group within the Human Factors and Ergonomics Society, and the publication of a newsletter by that technical group. These activities stand in contrast to the characteristic of guilds that tries to maintain exclusivity through secrecy. Issues of proprietary findings and secrecy have not been a problem to date among NDM researchers, and hopefully the open exchange of ideas will continue. Secrecy is more of an issue in working on applications for commercial clients who want to protect proprietary ideas.

A third requirement for a guild is to be able to bring trainees and students up to speed in the methods and concepts, and here we do not find many signs of progress. Because NDM researchers tend to work outside university settings, they are not in a position to develop graduate courses. It may take a fairly long time for such courses to appear. One of the limiting factors for graduate students interested in NDM is that the research paradigms are different from traditional theses and dissertations. NDM research requires different types of theories, methods, and materials. Few graduate students have the time or opportunity to learn enough about a specific domain to carry out Cognitive Task Analyses, although there have been some important exceptions. Another limiting factor here is that most of the models in NDM grow from applied projects, rather than laboratory-based efforts. Some university researchers are successful at working in field settings, and improved simulated tasks should permit an increasing amount of laboratory studies. However, the strength of NDM research—its focus on naturalistic settings—may also be a hindrance with regard to recruiting graduate students. Special steps such as cooperative arrangements between academia, government agencies, and private companies may be needed.

A fourth requirement for a guild is to develop a core of insights. Here, we continue to make important progress—U.S. NDM researchers continue to discover how people actually make decisions in a variety of different field settings. These descriptions are not at too abstract a level (e.g., "She used an RPD strategy," or "He used a noncompensatory strategy"), but are one level deeper, showing the details of how the strategy was carried out. By collecting these discoveries, and by disseminating them, the NDM community is obtaining a shared set of discoveries.

A fifth requirement for a guild is for the members of the NDM community to know each other well enough to be able to identify the relevant specialists for difficult tasks. A project team may need specialists in carrying out methods, or specialists in the domain or in a related domain, or specialists in related techniques such as expertise in human factors, or training development, or Artificial Intelligence, or classical decision making and probability models. A problem-oriented team must be able to readily call on different types of expertise. The new technical group on Cognitive Engineering and Decision Making could make it easier for the NDM community to become familiar with the skills and backgrounds of its members, in order to draw on necessary support.

A sixth requirement of a guild is to regulate the entry of new members and to maintain quality checks of the work of all members. Clearly, the NDM community will not be determining who is permitted to perform NDM research. However, there are some concerns. One concern is about the difficulty of carrying out some of the Cognitive Task Analysis methods. Untrained investigators may perform an inadequate job of implementing these methods, and conclude that the methods do not work very well. Another concern is to define what counts as research on naturalistic decision making. As the field becomes more popular, people may begin to frame traditional studies as examples of naturalistic research if any of the criteria described in the introductory chapter are even remotely satisfied. The definitions of NDM offered in chapter 1 provide a basis for defining the characteristics of studies of naturalistic decision making. Salas (1996) provides another attempt to define NDM. As we learn better how to resolve the question of what counts as an NDM study, we will be helping to form the identity of the field.

The concept of a guild may be useful in helping us anticipate the progress of the NDM movement in the years to come. Whether or not the NDM researchers develop guild-like organizations, what is important is to determine if the field is making progress or is losing momentum.

One of the indicators that a field is becoming stagnant is when it spends more time looking for applications of its existing methods than in constructing new ones. In basic research fields, that point is reached when researchers who had used a method to study a phenomenon become more interested in studying the method itself. In applied research fields, the point is reached when the practitioners complain that the users are not bringing them the right kind of problem to work on, to fit their techniques. These instances bring to mind the saying that "to a child with a hammer, the world is a nail." If the NDM movement reaches the point where practitioners are giving more emphasis to finding the right applications for its techniques than in finding or building the right techniques for user requirements, that will be a sign of impending decline. NDM researchers need to be tool makers, and need to keep attention centered on the cognitive performance problems encountered in operational settings. That sort of attitude, by itself, carries potential for progress.

REFERENCES

Beach, L. R. (1990). *Image theory: Decision making in personal and organizational contexts*. West Sussex, England: Wiley.

Chi, M. T. H., Glaser, R., & Farr, M. J. (1988). *The nature of expertise*. Hillsdale, NJ: Lawrence Erlbaum Associates.

Doherty, M. E. (1993). A laboratory scientist's view of naturalistic decision making. In G. A. Klein, J. Orasanu, R. Calderwood, & C. E. Zsambok (Eds.), *Decision making in action: Models and methods* (pp. 362–388). Norwood, NJ: Ablex.

Ericsson, K. A., & Smith, J. (1991). *Toward a general theory of expertise: Prospects and limits*. New York: Cambridge University Press.

Hammond, K. R. (1993). Naturalistic decision making from a Brunswikian viewpoint: Its past, present, future. In G. A. Klein, J. Orasanu, R. Calderwood, & C. E. Zsambok (Eds.), *Decision making in action: Models and methods* (pp. 205–227). Norwood, NJ: Ablex.

Hammond, K. R., Hamm, R. M., Grassia, J., & Pearson, T. (1987). Direct comparison of the efficacy of intuitive and analytical cognition in expert judgment. *IEEE Transactions on Systems, Man, and Cybernetics*, SMC–17(5), 753–770.

Hoffman, R. R., & Deffenbacher, K. A. (1993). An analysis of the relations between basic and applied psychology. *Ecological Psychology, 5*, 315–352.

Howard, G. S. (1994). Why do people say nasty things about self-reports? *Journal of Organizational Behavior, 15*, 399–404.

Kaempf, G. L., Klein, G. A., Thordsen, M. L., & Wolf, S. (1996). Decision making in complex command-and-control environments. *Human Factors 38(2)*, 220–231.

Kaempf, G., Klinger, D., & Wolf, S. (1994). *Development of decision-centered interventions for airport security checkpoints* (Final Tech. Rep.). Fairborn, OH: Klein Associates Inc. (Prepared under contract DTRS–57–93–C–00129 for the U.S. Department of Transportation, Cambridge, MA)

Klein, G. (in preparation). Nonlinear aspects of problem solving. *Human Performance*.

Klein, G., & Crandall, B. W. (1995). The role of mental simulation in problem solving and decision making. In P. Hancock, J. Flach, J. Caird, & K. Vicente (Eds.), *Local applications of the ecological approach to human–machine systems* (Vol. 2, pp. 324–358). Hillsdale, NJ: Lawrence Erlbaum Associates.

Klein, G. A., Orasanu, J., Calderwood, R., & Zsambok, C. E. (1993). *Decision making in action: Models and methods.* Norwood, NJ: Ablex.

Klein, G., Wolf, S., Militello, L., & Zsambok, C. (1995). Characteristics of skilled option generation in chess. *Organizational Behavior and Human Decision Processes, 62*(1), 63–69.

Lipshitz, R. (1993). Converging themes in the study of decision making in realistic settings. In G. A. Klein, J. Orasanu, R. Calderwood, & C. E. Zsambok (Eds.), *Decision making in action: Models and methods* (pp. 103–137) Norwood, NJ: Ablex.

Militello, L., & Lim, L. (1995). Patient assessment skills: Assessing early cues of necrotizing enterocolitis. *The Journal of Perinatal & Neonatal Nursing, 9*(2), 42–52.

Payne, J. W., Bettman, J. R., & Johnson, E. J. (1988). Adaptive strategy selection in decision making. *Journal of Experimental Psychology: Learning, Memory, and Cognition, 14*(3), 534–552.

Reason, J. (1990). *Human error.* Cambridge, England: Cambridge University Press.

Salas, E. (1996). Special section preface. *Human Factors, 38*(2), 191–192.

Salas, E. (1995, October). Discussant Comments from the *Symposium on Applying the Naturalistic Decision Making Perspective to Training,* Human Factors Ergonomics Society, San Diego, CA.

Shanteau, J. (1995, August). *Why do experts disagree?* Paper presented at the 15th Bi-Annual Conference on Subjective Probability, Utility, and Decision Making, Jerusalem.

Smith, P. J., Giffin, W., Rockwell, T., & Thomas, M. (1986). Modeling fault diagnosis as the activation and use of a frame system. *Human Factors, 28*(6), 703–716.

Woods, D. D., Johannesen, L. J., Cook, R. I., & Sarter, N. B. (1995). *Behind human error: Cognitive systems, computers, and hindsight* (State-of-the-Art Rep.). Wright-Patterson AFB, OH: CSERIAC.

Yates, J. F., & Lee, J. W. (1995). Chinese decision making. In M. H. Bond (Ed.), *Handbook of Chinese psychology* (pp. 338–351). Hong Kong: Oxford University Press.

Author Index

Subject Index